MANAGING GLOBAL GENETIC RESOURCES

Agricultural Crop
Issues and Policies

MANAGING GLOBAL GENETIC RESOURCES

Agricultural Crop Issues and Policies

Committee on Managing Global Genetic Resources:
Agricultural Imperatives

Board on Agriculture
National Research Council

NATIONAL ACADEMY PRESS
Washington, D.C.   1993

NATIONAL ACADEMY PRESS • 2101 Constitution Avenue, NW • Washington, D.C. 20418

NOTICE: The project that is the subject of this report was approved by the Governing Board of the National Research Council, whose members are drawn from the councils of the National Academy of Sciences, the National Academy of Engineering, and the Institute of Medicine. The members of the committee responsible for the report were chosen for their special competences and with regard for appropriate balance.

This report has been reviewed by a group other than the authors according to procedures approved by a Report Review Committee consisting of members of the National Academy of Sciences, the National Academy of Engineering, and the Institute of Medicine.

This material is based on work supported by the U.S. Department of Agriculture, Agricultural Research Service, under Agreement No. 59-32U4-6-75, and by the U.S. Agency for International Development under Grant No. DAN-1406-G-SS-6044-00. Additional funding was provided by Calgene, Inc.; Educational Foundation of America; the Kellogg Endowment Fund of the National Academy of Sciences and the Institute of Medicine; Monsanto Company; Pioneer Hi-Bred International, Inc.; Rockefeller Foundation; U.S. Forest Service; W. K. Kellogg Foundation; World Bank; and the Basic Science Fund of the National Academy of Sciences, the contributors to which include the Atlantic Richfield Foundation, AT&T Bell Laboratories, BP America, Inc., Dow Chemical Company, E.I. duPont de Nemours & Company, IBM Corporation, Merck & Co., Inc., Monsanto Company, and Shell Oil Company Foundation. In addition, dissemination of the report was supported by the W. K. Kellogg Foundation.

Committee on Managing Global Genetic Resources: Agricultural Imperatives

PETER R. DAY, *Chairman,* Rutgers University
ROBERT W. ALLARD, University of California, Davis
PAULO DE T. ALVIM, Comissão Executiva do Plano da Lavoura
   Cacaueira, Brasil[*]
JOHN H. BARTON, Stanford University
FREDERICK H. BUTTEL, University of Wisconsin
TE-TZU CHANG, International Rice Research Institute, The
   Philippines (Retired)
ROBERT E. EVENSON, Yale University
HENRY A. FITZHUGH, International Livestock Center for Africa,
   Ethiopia[†]
MAJOR M. GOODMAN, North Carolina State University
JAAP J. HARDON, Center for Genetic Resources, The Netherlands
DONALD R. MARSHALL, University of Sydney, Australia
SETIJATI SASTRAPRADJA, National Center for Biotechnology,
   Indonesia
CHARLES SMITH, University of Guelph, Canada
JOHN A. SPENCE, University of the West Indies, Trinidad and
   Tobago

Genetic Resources Staff

MICHAEL S. STRAUSS, *Project Director*
JOHN A. PINO, *Project Director*[‡]
BRENDA BALLACHEY, *Staff Officer*[§]
BARBARA J. RICE, *Project Associate and Editor*
ALWIN Y. PHILIPPA, *Senior Program Assistant*

---

[*]Executive Commission of the Program for Strengthening Cacao Production, Brazil.
[†]Winrock International, through January 1990.
[‡]Through June 1990.
[§]Through November 1989.

iii

Subcommittee on Plant Genetic Resources

---

*Formerly at Indian Council of Agricultural Research, New Delhi.

Subcommittee on Animal Genetic Resources

HENRY A. FITZHUGH, *Chairman,* International Livestock Center for
    Africa, Ethiopia
ELIZABETH L. HENSON, Cotswold Farm Park, England
JOHN HODGES, Mittersill, Austria
DAVID R. NOTTER, Virginia Polytechnic Institute and State
    University
DIETER PLASSE, Universidad Central de Venezuela (Retired)
LOUISE LETHOLA SETSHWAELO, Ministry of Agriculture, Botswana
THOMAS E. WAGNER, Ohio University, Athens
JAMES E. WOMACK, Texas A&M University

Board on Agriculture

The National Academy of Sciences is a private, nonprofit, self-perpetuating society of distinguished scholars engaged in scientific and engineering research, dedicated to the furtherance of science and technology and to their use for the general welfare. Upon the authority of the charter granted to it by the Congress in 1863, the Academy has a mandate that requires it to advise the federal government on scientific and technical matters. Dr. Bruce M. Alberts is president of the National Academy of Sciences.

The National Academy of Engineering was established in 1964, under the charter of the National Academy of Sciences, as a parallel organization of outstanding engineers. It is autonomous in its administration and in the selection of its members, sharing with the National Academy of Sciences the responsibility for advising the federal government. The National Academy of Engineering also sponsors engineering programs aimed at meeting national needs, encourages education and research, and recognizes the superior achievements of engineers. Dr. Robert M. White is president of the National Academy of Engineering.

The Institute of Medicine was established in 1970 by the National Academy of Sciences to secure the services of eminent members of appropriate professions in the examination of policy matters pertaining to the health of the public. The Institute acts under the responsibility given to the National Academy of Sciences by its congressional charter to be an adviser to the federal government and, upon its own initiative, to identify issues of medical care, research, and education. Dr. Kenneth I. Shine is president of the Institute of Medicine.

The National Research Council was organized by the National Academy of Sciences in 1916 to associate the broad community of science and technology with the Academy's purposes of furthering knowledge and advising the federal government. Functioning in accordance with general policies determined by the Academy, the Council has become the principal operating agency of both the National Academy of Sciences and the National Academy of Engineering in providing services to the government, the public, and the scientific and engineering communities. The Council is administered jointly by both Academies and the Institute of Medicine. Dr. Bruce M. Alberts and Dr. Robert M. White are chairman and vice-chairman, respectively, of the National Research Council.

Preface

The crucial role of genetic resources in supporting human society is frequently overlooked and greatly undervalued. They are of tremendous practical and historical significance for human life. They underpin both our daily survival and are responsible for generating a large part of the wealth of nations.

Germplasm is a resource that consists of the genetic materials that can perpetuate a species or a population of an organism. It can be used both to reproduce and, through hybridization and selection, to change or enhance organisms. Conserving genetic resources in the form of crop, livestock, microbial, and tree germplasm is a means of safeguarding the living materials now exploited by agriculture, industry, and forestry to provide food for humans and feed for livestock, fiber for clothing and furnishing, fuel for cooking and heating, and the food and industrial products of microbial origin.

Genetic conservation is also an integral part of a still broader activity concerned with protecting and maintaining the quality of air, water, and soil and the many plants, animals, microorganisms, and communities of organisms that help to mold and stabilize the global environment. Conservation ensures that future generations of humans will also benefit from earth's biological resources.

In 1985 the Board on Agriculture of the National Research Council, under the chairmanship of Dr. William L. Brown, concluded that an assessment of the status of global genetic resources important to agriculture was needed. Encouragement for this study came from several government and foundation officials and scientific associations. The National Research Council established the Committee on

Managing Global Genetic Resources: Agricultural Imperatives in November 1986. The scope of its study was largely restricted to plants, animals, and microorganisms used in commerce or having potential for such use.

Why is it necessary to preserve materials that were originally collected from undeveloped agricultures or from the wild? The success of modern high-yielding crop varieties is such that they tend to replace older peasant varieties even in remote parts of the world. Even though many of the varieties widely grown 20 years ago can still be found, they are also increasingly being replaced. These materials can be important sources of genetic variation. People have also destroyed or altered many natural habitats of wild crop relatives and made them unsuitable for the plants that once grew there.

There is widespread concern among many agricultural scientists about the status of conserved germplasm worldwide. Most collected materials, to be of use, must be evaluated and tested at the expense of considerable effort and resources. If properly conserved and catalogued such material is available to others who may wish to use it. Is enough being done? Is the material already conserved in seed stores and other facilities adequately documented, properly stored and managed, and freely available to anyone with a legitimate need? Does it include all the potentially important genetic information that can still be collected now but which may be disappearing and therefore not available much longer? Are sufficient resources being applied by national governments to their own and to global needs? What priorities have been established and are they correct? What mechanisms are in place for conserving genetic resources?

Conserving the genetic resources of exploited species first arose when humans saved individual or small groups of animals and part of their harvests of gathered seeds, roots, and tubers of plants. They were kept for herd increase and planting. Putting aside the better forms for future use began the long process of selection and improvement responsible for the development of agriculture. The first crop plants and livestock were undomesticated wild species that gave rise, thousands of years later, to modern varieties and breeds. The rediscovery of genetics at the end of the last century gave breeders an explanation of the mechanism of inheritance. The earlier cultivars and breeds and their closely related wild species were sources of useful variation that could be introduced by hybridization. Breeders assembled collections of useful materials that were described, catalogued, and tested and could be saved from year to year. These collections were the first forms of germplasm to be systematically conserved, at least during the working lifetime of the breeder.

The emergence of plant and animal breeding programs during the first half of this century created a demand for germplasm exchange among breeders and for collecting expeditions and explorations to satisfy the growing need for such crop plant characters as disease resistance, insect resistance, earliness, stiff stalks, and grain quality. At first much of the material collected was wasted because it was not properly stored and regenerated to avoid admixture, nor was it adequately described. Similarly livestock producers wishing to improve the productivity and adaptation of herds to local conditions imported exotic breeds from other parts of the world. When breeders retired, their stocks were often either discarded as useless or sold for slaughter by colleagues who did not appreciate their value.

In the United States, a nation almost wholly dependent on crop plants and livestock not native in North America, the elements of a national germplasm program evolved from early requests for U.S. travelers abroad to send back seeds or plants of promising trees or crops. From 1836 until 1862, when the U.S. Department of Agriculture (USDA) was established, the U.S. Patent Office distributed seeds and plants from overseas to U.S. farmers. In 1898 a Seed and Plant Introduction Section of USDA began to promote the collection and introduction of new crops. However, it became clear that it was necessary to minimize the risk of bringing in pests and diseases. In the United States, the first plant quarantines were initiated by individual states. California, in 1881, was the first to pass an act to prevent the spread of the grape gall louse from other states. Federal plant quarantine regulations were not adopted until 1912, although drastic legislation had been accepted in Europe and Australia by 1877 to restrict the introduction of the same grape pest from the United States.

Other nations in the developed world followed suit. National plant germplasm collections were established in Germany, France, the Netherlands, Scandinavia, the former Soviet Union, and the United Kingdom. Following World War II and the establishment of the World Bank to aid the economic development of poorer nations, a network of international agricultural research centers (IARCs) began to emerge. These centers were designed to provide both improved germplasm and agronomic research to help developing countries become self-sufficient and improve their agricultural exports to world markets. Among the most visible of the IARCs are those administered by the Consultative Group on International Agricultural Research (CGIAR), a body representing the principal donors that fund its activities. Centro Internacional de Mejoramiento de Maíz y Trigo (International Maize and Wheat Improvement Center) and the International Rice Research

Institute (IRRI), the first two centers, were established in Mexico and the Philippines, with support from the Ford and Rockefeller foundations. Both centers have seed banks. The IRRI seed bank, with more than 80,000 rice accessions, is the largest for any single crop, with almost complete representation and excellent documentation. The importance of germplasm to the developing world was recognized by the establishment of the International Board for Plant Genetic Resources as part of the CGIAR in 1974 to promote the development of international, regional, and national activities and programs for building a worldwide network of genetic resources centers.

The Committee on Managing Global Genetic Resources was assisted by two subcommittees and several working groups that gathered information or prepared specific reports. As the committee's work progressed it was decided to publish its reports in several volumes. This report reviews and comments on the scientific basis for germplasm conservation and the generic issues that apply broadly to the many different kinds of germplasm.

Much of the information the parent committee has drawn on for its review is based on the extensive literature on plant germplasm. Compared with other organisms, considerably more information on crop plants and related species exists. For this reason the report mostly uses plants to discuss the principles and legal, political, economic, and social issues surrounding global genetic resources management as they relate to agricultural imperatives. Other reports in the series *Managing Global Genetic Resources* are *Livestock* (1993), *Forest Trees* (1991), and *The U.S. National Plant Germplasm System* (1991). In addition a special working group was invited to review plans developed by the USDA for improving the National Seed Storage Laboratory at Fort Collins, Colorado. The committee issued its report, *Expansion of the U.S. National Seed Storage Laboratory: Program and Design Considerations,* in 1988.

The committee also saw the need for an assessment of the state of science and national and international policies regarding aquatic genetic resources. The diversity of fish and shellfish species worldwide presents a rich resource for development of food for future populations and a resource for germplasm for aquaculture. To date aquatic genetic resources has been a largely neglected area of focus in genetics and conservation management. Although the committee was unable to include a major report on aquatic genetic resources in its series, the committee does emphasize the importance of aquatic germplasm and the need for research and evaluation.

Because the subject of genetic resources is so wide in scope and so important to humankind the committee treated it at length. Even so, this report is limited to those resources that have identified economic value of importance to agriculture, forestry, and industry. However, it is likely that a wealth of unexplored and unknown materials of value to future generations are also threatened or are being lost. The concepts and methods discussed in this report will be useful for conserving these other materials. In the chapters of this volume the committee provides the background needed to understand its findings and recommendations. The report provides essential information for scientists, policymakers, staff of public and private institutions, development agencies, conservationists, educators, and students.

The report contains an Executive Summary, with the committee's major recommendations; an Overview, which introduces the subject for those readers without a background in genetic conservation; and two parts. Part One addresses basic science issues and entails Chapters 1 to 10. In Chapters 1 and 2 the arguments about the vulnerability of monocultures are reassessed and the adequacy of present systems to protect agriculture from risk in the next century are explored in light of events and experience since publication of the National Research Council's report *Genetic Vulnerability of Major Crops* in 1972. Chapter 3 discusses the role of in situ conservation and its relation to ex situ conservation. The principles and scientific basis of germplasm collection and management are discussed in Chapters 4 and 5, and ways of identifying useful variability and maximizing its availability to breeders are discussed in Chapter 6. Chapter 7 explores how developments in biotechnology may increase the efficiency of germplasm conservation and use and simplify some of the tasks. Documentation and data management, reviewed and discussed in Chapter 8, are essential if the contents of a germplasm bank are to be known and readily accessible to users. Chapter 9 describes the kinds of genetic stock collections that are globally important and discusses the special problems of safeguarding them. The conservation of microorganisms, discussed in Chapter 10, has all of the problems associated with conserving other kinds of germplasm including maintaining genetic stability in culture, protection of intellectual property rights, data banks, and the establishment of global networks.

Part Two addresses policy issues and entails Chapters 11 to 15. The principles of plant quarantine and the impact of quarantine regulations on exchange and movement of plant germplasm are described in Chapter 11. Chapter 12 reviews the mechanisms used for intellectual property protection of plant materials and their implications for international exchange of plant germplasm. The economic soundness of investing in the collection, preservation, and management of genetic resources is discuss in Chapter 13. It presents an original analysis of the contribution of the IRRI germplasm bank to breeding new rice cultivars for India to exemplify the value of properly managed and accessible germplasm for all crops. Chapter 14 traces the genesis of "the North-South debate," which features the industrialized nations of the northern hemisphere and the developing countries of the southern hemisphere. The genesis and operation of national and international germplasm conservation programs are discussed in Chapter 15.

Conserving genetic resources is an important part of agriculture. These resources, and the environments in which they are found, provide our daily bread. What stronger motivation could we possibly need to improve our management and support for them?

PETER R. DAY, *Chairman*
Committee on Managing
Global Genetic Resources:
Agricultural Imperatives

Acknowledgments

Many scientists and policymakers have contributed time, support, and information instrumental to the committee's analyses contained in this report. The committee thanks Robert Craig and Michael Lesnick for including several members of the committee and its staff in the Keystone Center's International Dialogues on Plant Genetic Resources; Henry Shands for support and valuable assistance throughout the project; Robert Weaver for early discussions on conservation and economics; Charles M. Rick for information on tomatoes; and Takuma Tsuchiya for information on barley.

The assistance of Dennis Allsopp, Ivan Bousefield, Les Breese, Paul D. Bridge, Stephen Brush, David L. Hawksworth, Douglas Gollin, William Grant, David A. John, Robert P. Kahn, Steven King, Barbara E. Kirsop, Jan Konopka, James M. Price, Peter H. A. Sneath, Judith Lyman Snow, and Theo van Hintum is gratefully acknowledged. The committee appreciated the contributions of these individuals throughout its deliberations.

Many other scientists and policymakers assisted by sharing their insights and expertise throughout the study. Numerous scientists and agriculturalists around the world gave us candid assessments of genetic resources activities in their countries. To all of these individuals the committee expresses its gratitude.

While the many subcommittee and work-group participants provided valuable information for this report, the conclusions and recommendations are those of the committee.

Administrative support during various stages of the development of this report was provided by Philomina Mammen, Carole Spalding,

Maryann Tully, Joseph Gagnier, and Mary Lou Sutton, and they are gratefully acknowledged. The committee also thanks Joi Brooks and Sherry Showell, interns sponsored by the Midwestern University Consortium for International Activities, for assisting in the development of the report, and Michael Hayes for his editorial work.

The committee also honors the memory of William L. Brown, whose leadership and early vision of the importance of crop genetic resources were crucial to the launching of this effort.

Contents

PART II
Managing Global Genetic Resources: Policy Issues

Other International Centers, 366
Constraints to Effective International Programs, 374
The Future, 375
Recommendations, 379

Agricultural Crop
Issues and Policies

Executive Summary

From the early beginnings of agriculture, germplasm resources have been selected, exchanged, collected, and preserved. Over time, cultivation of crops and farm animals has yielded landraces and breeds that are specific to various regions of the globe. Artificial selection by the farmer and natural selection by the elements developed a combination of genetic traits to meet the agricultural needs and climate and soil conditions that characterized each region. While many landraces of crops and breeds of animals can be traced back to their wild relatives, and even to their geographic centers of origin, there are others where the path of migration or genetic refinement to an agriculturally important genotype remains speculative.

In the past as well as today, the management of germplasm resources reflected a combination of the technology available and the particular uses of these resources. The age of exploration that erupted in the fifteenth century with the discovery of the New World resulted in exchanges of agriculturally valuable germplasm to and from every travelled region of the globe. Many of the exchanges during this great age of germplasm explorers were pivotal to the course of human history. The linkage of human slavery in the Americas to sugarcane and cotton and the role of the potato in the population explosion in Europe and consequently the industrial revolution are well-known examples (Harlan, 1992; Heiser, 1981). In addition, the accidental introduction of old pests, which were carried along with exchanged crops, and the appearance of new diseases or pests, which attacked introduced crops, resulted in quarantine measures.

1

Since the beginning of the twentieth century, the science of genetics has greatly escalated the speed and accuracy of breeding new genotypes. Earlier, in the 1860s, the Austrian Abbott Gregor Mendel, working with peas in his garden, discovered that most genetic traits follow mathematically predictable patterns of inheritance. His discovery permitted a new level of precision in genetic selection. More recent developments, since the 1950s and the discovery of deoxyribonucleic acid (DNA), allow scientists to identify, isolate, and manipulate individual genes at the molecular level. These developments continue to drive major advances in the technologies available to cull and manage germplasm resources. Genetic tools have produced an explosion in the rate of artificial selection and breeding for crop improvement, which is largely responsible for the manyfold increases in agricultural yields during this century. For example, U.S. corn yields have increased from 20 bushels per acre in the 1930s to 120 bushels per acre in the 1980s. About 50 percent of the increase in yield is due to genetic gain (Duvick, 1992). The ever-increasing pace of crop breeding has greatly refined and enhanced the value of germplasm collections. Yet as the high-yielding, highly selected varieties become utilized in agriculture around the globe, they threaten to displace their ancestral landraces and breeds.

AN AGRICULTURAL ENDOWMENT

Selecting, collecting, exchanging, and preserving germplasm resources are not new activities or issues. They are in fact as old as agriculture itself, as old as our knowledge of growing crops for food. But what is new, and pressing, is how to make national and international decisions about managing these activities for the future.

Circumstances throughout the world are changing at a rapid rate. The pace of scientific and technological developments has increased and, with it, the political and material demands of societies worldwide. People's needs for food, medicines, housing, land to cultivate, and the other natural resources of the earth—and their concern for having a voice in decision making about their needs—are increasing.

The natural environments of the earth are rapidly becoming managed—or mismanaged—ecosystems. The threats to biological diversity and the challenge to preserve and manage these natural endowments are well recognized (Wilson and Peter, 1988). The issue of global genetic resources is not merely the more clearly defined processes of collection, exchange, and preservation. The issue involves a dynamic system, an endowment, that requires management. And at

The viability of U.S. agricultural production depends on developing new and improved crop varieties. Credit: Pioneer Hi-Bred International, Inc.

this time, it is a system that is experiencing scientific and social pressures that often push from opposite directions.

This genetic resources endowment to a small extent has been captured as stored seed, in a variety of institutions throughout the world, that can be planted and grown out. But more broadly, a number of tangible and intangible components of this endowment require persistent attention over time. They include seeds and propagules of plants or sperm and embryos of animals that cannot be held in cold or long-term storage; the special collections and experience represented by individuals throughout the world who have devoted their careers to the germplasm of particular species; the exchange of knowledge as well as germplasm resources that takes place formally and informally; and the continual natural selection in an Amazonian forest, grassland reserve, or cultivated area that persistently reshapes genotypes. Similarly the continuous artificial selection practiced by breeders to create new cultivars or by farmers to emphasize special traits in a landrace or breed further refines germplasm resources.

Global genetic resources are not static. They constitute a living, changing, diverse system that can best benefit humankind when thoughtfully managed. Without proper management of these resources, many of their riches—the keystones to our ability to produce food—will be lost. Once lost, genetic resources cannot be regained.

Plant and animal germplasm is an agricultural endowment that offers the dividends of improved varieties and breeds of food crops and animals and increases in yield. Sound management practices can ensure that the "capital" derived from this endowment will not be squandered.

In this report, the Committee on Managing Global Genetic Resources: Agricultural Imperatives summarizes the status of global genetic resources. It assesses genetic vulnerability, the condition that results when a crop is uniformly susceptible to a pest, pathogen, or environmental hazard as a result of its genetic constitution, thereby creating a potential for vulnerability or even disaster. It describes the importance of in situ conservation of genetic resources, and details the science of collection, use, and management of these resources. Further, the report discusses policy issues relating to quarantine; proprietary rights; and conflicts over ownership, management, and use. It also presents a basis for assessing the economic value of genetic resources and evaluates existing national and international genetic resources programs.

This report structures the foundation for managing global resources for the future. The challenge to scientists and farmers alike is to be able to respond to the continuously changing landscape of modern agriculture. The challenge to policymakers and national leaders is to ensure that the genetic resources endowment is managed wisely to serve society, now and in the future.

THE NEED FOR GENETIC RESOURCES

The genetic components of crops today are based on landraces, slightly improved and obsolete varieties, recently developed breeders' lines, and wild species. As biotechnology enhances the ability to move genes among distantly related species, the breadth of species diversity that constitute a crop's genetic resources will increase. For livestock, genetic resources encompass the wide variety of breeds and types found around the world.

The rapid spread of improved crop varieties throughout the world has replaced many of the genetic resources essential to their continued improvement. Landraces, selected over decades in the presence of a wide range of pathogens, have proved to be important sources of genes for such traits as disease and pest resistance or drought tolerance. Thus, the conservation of traditional landraces as part of a crop's genetic resources has become imperative.

Nonetheless, landrace replacement is only one source of risk to the world's genetic resources. Population expansion, urbanization, deforestation, and pollution and destruction of the environment also threaten much of the world's biological resources and their diversity. As a countermeasure, many national and international collections have been established with the intention of rescuing and conserving them for future use. Some, such as the rice collection of the International Rice Research Institute (IRRI), have captured a major portion of the total genetic diversity of a particular crop. Other notable collections include those for wheat, maize, soybean, potato, tomato, sorghum, legumes, and many of the world's grain, fruit, vegetable, fiber, forage, and industrial crops. However, despite efforts in germplasm collection and maintenance by the commodity-oriented centers of the Consultative Group for International Agricultural Research (CGIAR), CGIAR's International Board for Plant Genetic Resources (IBPGR), the Food and Agriculture Organization (FAO) of the United Nations, and many national genetic resources programs, much work remains for this broad and difficult mission.

The better managed and more comprehensive a collection, the more valuable it is to researchers who must locate needed materials within it in an efficient manner. The likelihood that any single accession of a major collection has an essential gene or gene combination is low and, thus, it may have little value by itself. The value is not in individual accessions, but in the entire collection as a source of the crop's genetic diversity. The value of genetic resources collections is illustrated by increases in yields in rice productivity in India between 1972 and 1984. Varieties that were improved by means of genetic resources collections contributed more than one-third of these gains during this period.

An accession of unimproved germplasm of unrecognized potential is of value only when it is useful in an improvement program. Accessions that have no value today may suddenly have strategic value tomorrow. Unimproved genetic resources are not traded in markets in a manner similar to improved breeding lines or livestock breeds. Thus investment in a conservation program must be based on its potential to yield a benefit in the future. Whereas the developed nations are most able to support such conservation investments, the developing countries are where the greatest diversity of crop ge-netic resources exists. This disparity, given inequality among nations and the limited financial resources of most developing countries, has led to calls by many in the developing world for compensation from those who have collected genetic resources from another country and used these resources to establish improved varieties.

ACCESSING GENETIC RESOURCES

In a broad sense, conserving genetic resources, either in collec-tions or within reserves or other natural areas, preserves access to them. Efforts to conserve the biological diversity of the world's tropical forests, for example, save many as yet unidentified species that may have potential utility. Access to well-maintained stocks of germplasm resources of agricultural species is required by researchers and others who may need specific resistance factors, breeding lines, species, or populations for use in study or breeding. Ease of access depends on both the technology and practice of germplasm use and the policies and laws that affect exchange of genetic resources between research-ers, institutes, or nations.

The Role of Science and Technology

The most extensive systems for managing genetic resources exist for agricultural crops. They are supported by a considerable body of scientific information and technology. Emerging biotechnologies are also expected to lead to more information on existing collections and to increase the efficiency and precision of germplasm use.

In Situ or Ex Situ Conservation?

Many wild plant genetic resources can be conserved where they occur naturally (in situ). Although domesticated materials can some-times be maintained where they were originally selected, unless there is an incentive to continue planting them, they tend to be replaced by modern cultivars or lost due to human neglect.

Ex situ conservation occurs outside the species' natural environ-ment. It can include field plantings; seeds or pollen held in cold storage; tissue cultures; or seed, pollen, or tissues held under cryo-genic storage (–150° to –196°C). Most major collections of crop ge-netic resources in the world are ex situ collections of seed. Genetic resources in ex situ collections are often readily accessible to breeders and scientists. They are relatively secure from loss through future displacement, environmental damage, pests, or disease, although they are separated from the dynamic evolutionary forces that shaped them.

In situ conservation also preserves germplasm resources against loss, but in situ materials are often less accessible for breeding or research because they are usually not well characterized and studied, and because seeds or propagules may not be generally available. However, in situ collections may preserve as yet unrecognized or rare genetic traits that could be lost if only selected samples were preserved. For agricultural crops, ex situ methods are the primary means of conservation. Ex situ collections provide breeders or other researchers more immediate access to characterized accessions, which may include improved varieties and breeding lines.

Microorganisms are conserved almost exclusively ex situ. Some 18,500 species are estimated to be held in world culture collections that are part of an international network for their management, distribution, and utilization. These collections are important to the agriculture, food and beverage, and pharmaceutical industries and as research tools.

Managing the Numbers

The growth of genetic resources efforts has led to several large ex situ collections (Tables 1 and 2). However, the extent of genetic coverage within a crop and the level of care to preserve the collection vary markedly. Although large collections can improve the chance that most genes are captured, their sizes make it difficult to maintain viability, regenerate aged seed, document accessions, and screen for particular traits. Furthermore, long-term conservation, evaluation, and characterization become increasingly difficult as large collections increase in size. Data storage and retrieval methods and management tools, such as the use of core subsets, can give users more direct and selective access to large collections. An ideal core subset within a collection would contain a range of accessions that represents, with an acceptable level of probability and minimum redundancy, much of the genetic diversity of the crop species. Typically, it would consist of no more than 10 percent of a whole collection.

Selecting too few samples for a collection can leave important alleles uncollected. Considerations for determining how much to collect should include the mating system of the population (whether it is outcrossing or inbreeding) and its mobility (by dispersal of pollen or seed and the nature of the samples). In general, the less frequently a particular allele occurs in a population, the greater the number of plants that must be sampled to collect it. Samples from no

TABLE 1   Conservation Status of Major Crops

Crop	Total Accessions in Germplasm Banks	Distinct Accessions	Wild Accessions	Cultivars Uncollected (percent)	Major Needs
Wheat	410,000	125,000	10,000	10	E, M
Grain, oil legumes	260,000	132,000	10,000+	30-50	C, E, M for peanuts
Rice	250,000	120,000	5,000	10	C (wild), E, M
Sorghum	95,000	30,000	9,500	20	E, M
Maize	100,000	50,000	15,000	5	M, E
Soybean	100,000	30,000	7,500	30	C (wild), E
Common potato	42,000	30,000	15,000	10-20	C, E
Yams	8,200	3,000	60	High	C
Sweet potato	8,000	5,000	550	>50	C, E

NOTE: C, collection; E, evaluation; M, maintenance.

SOURCE: Chang, T. T. 1992. Availability of plant germplasm for use in crop improvement. Pp. 17–35 in Plant Breeding in the 1990s, H. T. Stalker and J. P. Murphy, eds. Wallingford, U.K.: CAB International. Reprinted with permission, ©1992 by CAB International.

more than 5 plants are adequate to collect alleles present locally at frequencies of 75 percent, whereas 200 or more plants must be sampled to secure alleles present at frequencies of 1 percent. Generally, sampling more sites is more effective at capturing genetic variants than is sampling more plants within a site.

Genetic diversity is usually not dispersed uniformly. Empirical studies of the slender wild oat revealed that genetic variation is correlated with environmental factors such as rainfall, temperature, slope, exposure, and soil type. Studies of variation in maize, a domesticated outbreeding species, show similar variability. Collection sites must therefore represent as many environments as possible and, in the case of domestic species, as many cultural environments as possible.

Describing Collections

Predicting which accessions in a collection are likely to contain particular genetic traits is aided by information on the collection. The more information available, the more easily a breeder or researcher can locate likely accessions. Three kinds of information about accessions are required: passport data, which relate to the ecogeographic origin of the accession; characterization data, which describe appearance and structural traits that are highly heritable and generally unaffected by differing environments; and evaluation data, which involve the more variable qualities of an accession, such as relative yields, disease resistance, or fruit quality. The latter data are of most interest to a breeder; passport data are essential for management,

TABLE 2   Estimates of Germplasm Holdings in the Five Largest National Plant Germplasm Systems and Major International Centers

Country / Center	Categories Concerned	Total
United States	All crops	557,000
China	All crops	400,000
Former Soviet Union	All crops	325,000
IRRI	Rice	86,000
ICRISAT	Sorghum, millet, chickpea peanut, pigeon pea	86,000
ICARDA	Cereals, legumes, forages	77,000
India	All crops	76,800
CIMMYT	Wheat, maize	75,000
CIAT	Common bean, cassava forages	66,000
Japan	All crops	60,000
IITA	Cowpea, rice, root crops	40,000
AVRDC	Alliums (onion, garlic, shallot), Chinese cabbage, common cabbage, eggplant, mungbean, pepper, soybean, tomato, other vegetables of regional importance	38,500
CIP	Potato, sweet potato	12,000

NOTES: IRRI, International Rice Research Institute; ICRISAT, International Crops Research Institute for the Semi-Arid Tropics; ICARDA, International Center for Agricultural Research in the Dry Areas; CIMMYT, Centro Internacional de Mejoramiento de Maíz y Trigo; CIAT, Centro Internacional de Agricultura Tropical; IITA, International Institute for Tropical Agriculture; AVRDC, Asian Vegetable Research and Development Center; CIP, Centro Internacional de la Papa.

SOURCES: Asian Vegetable Research and Development Center. 1992. 1991 Progress Report. Shanhua, Taiwan: Asian Vegetable Research and Development Center; Chang, T. T. 1992. Availability of plant germplasm for use in crop improvement. Pp. 17–35 in Plant Breeding in the 1990s, H. T. Stalker and J. P. Murphy, eds. Wallingford, U.K.: CAB International; Vitkovskij, V. L., and S. V. Kuznetsov. 1990. The N. I. Vavilov All Union Research Institute of Plant Industry. Diversity 6(1):15–18.

use, and assessment of a collection's quality. In general, gathering passport and characterization data is part of collection management. Evaluation is usually done by breeders or specialists, and, although location specific, this information should also become a part of the accession's record.

Unfortunately, capturing passport, characterization, and evaluation data has often received little attention. Detailed evaluation data are lacking for most of the world's collections, and even passport information is often surprisingly incomplete and unusable.

Regeneration: Essential but Overlooked

The growing out of collection samples to increase seed and maintain their viability is another important activity. Plants grown from seed accessions are vulnerable to environmental stresses that can produce unique selective pressures. Strategies for regeneration must preserve the original integrity of the sample to keep the regeneration process itself from becoming a source of genetic loss.

Too often regeneration is poorly considered in the planning of germplasm management activities. Problems can arise when the magnitude of the task exceeds available resources, as well as from inadequate consideration of the potential biological difficulties. Growing accessions too near one another, or without using isolation methods such as cages or bags to prevent cross-pollination, can result in unintended hybridization and loss of genetic integrity. If the accession is self-pollinating, as are many beans, there may be little concern. However, species pollinated by insects, which may carry pollen between widely separated plants, can present serious difficulties. Typically it is regeneration capability that ultimately limits the effective size of the collection. There is little future in amassing large collections that die in storage due to the inability to regenerate. Thus, planning should begin with regeneration requirements.

The Role of Biotechnology

Biotechnology influences germplasm conservation in at least four ways. First, it provides alternatives, such as plant tissue cultures, for conserving organisms. Second, modern techniques may simplify quarantine requirements by providing better tools, such as DNA tests for the presence of viruses in plants or seed stocks. Third, the application of molecular techniques may eventually address problems of using germplasm. For example, DNA probes can be used to screen germplasm accessions for the presence of particular genetic sequences

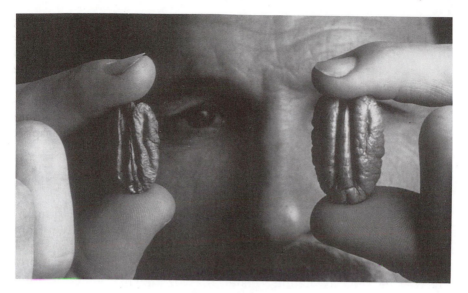

A plant geneticist at the W. R. Poage Pecan Field Station of the U.S. Department of Agriculture compares an early unimproved pecan with a newer variety. Credit: U.S. Department of Agriculture, Agricultural Research Service.

that encode specific traits. Dot blot tests can use DNA probes to detect the presence of genetic information from rye in wheat breeding lines. Fourth, biotechnology itself can lead to increased demand for germplasm and conservation services because it makes them easier to use. For example, the numbers of genes that have been isolated, cloned, and sequenced increase daily and represent a resource of considerable scientific and commercial value.

A direct application of biotechnology to germplasm use is the transformation of DNA to bring about directed genetic changes. Transformation has been carried out in bacteria, fungi, algae, and more than 30 species of plants and in nematodes, insects, fish, and mammals including several species of livestock. For the past several years field trials of genetically engineered crop plants, carrying foreign genes for resistance to herbicides, viruses, and insects, have shown the potential of transformation as a breeding tool.

Exchange and Ownership

The larger portion of accessions in collections today comes through exchange activities. International and national collections have been used to restore genetic resources lost through war, neglect, or natural disaster. There are, however, policy and legal constraints to exchanging genetic resources. For example, in an effort to protect national agriculture and prevent the spread of pests and diseases, nations may limit exchange through the use of quarantine restrictions. Restrictions have sometimes been used to limit trade rather than to protect crops; such economic or political restrictions can be major limitations for germplasm exchange. More controversial has been the application of ownership rights to germplasm.

Quarantine

When animal or plant genetic resources are brought into a country they may carry pests or pathogens that could damage agriculture. Cassava bacterial blight (*Xanthomonas manihotis*), for example, is believed to have been introduced to Africa and Asia from tropical America by way of infected planting material. For animals, rinderpest, an acute infectious disease and a significant problem for some countries of sub-Saharan Africa, was introduced from Central Asia near the end of the last century. In rinderpest epidemics, mortality ranges up to 90 percent.

Quarantine is designed to exclude potential sources of pests or pathogens from entry. It is typically a government responsibility, and nations differ as to how it is implemented. Quarantine practices generally prohibit exotic pathogens, pests, or parasites. They attempt to contain, suppress, or eradicate pests and pathogens. Government quarantine agencies may also assist exporters in meeting the quarantine requirements of importing countries.

When quarantine regulations require delay of importation or exclusion of imported germplasm, they may be viewed as restricting access to genetic resources. In the United States importation of *Prunus* (stone fruits such as apricots, plums, and almonds) germplasm can be delayed 10 years or more while samples are tested for the presence of viral pathogens. Some nations protect themselves by forbidding the entry of designated species. Biotechnology promises powerful tools for rapid disease screening of imported materials and may help to reduce these delays. Regional and international cooperative efforts to assemble and disseminate information about the distribution of pests and diseases are also important to international quarantine. Close cooperation between germplasm experts and those charged with implementing quarantine regulations can greatly improve access to genetic resources.

Ownership

Proprietary protection, in the form of patents, plant variety certificates, or copyright, for plant genetic resources has become a divisive issue that some fear may restrict access to plant genetic resources. Some developing countries are concerned that they have reaped few of the monetary benefits that have resulted from the use of their germplasm to produce modern cultivars. However, the rise of biotechnology has engendered the expectation that the developers of new crop varieties—whether breeders, biotechnologists, or industry—must be assured of a fair return for their investment. These issues have been debated in forums such as the FAO Commission on Plant Genetic Resources, the Keystone International Dialogue on Plant Genetic Resources, the Uruguay round of the General Agreement on Tariffs and Trade negotiations, and the 1992 United Nations Conference on Environment and Development (UNCED) in Rio de Janeiro.

Apart from proprietary protection, several countries have limited access to germplasm that may be of particular national value. For example the export of germplasm of coffee, black pepper, rubber (*Hevea*), soybean, and tobacco is restricted by selected countries. As debate over the value of unimproved genetic resources and wild species has escalated, some nations have restricted collectors' access to germplasm.

The geopolitical debate over plant genetic resources can be illustrated by the extreme positions. One, widely held in the developed world, is that all undeveloped genetic resources, including wild species, primitive landraces, and traditional cultivars, are public goods and should be freely available to all without restriction. Inbred lines, new cultivars, and other forms developed from these materials and protected by patents or plant breeders' rights should not be freely available to ensure that those who invested in their development receive a fair return. The contrasting view, held in parts of the developing world, is that these latter materials should be freely available, without proprietary restrictions, since they were largely developed from materials that originated in the developing world. This view has been reinforced by the suggestion of some form of "farmers' rights" to reward and recognize the contribution of developing world farmers toward the development and preservation of crop plant landraces. Discussions on international genetic resources, such as those sponsored by the Keystone Foundation, Keystone, Colorado, have led to agreement that compensation should come from an international fund, but the mechanisms for supporting it or distributing its proceeds remain to be developed.

Until the early 1990s, there was minimal impact of intellectual property rights—or, more fundamentally, of the increased value that biotechnology had given to genetic resources—on access to these resources. In the early 1990s, however, some developing nations saw these resources as one of their strongest bargaining chips in dealing with the developed world. Representatives at UNCED adopted the Convention on Biodiversity on June 5, 1992; the United States signed the convention on June 4, 1993. This convention, which will be a binding legal document, gives national governments the authority to control access to genetic resources. It is, therefore, likely that the era of free and open exchange of agricultural germplasm will soon be over.

BUILDING COOPERATION

Germplasm is one of the world's fundamental resources. Management of genetic resources has moved from being the concern of a few to becoming a vital interest for all nations, since it is vital for the support and development of every nation's agriculture. Neither genetic resources nor the means to conserve them are spread evenly over the globe. Wise and effective management requires broad international cooperation.

National Programs

Effective programs for managing and using genetic resources are essential for national agricultural health and food security. However, their cost is clearly beyond the capacity of many nations. It is both impractical and unnecessary for each nation, including the United States, to hold complete collections of all important crops and species. Regional associations that pool financial and technical resources can provide the benefits of genetic resources programs while making the best use of limited funds. The international agricultural research centers hold large global collections that can service many national and regional requirements. Adequate financing and suitably trained professional and technical staff are also needed.

International Cooperation

While the primary task of using genetic resources rests with individual nations, there are compelling reasons for developing international cooperation in management. Disparities in both the distribution of genetic resources and the financial and technical resources necessary for their management can be addressed through a global system of cooperation that provides benefits to all.

To be effective, a global system must ensure the security of its collections by providing financial support and enforcing rigorous standards. Although many national and international organizations promote the conservation of food crops, no coherent international program has yet emerged to provide a system of funding and coordination that ensures continuity.

International cooperation has been developing through the joint and separate actions of the FAO, most recently through its Commission on Plant Genetic Resources, the IBPGR, and the Keystone germplasm conferences. Questions of germplasm ownership and the distribution of support for national and international efforts have complicated efforts toward more cooperation. Another issue is the degree of authority that an international organization can exercise over the collections of cooperating nations.

International agreements and national policies are major hurdles for developing strong global cooperation. While the debate goes on about legal and effective ownership, genetic erosion continues to take place inside and outside germplasm banks and nature preserves, notably in the developing world, which harbors most of the world's genetic resources. It is in the interest of all that these vital natural resources are adequately conserved and made accessible.

LOOKING TO THE FUTURE: RECOMMENDATIONS

Genetic diversity provides the raw materials for the technology of plant breeding and crop improvement. It is also one of the first lines of defense against serious crop loss in agriculture. Much of the germplasm used to develop today's crop plants would have been lost forever had it not been for the collection efforts of the past, safekeeping of germplasm in public facilities, and availability to any investigator.

The following paragraphs include the committee's principal recommendations. More details and additional recommendations are given in the chapters of this report.

Managing Crop Vulnerability

Crop vulnerability is a measure of the susceptibility of a crop species to catastrophic decline in its production. Although losses often result from drought, severe storms, floods, or other environmental stresses, they can also be greatly aggravated by biologic stress, such as insect infestation or pathogen infection. Genetic vulnerability results when an organism lacks genes that enable it to withstand an environmental or biological stress or when its genes confer susceptibility to a pest, pathogen, or natural stress.

The peasant farms of traditional agriculture are less vulnerable to catastrophic loss because they grow a wide variety of plants under diversified farming. Many of these plants are landraces grown from seed passed down from generation to generation and selected over the years to produce desired production characteristics. Landraces are genetically more heterogeneous than modern cultivars and can offer a variety of defenses against vulnerability. By contrast, a pest or pathogen has a much less difficult barrier to breech when it encounters a genetically uniform modern cultivar grown under continuous monoculture over wide areas. Consequently, today an entire crop could be attacked and seriously damaged or even destroyed.

Crop vulnerability is not a problem that once addressed is no longer a concern. Rice varieties released by IRRI illustrate the potentially transient nature of breeding success. In 1966, IRRI released the first of its new, short stature, high-yielding varieties, IR-8. Only 3 years later, in 1969, a virus epidemic seriously reduced yields and a newer variety, IR-20, was released. This fell prey to yet another virus and an insect, the brown planthopper (*Nilaparvata lugens*). It was replaced in 1973 by IR-26. A new genetic variant (biotype) of the brown planthopper that could successfully attack IR-26 subsequently arose and so IR-36—resistant to this new biotype—was released. Rice production in Asia once again rose dramatically. Notwithstanding, yet another biotype of the insect necessitated the release of IR-56 by 1980. And so it goes. Each new release is the result of years of research and breeding.

In the early 1970s, a major epidemic of southern corn leaf blight occurred in the United States and prompted a report of the National Academy of Sciences. This report recognized the potential for the rapid spread of pests or pathogens through the highly uniform, genetically vulnerable fields of modern U.S. agriculture (National Research Council, 1972). Since that time many highly developed crop varieties and livestock species have spread around the world. Despite remarkable increases in agricultural production and the growth of programs for collecting, managing, and using genetic resources, significant tasks still confront the world community. For example, researchers need to assess the genetic vulnerability of currently grown crop varieties.

Diversity in wheat and maize has increased in the United States since 1970, but diversity in rice, wheat, and maize in developing countries has decreased. Landraces are being replaced by modern high-yielding varieties. Breeding and extension programs in many developing countries are not prepared to react rapidly to major epidemics. Thus, when yields of varieties decline, farmers may find that replacements are not available or transient in durability.

The challenge to scientists and farmers alike is to be able to respond to the continuously changing landscape of modern agriculture. To do so requires a long-range effort that constantly seeks new genetic resources and incorporates new genes or cytoplasms for developing future livestock breeds and crop varieties. Modern biotechnology promises even greater precision to scientists in manipulating crop and livestock genomes. However, the need for germplasm collections as resources for genetic diversity will continue to be of crucial importance for plant improvement.

The potential for crop vulnerability must be nationally and globally monitored.

International centers that distribute breeding materials to national programs may disseminate germplasm possessing similar genetic profiles to many scientists. In Asia, many rice varieties are tailored to local conditions, but a large majority share a common semidwarfing and maternal parentage. Thus the potential exists for all of them to possess genes that might confer susceptibility to an as-yet unimportant pest or pathogen.

The extent of food crop vulnerability can most easily be estimated by correlating data on the acreage of major cultivars, the races of pests and pathogens, and the extent of genetic uniformity for resistance or susceptibility. However, reductions in public funding for agricultural research have severely diminished capabilities for monitoring crop vulnerability, even in the developed world. As a consequence important survey information is no longer gathered.

A wide range of genetic and agronomic strategies should be employed to minimize crop uniformity and consequent susceptibility.

Genetic strategies that enhance stability include the planting of multilines, varietal mixtures, and relay plantings of distinctly different genotypes. A multiline is a mixture of lines with different genes for pest or disease resistance that have a common varietal or genetic background. A varietal mixture contains several different varieties, usually with dissimilar genetic backgrounds, that are selected to ripen for harvest at the same time. They may differ not only in pest and disease resistance but also in their tolerance of extremes of temperature and humidity. Relay plantings impose deliberate local or regional diversity of varieties, both in space and time, to avoid widespread genetic uniformity. Thus adjacent fields, or regions, are planted to varieties with different resistance genes. Spring and winter forms of the same crop are chosen to avoid using the same resistance genes in order to prevent the development of pathogen races that survive on one crop and attack the following one (the green bridge effect). These strategies require a diversity of genetic resources for their implementation.

Modern agricultural practices include monoculture in combination with fertilizers and herbicides, and often with fungicides and insecticides, to achieve high yields. The stability of traditional farming systems is due in part to the mix of crops they contain, cropping patterns, and times at which crops are planted. In recent years, there has been a growing interest in agricultural systems that mimic natural ecosystems. Alternative agricultural strategies, such as crop rotation, modified tillage practices, and integrated pest management can help provide stability of production over time.

Management and Use

Agricultural production has increased dramatically since the 1960s. India, a country many in the 1960s advised was beyond help, not only produces food sufficient for its population, but became a net exporter of grain. An essential element of such successes was the development of new, high-yielding crop varieties that enabled dramatic increases in agricultural production. Central to the development of these and other high-yielding varieties are breeding programs supplied with broad arrays of genetic diversity. The magnitude and importance of these programs, coupled with the globally increasing need for food, underlines the urgent need for effective conservation, management, and use of the world's genetic diversity.

Unique, endangered wild populations that have present or potential value as crop genetic resources should be conserved in situ.

In situ conservation preserves evolutionary and adaptive processes. While many calls have been made for increased in situ conservation, few efforts are specifically directed toward crop genetic resources. Conservation of the maize relative, teosinte, in Mexico and a study of in situ genetic relationships in a wheat relative, *Triticum dicoccoides*, in Israel are notable exceptions.

The capacity to regenerate seed accessions when needed should be a major determinant for limiting the size of germplasm collections.

The timing and extent of regeneration efforts may be determined more by the availability of resources than by the need to preserve genetic integrity. If not performed with adequate scientific oversight, regeneration can lead to changes in the genetic makeup of the accession (genetic drift), contamination, or loss. These effects are compounded over successive cycles. The need to minimize the number of times accessions in collections are regenerated has prompted the use of techniques such as long-term cryogenic storage. Unfortunately, the capacity of collections to support genetically sound regeneration of their holdings is rarely considered when they are assembled.

Core subsets should be identified for large collections to aid management, evaluation, and use of accessions.

The aim of identifying a core subset is to facilitate collection use. A breeder may begin a search for a particular genetic trait, such as disease resistance, by first examining and testing the core subset. If the trait desired is found in a part of the core, then other accessions from the same ecogeographic region or with similar characteristics could be examined.

Core subsets also have advantages for curators. Accessions in the core subset would receive priority in evaluation and characterization, so that in time many more traits would be evaluated for all of the core samples than would be evaluated for other samples. In this way, curators could better use a limited budget to promote the distribution of information and material and, hence, facilitate use of the entire collection.

The principal disadvantage of the core concept is the possibility—or indeed the probability—that the remainder of the base collection would erode away and disappear from neglect or, alternatively, would be seen by some administrators as being of less value and therefore dispensable in the interests of economy. Under the core concept, the entire base collection serves at least two important functions: (1) as a back-up collection to be screened if needed variation is not found in the core subset, and (2) as a source of additional diversity when many different genes are required for the same trait, as in breeding strategies for disease and pest resistance.

Several practical problems must be overcome before the core concept can be fully implemented. One is establishing the appropriate size of a core subset and defining its composition so that it contains a representative sample. Although they should be selected to represent the diversity of the main collection, core subsets are unlikely to capture all of the genetic diversity within all accessions. It is probable that if 10 percent of the accessions in a collection are assigned to a subset, no more than 70 percent of all the available alleles will be included (National Research Council, 1991a). Thus, the rest of the collection will continue to serve as a source of diversity similar to that in the core, but which results from different genes or alleles. This larger portion may also serve as a resource for genetic traits not captured in the core subset.

Core collections can be used to develop priorities for evaluation and acquisition, but without the construction or establishment of separate collection facilities. The identification of cores does, however, require a basic amount of information about accessions. Minimally this should be the passport data regarding the origins and environment of the accession. Such information is essential to linking core subsets to the potentially wider genetic diversity of the entire collection. A first step in identifying cores, therefore, is the recording of basic passport information about each accession in a collection.

Redundancies within global collections should be minimized.

The costs of eliminating duplicates within a collection usually exceeds the benefits derived from such an effort. Characterization technologies, such as isozyme or restriction fragment length polymorphism analysis, are still too costly for routine germplasm bank use. For many collections, the gains in efficiency from a reduction in duplicate accessions would be modest.

Greater gains in efficiency could be achieved by reducing duplication between collections to that necessary for insurance against catastrophic loss and allow a reasonable degree of accessibility. This is difficult to achieve, however, because it requires a high level of cooperation and coordination among national and international germplasm banks. For countries with economic dependence on particular crops, germplasm collections have strategic importance, and large collections are seen as protection against an uncertain future. This insurance is especially relevant in light of controversies surrounding ownership and control of crop genetic resources.

The limited scope for reducing collection size by eliminating redundancy within and among germplasm banks has led to proposals for improving the management, characterization, evaluation, and use of collections. Use of the new tools from biotechnology and information management will help to minimize duplication. Many problems of duplication would be solved if passport data were adequately obtained and maintained.

Programs of genetic enhancement should be developed to make diverse germplasm resources useful to crop breeders.

Genes are the building blocks of crop and livestock development. Genetic alterations to wheat and rice have improved harvest indices by creating shorter plants that devote more of their photosynthetic products to grain production. A particular gene may make a plant resistant to a particular pest or disease. Breeders use such plants as sources of traits, such as disease resistance, for the crops they study. Germplasm resources also enable breeders to introduce substantial amounts of new genetic information from untapped sources through recombination.

Within the spectrum of variants comprising the genetic diversity of a crop are accessions that may contain a useful genetic trait, such as resistance to a particular disease or insect. However, such accessions are typically unimproved landraces or wild species that exhibit lower yields and poorer quality along with traits of interest. Breeders may be reluctant to use such an accession in breeding programs because it will require extensive backcrossing to more desirable cultivated forms, and such crosses may only be achieved with difficulty. Thus, prebreeding, or germplasm enhancement, moves desired traits into a more suitable genetic background and is frequently essential to using genetic diversity.

For example, resistance to eyespot disease (*Pseudocercosporella herpotrichoides*) in wheat is available from wild goat grass (*Aegilops ventricosa*). However, this species cannot be crossed directly with cultivated wheat (*Triticum aestivum*) but must be crossed with another species compatible with both. The process is time consuming and difficult. Also goat grass seed heads shatter and scatter their seeds as soon as they are ripe. In such cases prebreeding is essential to place the resistance gene in a manageable and useful background.

Germplasm collections must at a minimum make available passport information for the materials they hold.

Collections must be well documented to allow rational and efficient searches for accessions with specific genetic traits. Unfortunately, even passport information, which includes the location and general environment where the sample originated, is often unavailable. This basic information would allow breeders to identify those accessions most likely to possess selected genes. Characterization data as well as passport information must receive greater emphasis from plant explorers, germplasm bank managers, and curators. Conversely, plant breeders can make an important contribution by supplying evaluation data to the accession records.

Technology Development and Research

Research has led to technologies for the long-term storage of germplasm, as well as for its rapid and efficient use. In situ conservation, a method widely recognized as valuable for wild species, could benefit from better understanding of the biological and ecological processes affecting populations. Several important crops cannot be held in long-term storage because of their incompatibility with available methods. Finally, the emerging molecular technologies promise to provide powerful new tools for managing, using, and exchanging germplasm.

Research is needed to elucidate the components for establishing viable and genetically diverse populations of wild species.

In situ conservation areas must preserve the genetic structure of the target species. This preservation requires consideration of the extent of genetic variation within the species that is present in an area. Multiple sites may be required to capture genetic variation associated with adaptation to different soil types, humidity, exposure, and so on. However, effective in situ conservation of this genetic diversity is only possible when there is information on the genetic structures of populations and species. Populations conserved in situ should also be large enough to be self-regenerating and to minimize loss of rare alleles. Since the costs of in situ conservation of large populations can be high, some international cost-sharing may be required.

Research is needed to apply in vitro culture and cryogenic storage methods to a broad range of plant and animal germplasm.

In vitro tissue cultures could be used to preserve species that cannot easily be kept as seeds. For some species, especially those that are propagated as clones, this technique may be more efficient and less expensive than maintaining whole plants in field culture. Sterile cultures also are protected from the diseases and pests that threaten field-grown plants. They also can be useful for exchanging and distributing some clonally propagated germplasm.

The storage of seeds or tissue cultures at cryogenic temperatures in liquid nitrogen could preserve plant germplasm for far more than 25 years. The method is not as well developed for plants as it is for semen and embryos of some livestock species. In general, small seeds that are known to withstand freezing temperatures and long-term storage (orthodox seed) can be maintained in liquid nitrogen.

Biotechnology research efforts should focus on developing enhanced methods for characterizing, managing, and using genetic resources.

Rather than obviating the need for germplasm collections, biotechnological innovations heighten their utility. For the molecular biologist, as well as the breeder, these collections are the raw material from which to draw diverse new genes. The rapid development of DNA sequence data banks, plasmid libraries, and cloned DNA fragments have in turn created a new genetic resource of growing importance. Managers and researchers urgently need more effective data handling systems for storage, retrieval, and sequence comparisons; some may be by-products of the considerable investment in sequencing the human genome. But with these increased capabilities and resources will come an even greater need for well-documented and managed germplasm collections.

As biotechnology becomes more affordable, it could improve access to collections by providing tools for characterizing their accessions. Information on the genetic similarity between accessions could aid in developing priorities for acquisition. Molecular techniques for characterizing genetic material, such as restriction fragment length polymorphism analysis, appear likely to provide the breeder with greater efficiency in selecting and developing new breeding lines and varieties. As a result, breeders may become more willing to use germplasm resources and hence to make greater demands on germplasm resource systems.

Policy and Politics

Awareness is growing among scientists, policymakers, and international organizations that germplasm is an important natural resource. In an international environment with conflicting economic systems and a skewed distribution of wealth and control over agricultural production and markets, the problems of germplasm conservation and management extend beyond purely technical issues. Plant and animal breeders, molecular biologists, and microbiologists must consider the broader social and economic ramifications of their work. If they fail to do so and isolate themselves in their laboratories in pursuit of research, the future of germplasm exchange will be determined by politicians, lawyers, and economists.

Quarantine should not be used to promote economic or political objectives.

The legal foundation that supports national quarantine regulations and actions is usually either legislation passed by national governments as acts, statutes, orders, decrees, or directives or enabling legislation that authorizes a minister or secretary of agriculture to issue regulations. Some countries are also bound by quarantine regulations that are promulgated by regional parliaments, such as the

European Community or Andean Pact. Most international plant quarantine activities take place under the legal umbrella of the International Plant Protection Convention of 1951.

Quarantine policies and practices should be biologically based
and in accordance with known or potential risk that a hazardous pest
or pathogen will gain entry through imported germplasm. Quarantine is essential for protecting a nation's agricultural system. However, regulations and practices must continually balance the potential
for release of harmful pests or pathogens with the needs of germplasm
scientists, research efforts, and breeding programs.

*Intellectual property protection systems should be designed to minimize
the potential for restricting the free exchange of germplasm among nations.*

Plant variety protection continues to be a controversial issue in
the debates surrounding the free international flow of germplasm.
Other forms of proprietary rights are likely to become equally controversial as plants and animals as well as individual genes are given
protection under the regular patent system. International political
concern has been expressed over the growth of proprietary rights in
the biological area, and particularly over the possibility that such
rights benefit the developed world at the expense of the developing
world.

In both the public and the private sectors, the increased value of
germplasm (a value derived from commercial opportunities created
through biotechnology) is leading to hesitation in sharing the material. The patent system plays a mixed role in this hesitation. In
general, although the patent system appears to assist in increasing
the rate at which technology advances, it sometimes makes it more
difficult to use or commercialize any specific technological item (including new plant varieties), especially those developed privately.

Germplasm material held in collections, especially that originat-
ing in other nations, must continue to be broadly available. Intellectual property rights must allow protected materials to be used for
research and breeding purposes.

*Public institutions should not respond to the commercialization of germplasm
by enacting restrictions that could limit the use of genetic resources by
developing nations.*

For many plant species, advanced traditional breeding is conducted primarily in the public sector, for example, in land-grant universities in the United States, in the more advanced research universities and publicly funded research centers in developed nations, and
in a variety of international research institutions, such as those oper-

ated by CGIAR. Although these institutions sometimes obtain legal protection for their working materials and research products, they have made such materials and products freely available throughout the world. However, the public sector has more recently used confidentiality or patents to support research or as a defensive step to prevent ideas or material from being patented or misappropriated by others. Public institutions should provide royalty-free licenses for the benefit of developing nations. In this way, they could use the patent system to assure continued broad availability of germplasm resources.

Disagreements over the ownership, control, and use of genetic resources must be resolved.

Although political will and cooperative relations among scientists are important to germplasm exchange, funding and technical resources for collection and management are major influences on distribution. Barriers to free exchange remain because of the lack of funding and staff to maintain and distribute materials on request. Governments and institutions that host international collections must be committed to maintaining and distributing them.

Government officials, agricultural scientists, and germplasm conservation personnel in developed and developing countries must recognize the necessity of cooperation in genetic resources conservation, exchange, and use. Cooperation is needed between the developing countries, which serve as centers of diversity, locations for rejuvenation of seed, and sites for in situ conservation, and the developed countries, which serve as sources of funding, preservation and distribution sites, and training centers. International cooperation and exchange of germplasm can be substantially strengthened by agencies, such as FAO, IBPGR, and the international agricultural research centers. Cooperation cannot be secured by statutory means alone, however. It must begin with scientists, environmentalists, and other interested parties who share a common concern for the world's genetic resources.

Institutions

As the debates on freedom of access, ownership, and international accountability continue, genetic erosion also continues. In the meantime a practical and effective worldwide system of genetic conservation must be established. This will depend on strong national participation, particularly in the developing world, and the provision of stable, long-term, financial support.

The capacities for plant breeding, seed production, and biotechnology in developing countries should be strengthened. Plant breeding programs should have regional or international sources of reliable financial and technical support. Innovative methods of improving seed production and distribution in developing countries should be encouraged.

The governments of developing nations could strengthen biotechnology sectors by establishing appropriate economic incentives, regulatory requirements, and institutions. Public sector and extension activities should be structured to ensure that subsistence farmers participate in the benefits provided by these technologies. Accomplishing these goals will require bilateral and multilateral assistance in the form of training, facilities, and operating support.

Until recently, it has been relatively easy politically to conduct a large-scale international public-sector agricultural technology development program. That type of program, which was responsible for the so-called green revolution, was often quite successful. Today, multilateral programs are often the objects of political criticism. Agricultural technology is also increasingly being developed in the private sector rather than in the public sector. These transitions underlie the germplasm controversy; they reflect the difficulty of ensuring that private-sector agricultural biotechnology, including breeding and seed production, provides expanded opportunities for developing countries.

International responsibility for conserving, managing, and using genetic resources must be translated into a form of funding that satisfies the basic principles of the International Undertaking on Plant Genetic Resources of the Food and Agriculture Organization.

Any strategy should build on existing framework and activities, stressing national involvement and the cooperation of FAO, IBPGR, and other CGIAR centers. A major criticism of the CGIAR has been that it has no formal legal basis for action among governments. Nevertheless, several of the CGIAR institutions have been leaders in conserving plant genetic resources. The overall CGIAR program in genetic resources might therefore be modified to include some form of governance as suggested by the FAO Commission on Genetic Resources. This might be done, for example, by an agreement between the CGIAR or its individual institutions and the FAO that obligates the institutions to consult with the FAO with respect to important policy issues.

It is also important to maintain existing policies with respect to free exchange of existing germplasm collections at the national and international level; the Convention on Biodiversity leaves this material subject to free access. The convention will allow nations to place restrictions on the flow of later-collected germplasm. It will also be important to develop reasonable policies for respecting such restrictions.

Although the debate on legal and effective ownership continues, genetic erosion also continues, notably in developing countries (and in some germplasm banks), where most of the world's untapped genetic resources exist. It is in the interest of world agriculture that countries in regions with major genetic diversity be provided with the means to participate more fully in genetic resources conservation and use of the world's biological resources.

Genetic erosion is also prevalent in many germplasm banks. Sustained support of the banks and vigilance over their management are crucial to the security of germplasm collections.

An adequate and appropriate funding mechanism must be established to support national and international conservation, management, and use of genetic resources.

The world scientific and technical community has put in place the beginning of a functional network of both national and international genetic resources programs that have the potential to safeguard germplasm. Present knowledge and rapid developments in several fields related to genetic conservation, including modern biotechnologies, provide a good basis for rational and selective conservation. Adequate operational funding is a key constraint.

Much has been achieved in collection efforts, training, the establishment of germplasm banks, documentation, and research support. However, the funding available for genetic resources activities at national and international levels does not reflect the substantially increased public and political awareness of the importance of genetic resources. The CGIAR has been a major provider of funds since the early 1970s. Funds from most other international sources have been smaller and, the debates of recent years notwithstanding, have not provided the substantial support that is needed. The Rio negotiations have provided a disappointingly small additional contribution to developing-nation environmental funding, including genetic resources funding, but they have affirmed the possibility of using the Global Environmental Facility, sponsored by the United Nations Environment Program, United Nations Development Program, and World Bank. Should that approach not yield an adequate and sustainable long-term funding base, stronger approaches should be sought, such as national contributions tied to domestic seed sales or a tax on worldwide seed sales.

Multilevel Collaborations on Genetic Conservation

All nations and international agencies need to pool their limited resources and collaborate on the myriad facets of genetic conservation. Worldwide concern demands that periodic assessment and monitoring of collaborative activities be required in the future to ensure maintenance and use of genetic resources, our common biological heritage.

An Overview of Genetic Resources Management

This chapter is intended as a general introduction to some of the principal issues involved in germplasm conservation and management. It introduces the subject for those readers without a background in genetic conservation.

Germplasm consists of the genetic materials that can perpetuate a species or a population of an organism. Germplasm is thus a genetic resource that can be used both to reproduce and, through hybridization and selection, to change or enhance the organism. Conserving genetic resources is a means of safeguarding the living materials exploited by agriculture, industry, forestry, and aquaculture to provide food, feed, medicinals, fiber for clothing and furnishing, fuel for cooking and heating, and the food and industrial products of microbial activity.

Genetic resources are of tremendous practical and historical significance for human life from daily survival to generating the wealth of nations, yet their crucial role in supporting human society is frequently overlooked and undervalued. Genetic conservation is an integral part of a much broader activity concerned with protecting the many plants, animals, microorganisms, and communities of organisms that help to mold and stabilize the environment and maintain the quality of air, water, and soil. Conservation ensures that future generations will benefit from earth's biological resources.

The direct connection between genetic resources and humanity's food supply provided the motivation for this study which focuses on

29

improving the management and utility of genetic resources important to agriculture. This part of the report is intended as an introduction to the subject of genetic resources management for the nonspecialist.

The charge to the Committee on Managing Global Genetic Resources: Agricultural Imperatives in undertaking this study was to examine the effectiveness of germplasm conservation and management on a global scale and to make recommendations on how they might be improved. At the same time the committee was expected to

---

THE NATURE OF PLANT GENETIC RESOURCES

In this report the terms *variety* and *cultivar* are used interchangeably to mean cultivated forms within a species. The cultivated plant materials that are conserved are of five general kinds: landrace, folk, or primitive varieties; obsolete varieties; commercial varieties or cultivars; plant breeders' lines; and genetic stocks.

Landrace, folk, or primitive varieties are local varieties developed by indigenous farmers in traditional agricultural systems, over hundreds of years. By modern standards such varieties are often highly variable. Landrace cultivars have generally been replaced by modern scientifically bred cultivars in most crops but may still be locally important in some farming systems.

Obsolete varieties are varieties developed since the advent of scientific agriculture in the late nineteenth century and that are no longer cultivated. Although no longer grown commercially, such varieties are usually maintained in collections for use in current and future breeding programs.

Commercial varieties or cultivars are elite high-yielding lines in current use, developed by scientific plant breeding for modern intensive agriculture. The average life of modern varieties is relatively short (5 to 10 years) when they are replaced by more recent products of breeding programs.

Plant breeders' lines are as-yet unreleased lines, mutations, or parents of hybrids maintained by breeders as part of their working stocks. Breeders usually develop and carry many lines in their programs, of which only a very small number are ever released into commercial production.

Genetic stocks are genetically characterized lines of a species principally used in genetics and plant breeding research. They rarely have any commercial value but are nevertheless an important germplasm resource because of their usefulness in basic and applied research. Genetic stocks can be conveniently divided into the following three classes:

examine sociopolitical, economic, and other issues that affect germplasm conservation and the free exchange of germplasm among nations.

For the most part, this report concerns crop plant germplasm with some reference to the germplasm of agriculturally important livestock and microbes. For some readers, this focus will represent an unduly narrow and restricted view of the crisis now faced in conserving global biological or genetic diversity. Despite the truth of the observation "there are no useless plants, only plants we have not yet found a use for" (Falk, 1993), this report was limited to the staple

---

• Single gene or single trait variants are lines carrying artificially generated or naturally occurring variants for qualitative characters. Examples include differences in morphological and physiological characters, electrophoretic variation in proteins, and fragment length variation generated by restriction enzymes in DNA.

• Multi-allelic stocks include multiple gene marker stocks useful for linkage studies and stocks with special combinations of loci necessary for the expression of a single trait.

• Cytogenetic stocks include all variants whose chromosome structures differ from the norm because of the deletion, duplication, inversion, or translocation from one chromosome to another of specific DNA segments. These stocks are important in determining the chromosomal location of specific genes.

Genetic stock collections are the province of the specialist because they require skill and experience in regeneration to ensure that they remain viable and true to type.

Wild and weedy relatives are species that share common ancestry with crops, but that were not domesticated. Most crops have wild or weedy relatives which differ in their degree of relationship to the crop. The ease with which genes can be transferred from them to the crop varies. They may be classified into the following three gene pools:

• A primary gene pool consisting of the cultivated species and wild relatives, which are readily intercrossed so that gene transfer is relatively simple.

• A secondary gene pool composed of all the biological species that can be crossed with the crop but where hybrids are usually sterile. Gene transfer is difficult but not impossible.

• A tertiary gene pool made up of those species that can be crossed with the crop only with difficulty and where gene transfer is usually only possible with radical techniques. Biotechnology has, at least in theory, greatly enlarged this pool because transformation (a radical technique) makes possible the introduction of DNA from any species.

food crops and their wild relatives. A recent study concluded that 103 species provide about 90 percent of the world's food plant needs (Prescott-Allen and Prescott-Allen, 1990). The conservation of the genetic resources of these species is vital to agriculture and humanity. However, for long-term survival of the world's ecosystems, there is no substantial difference in methods used to protect taxa of proven worth and those that are used to protect others that may be unexplored and in danger of extinction in coming decades (Heywood, 1993). Simply put, cultivars have been selected for traits that fit a particular agronomic technology; wild species undergo natural selection to acquire traits that maintain their ability to survive in an ecosystem.

THE IMPORTANCE OF GERMPLASM

Why is it necessary or desirable to preserve materials collected from undeveloped agricultures or the wild? The obsolete crop varieties and livestock breeds and the unadapted wild relatives of crop plants frequently carry useful genes or alleles that, if not preserved, may no longer be available in the future. Modern, high-yielding crop varieties have largely replaced older landrace varieties even in many remote parts of the world. For many crops sown as seed, rather than as roots or tubers, few of the varieties that were widely grown 40 to 60 years or more ago can still be found in areas of commercial agricultural production. In some cases people have destroyed or altered many natural habitats of wild relatives of crop plants so that it is no longer possible to collect them at the original sites. Even if collection is still possible at undisturbed sites, the range of genetic variation present in the original collections may no longer be available. Before they can be used by breeders, the collected materials must be evaluated, tested, catalogued and properly conserved. They can then be made available to others who may wish to use them.

There is widespread concern about the status of conserved germplasm worldwide. Is enough being done? Are the materials already conserved in germplasm banks and other facilities adequately described, documented, properly managed and stored under conditions to safeguard their viability, and freely available to anyone who needs them? Do they include enough of the potentially important germplasm that can still be collected now but that may not be available much longer? Are sufficient financial and other resources being applied by national governments to their own and to global needs? What priorities have been established and are they correct?

THE ORIGINS OF CONSERVATION

Humans first began conserving genetic resources of exploited species when their primitive ancestors saved part of their harvest of gathered seeds, roots and tubers of plants, and individual or small groups of animals. The selected materials were not used for food but were kept for planting or herd increase when conditions improved or were carried from place to place during human migrations. In setting aside the better forms for future use, humanity began the long process of selection and improvement responsible for the development of agriculture. The first crop plants and livestock were undomesticated wild species that gave rise, thousands of years later, to modern varieties and breeds.

Among the first plants to be domesticated were several wild species that had been most useful to hunter-gatherers as energy food plants, probably annual, large-seeded grasses; legumes; and starchy tubers. In Southwest Asia these were barley, tetraploid wheat, peas and lentils; rice and soybeans in Southeast Asia; sorghum in Africa; and corn, common beans, and lima beans in South and Central America. This evidence also suggests that the domestication of cereals and legumes began about 10,000 years ago and that the domestication of each species was not a single event but a continuing process at many ecologically different places over many centuries.

The two main evolutionary processes during the earliest stages of domestication were almost certainly: (1) identification of the wild species that were most useful to cultivators, and (2) selection from within those species of the particular individuals that were best suited to the conditions of cultivation. As agriculture expanded and increasing numbers of plants were grown under cultivation, there was a corresponding increase in the number of novel mutations observed, and those favored under agricultural conditions were, as noted by Darwin (1859), preserved by selection, greatly increasing the accumulated store of economically useful variability.

By the dawn of recorded history nearly all of the world's present major crop species had been cultivated for several thousand years. The earliest written records show that major evolutionary changes had already taken place. For example, the many cultivars of wheats and pulses described by Theophrastus, who lived from 372 to 287 B.C.E., were all of nonshattering habit, in contrast to their wild ancestors. The apples he described were very different from those of their small-fruited, acidic, and astringent ancestors, which are still found growing wild in Southwest Asia.

The cultivars described by Theophrastus and his contemporaries provided the basic germplasm that was transported along ancient trade routes over the European, African, and Asian land masses and that continued to evolve under different natural and human selection pressures. Rates of evolutionary change accelerated in the eighteenth and nineteenth centuries. Seed growing developed as a business, and competition provided an incentive for selecting and rapidly distributing distinctive types of field and vegetable crops and ornamentals. The work of Knight (1799), who was one of the first to observe differences in disease resistance among wheat hybrids, stimulated the deliberate crossing of different types, promoting genetic recombination and the production of rich arrays of segregants from which to select.

THE USEFULNESS OF GERMPLASM

The practical development of genetics at the beginning of this century gave breeders an explanation for the mechanism of inheritance. Obsolete varieties and breeds and closely related wild species were sources of useful variation that could be introduced into new forms by hybridization. Breeders assembled collections of useful and new materials that were described, cataloged, and tested and that

---

THE U.S. GERMPLASM SYSTEM

In the United States, a nation almost wholly dependent on crop plants and livestock that were not native to North America, the elements of a national germplasm program evolved from early requests for U.S. travelers abroad to send back seeds or plants of promising trees or crops. From 1836 until 1862, when the U.S. Department of Agriculture (USDA) was established, the U.S. Patent Office distributed seeds and plants from overseas to U.S. farmers. In 1898 a Seed and Plant Introduction Section of USDA began to promote the collection and introduction of new crops. However, it became necessary to introduce plants and animals in such a way as to minimize the risk of bringing in pests and diseases at the same time. In the United States the first plant quarantines were the initiative of individual states. California, in 1881, was the first to pass an act to prevent distribution of the grape gall louse. Federal plant quarantine regulations were not adopted until 1912, although drastic legislation had been accepted in Europe and Australia by 1877 to restrict the introduction of the same grape pest from the United States.

Plant materials collected by the USDA that entered the United States were given numbers to create an inventory of the plant introductions.

could be saved from year to year. These collections were the first to be systematically conserved, although often only during the working lifetime of the breeder.

The emergence of plant and animal breeding programs during the first half of this century created a demand for germplasm exchange among breeders and for collecting expeditions and explorations to satisfy the growing need for such crop plant characters as earliness, stiff stalks, grain quality and resistance to diseases and insects. Foremost among the plant explorers was the Russian scientist, N. I. Vavilov whose work in the 1920s defined the centers of origin (now known as centers of diversity) for many important crop plants. In 1935 Vavilov (1951:53) wrote the following:

> We have barely begun the systematic study of the world's plant resources and have discovered enormous untouched resources, unknown to scientific breeders in the past. The tremendous potential source of species and varieties requires thorough investigation employing all the newest methods. The problem of the immediate future is a classification of the enormous diversity of the most important cultivated crops not only on the basis of their botanical and agronomic characters but also with the use of physiological, biochemical, and technological methods.

Catalogues of the properties of accessions, listed by their plant inventory numbers, were useful to breeders who could obtain seeds or cuttings from the U.S. regional center responsible for the evaluation and regeneration of the species concerned. In this way the U.S. National Plant Germplasm System evolved around a base collection, established in the 1950s at the National Seed Storage Laboratory in Fort Collins, Colorado (James, 1972).

In a report of the National Research Council (1991a), the present structure of the U.S. National Plant Germplasm System is analyzed, and recommendations are offered to make it a more effective national organization, rather than the collection of loosely coordinated regional and sectional programs that it is today.

Although germplasm conservation would appear to be designed to serve the interests of plant breeders, most modern breeders are interested primarily in highly adapted breeding lines that can be directly introduced into their crossing programs. Poorly adapted materials are less useful to them but can be a source of variation to remedy the limitations of their better lines. Breeders are rarely interested in using wild relatives directly because of the effort and expense of eliminating their many undesirable traits. Biotechnology may well change this by making it easier to transfer only the gene or genes of interest.

At first much of the plant material collected by the early explorers was lost because it was not properly stored under conditions of low temperature and low humidity and free from vermin, or it was mixed or contaminated when regenerated. Much of it was not adequately described or catalogued. However, the early expeditions laid the groundwork for many important collections in the United States and elsewhere.

THE THREAT OF GENETIC VULNERABILITY

As progress in selecting successful crop varieties continued, some scientists (Harlan and Martini, 1936; Sauer, 1938) expressed concern about increasing genetic uniformity and the loss of genetic variation associated with the disappearance of natural habitats of wild relatives. Although earlier plant disease epidemics, such as the Irish potato famine in 1845 and 1846, had caused widespread suffering and death, the impact of genetic vulnerability on modern agriculture was dramatically exemplified by the southern corn leaf blight epidemic of 1970, which destroyed more than 15 percent of the U.S. corn crop (National Research Council, 1972). The loss resulted from the widespread use of a form of cytoplasmic male sterility that simplified the production of hybrid seed. Hybrids with this cytoplasm accounted for more than 70 percent of the corn varieties grown in the United States at the time. They were very susceptible to a new race of the fungal pathogen (*Bipolaris maydis*) first described in the Philippines in 1963.

The 1972 report of the National Academy of Sciences questioned the extent to which corn and other food crops are vulnerable to similar unanticipated epidemics. It pointed to the modern tendency of farmers to use a few high-yielding varieties, replacing the many varieties of earlier times, and emphasized the risks implicit in the relative homogeneity of the leading varieties of the major crops.

The vineyards of Napa and Sonoma counties in California have been invaded by a new biotype of phylloxera—the aphid relative that attacks grape roots and devastated the European wine industry in the nineteenth century (4 million acres destroyed in France alone). Because 70 percent of the wine grapes of this district are grafted on a susceptible rootstock, they are seriously threatened by this pathogen, and vines have already been removed from certain vineyards. Spread of this pest into other grape districts is considered likely (Granett et al., 1991). Since the performance of this rootstock has been satisfactory for many years, research lagged on the breeding and ecology of rootstocks. Even though investigations have now been initiated, much time will be required before resistant stocks have been generated, tested, and used for replanting. This serious problem constitutes another example of the vulnerability of planting large areas in a monoculture fashion. The risk in this instance is much greater that for annual seed-propagated crops, because more time is required for researching and replanting woody crops.

EMERGENCE OF GLOBAL CONCERNS

Other nations had recognized the importance of plant germplasm and collections were established in Brazil, Colombia, the Commonwealth of Independent States, France, Germany, Japan, Mexico, the Netherlands, Scandinavia, and the United Kingdom. In the 1950s, germplasm conservation gained more visibility in the international community. In 1961 in Rome, the Food and Agriculture Organization (FAO) of the United Nations sponsored the first of a series of technical conferences (Whyte and Julén, 1963). A second technical conference, on plant exploration and genetic conservation, was planned and held jointly with the International Biological Program (IBP) in 1967, as was the third, held in 1973, both at FAO in Rome. These two conferences highlighted and developed the scientific issues involved in all phases of work related to genetic resources—collection, evaluation, conservation, documentation, and use. The tangible results were two books on genetic resources (Frankel and Bennett, 1970; Frankel and Hawkes, 1975a), which for some years served as reference sources on science and technology relating to genetic resources. In 1965, FAO had established the Panel of Experts on Plant Exploration and Introduction. With added IBP representatives, this panel had a major part in planning the technical conferences and in the discussions that led to the formation of the International Board for Plant Genetic Resources (IBPGR).

Following World War II and the establishment of the World Bank to aid the economic development of poorer nations, a network of international agricultural research centers (IARCs) was established in the 1960s and 1970s. The centers were designed to carry out scientific research and development to combat hunger in the developing world. They provide improved germplasm and carry out agronomic research to help developing countries become self-sufficient and improve their agricultural exports to world markets. The IARCs are administered by the Consultative Group on International Agricultural Research (CGIAR), which represents the principal donors that fund its activities. Warren Baum (1986) has described the establishment and operation of this international system during its first 15 years. The Centro Internacional de Mejoramiento de Maíz y Trigo (CIMMYT, International Maize and Wheat Improvement Center) and the International Rice Research Institute (IRRI) were the first two centers, begun in Mexico and the Philippines with support from the Rockefeller and Ford Foundations. CIMMYT conducts research on maize and wheat, and IRRI focuses on rice. Both centers have seed banks. The IRRI seed bank, with more than 80,000 rice accessions, is

GERMPLASM CONSERVATION AND BIOTECHNOLOGY

Much of the germplasm conserved today is kept as seeds in germplasm banks, planted in nurseries and gardens, or kept in refrigerators as planting stock or tissue cultures. These materials are conserved ex situ, that is, away from their natural place of origin or where they were originally selected. In situ conservation is designed to maintain viable populations in the places where they occur naturally. Where practical, in situ conservation can be cheaper and less laborious than ex situ conservation. However, once evaluation and utilization are undertaken, the materials involved usually move into an ex situ conservation mode.

Well-established principles and standards have been developed for collecting germplasm for ex situ conservation; they cover sample size and number, record keeping at the collection site, and sampling procedures. Once established, a germplasm bank should be properly managed by following tested guidelines for cataloging, evaluating, regenerating, and distributing the accessions. Documentation and data management are essential if the contents of the germplasm bank are to be known and readily accessible to users.

Rapid advances in understanding the biochemical and molecular basis of inheritance and development are having profound effects on the life sciences. What contributions are these and other developments in biotechnology likely to make to germplasm conservation? Plant tissue culture and in vitro propagation have already changed how the germplasm of some clonally propagated crops is conserved and distributed. Plant transformation has, in theory, made the gene pool that breeders may draw on limitless.

Biotechnology has introduced an information explosion at the molecular level. The rapid accumulation of DNA sequence information in computer data bases is itself an immensely valuable, although nonliving, genetic resource because it is the key both to describing genes and to synthesizing them from their component nucleotides. A sensitive method for recovering specific stretches of DNA sequence from dead the largest for any single crop, with almost complete representation, excellent documentation, and large-scale seed distribution.

Of the 17 IARCs, 12 of them (Centro Internacional de Agricultura Tropical [CIAT], Centro Internacional de la Papa [CIP, International Potato Center], CIMMYT, IBPGR, International Center for Agricultural Research in the Dry Areas [ICARDA], International Center for Research in Agroforestry [ICRAF], International Crops Research In-

---

materials such as herbarium specimens, inviable seeds, and fossils suggests that these materials may become useful genetic resources. Developments in biotechnology may increase the efficiency of germplasm conservation and use.

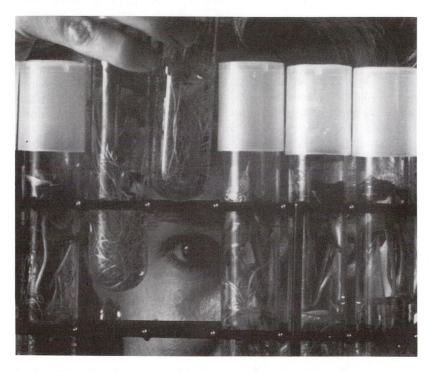

Many plants can be conserved by tissue culture storage. However, some plants quickly outgrow their test tubes or must be recultured every 90 days. Here a technician of the Agricultural Research Service checks sweet potato plants in experiments to slow their growth and stretch the time between regenerations. Credit: U.S. Department of Agriculture, Agricultural Research Service.

stitute for the Semi-Arid Tropics [ICRISAT], International Institute for Tropical Agriculture [IITA], International Livestock Center for Africa [ILCA], International Network for the Improvement of Banana and Plantain [INIBAP], IRRI, and West Africa Rice Development Association [WARDA]) are involved in conserving and using major global crops and forages. A recent estimate is that together they maintain some 600,000 germplasm samples (Alexander von der Osten, personal communication, CGIAR Secretariat, October 28, 1992). Hawkes (1985) analyzed the IARCs' effectiveness in conserving crop plant germplasm, pointing out that some have little interest in, or ambition to, becoming major international germplasm repositories.

The importance of germplasm to the developing world was emphasized in 1972 in a proposal prepared by FAO at the request of the Technical Advisory Committee (TAC) of the CGIAR. This document proposed the establishment of the IBPGR to encourage, coordinate, and support action to conserve genetic resources and make them available for use. After much debate about its relationship to FAO and the CGIAR, the IBPGR was established in 1974 with a secretariat provided by FAO and funding from CGIAR donors (Baum, 1986). The task of IBPGR was to promote the development of international, regional, and national activities and programs to build a worldwide network of genetic resource centers (Williams, 1988). When IBPGR was founded, only six long-term germplasm banks existed, most of which were in industrial countries. By 1986 some 50 banks were operating, with more under construction (Plucknett et al., 1987). A series of collecting expeditions was mounted for crops considered to be of the highest priority. IBPGR policy has been to seek permission for collecting from the countries visited and to share its collections with them.

In 1990, CGIAR initiated a process to establish a fully independent institute for plant genetic resources that would programmatically collaborate with FAO. The IBPGR will continue to operate until the International Plant Genetic Resources Institute can assume its duties.

DISSENTING VIEWS

Although many of the plant scientists involved in germplasm conservation believed that the mechanisms being put in place for protecting global genetic resources were both sound and fair, their view was strongly challenged by sociopolitical activist groups who regarded seeds and genetic resources as a common heritage of humankind (Mooney, 1979, 1983). Critics claimed that IBPGR enabled the industrialized, "gene-poor" countries to control genetic resources that properly belonged to the "gene-rich" countries where these materials were originally collected. Their recommendations included major increases in funding to launch a global campaign to collect landraces and wild species, codes of conduct that would make trade in seeds a national issue and an inappropriate activity for international firms, and a guaranteed right of nations to protect their botanical heritage from commercial exploitation.

At the 1981 FAO biennial conference, the director general was asked to prepare an international document to ensure that global plant genetic resources of agricultural interest will be conserved and used for the benefit of all human beings, in this and future generations, without restrictive practices that limit their availability or exchange, whatever the source of such practices (Food and Agriculture Organization, 1981). Further discussions led to the formation of the FAO Commission on Plant Genetic Resources in 1983 and the adoption of an International Undertaking on Plant Genetic Resources (Food and Agriculture Organization, 1983a), which attempted to right some of the perceived wrongs suffered by developing countries. These discussions came to be known as "the North-South debate" because they mainly featured the industrialized nations of the Northern Hemisphere and the developing countries of the Southern Hemisphere. The industrialized nations sought to protect and capitalize on their investments in crop plant improvement, while the developing countries supplied the raw germplasm material but felt exploited when it was sold back to them, albeit in an improved form. Similar discussions have taken place among developing countries, some of which have placed restrictions on germplasm exchange among themselves, giving rise to the so-called "South-South" debate.

In 1981, FAO, in cooperation with the United Nations Environment Program and the IBPGR, organized a technical conference on crop genetic resources that explored some of the successes and failures of the IBPGR global program for the conservation of genetic resources. Participants noted several trends that continued to limit the effectiveness of programs for collecting, conserving, and using plant germplasm. These trends included slowing the collection of seed crops and placing greater emphasis on characterization, documentation, and information exchange; focusing collection efforts on meeting emergency situations and filling important gaps; accelerating work on clonal crops and research on in vitro conservation; giving high priority to data-base development; and paying more attention to wild species so that collections are more representative of gene pools. Concern was expressed for the first time about the continuing rapid growth in the number and size of germplasm banks (Holden and Williams, 1984).

Since World War II the private sector in the industrialized world has made major and increasingly important contributions to plant breeding. A variety of methods are used to protect the industry's investment in breeding and to ensure that it receives a fair return. For example, proprietary rights in hybrid corn are often protected by not releasing the parents used to make hybrid seed. The hybrid seed is sold without restrictions and produces a high-yielding, uniform crop. However, when seed of the next generation is sown it does not reproduce the hybrid because of genetic segregation and generally results in an inferior crop. Patents and plant variety protection procedures were used to protect "plant breeders' rights."

In the course of discussions in FAO commission meetings about the commercial exploitation of germplasm, "farmers' rights" emerged as a counterpart to plant breeders' rights. Widely misinterpreted as a system for financially rewarding farmers from developing countries for selecting landraces, the concept now focuses on the importance of the past, present, and future contribution of farmers in general to the development of agriculturally important germplasm. In August 1988 a meeting in Keystone, Colorado (Keystone, 1988) attempted an operational definition of farmers' rights. A second meeting in Madras, India (Keystone, 1990) considered how intellectual property rights affected the conservation and exchange of germplasm. Both meetings brought together people with diverse interests for nonconfrontational discussions. A third meeting near Oslo in the summer of 1991 called for establishing a new Global Plant Genetic Resources (PGR) initiative. First priority was given to: emergency attention to the security of PGR collections in eastern Europe and Ethiopia; developing a special PGR program for promoting sustainable advances in crop productivity; and developing a global network of conservation centers dealing with germplasm relevant to the potential challenges introduced by global climatic change.

Commentaries on germplasm conservation usually conclude that both national and international expenditures on all aspects of these activities are inadequate. There is special concern about germplasm banks established during the past few decades in developing countries with external assistance. Several proposals have suggested the development of an international fund to finance global genetic conservation. One scheme would be to collect a small part of the royalties from seed sales around the world and use it exclusively for conservation.

THE PROBLEM OF ECONOMIC ANALYSIS

Many examples exist of the value and usefulness of individual germplasm accessions to animal and plant breeders. However, it is difficult to place an accurate economic value on the material conserved in the world's germplasm collections. Much germplasm has been collected in case of need but will never be used. Also, it is difficult to predict what crop plant characteristics or features may be sought in the future. Therefore, conservation, like research, must be well organized and thorough, can be expensive, and the usefulness of any particular element is unpredictable. Information on resistance to known pests and diseases is useful, but germplasm managers cannot screen for resistance to agents that have not yet been recognized as a threat. Germplasm collections may be the only readily accessible source of resistance to new pests and diseases and other characters of interest to breeders. They are an insurance against future threats. They are also repositories of materials that can no longer be collected.

There are few detailed economic analyses of the value of germplasm in the literature. This report presents an original analysis of the contribution of rice germplasm, mostly from the IRRI germplasm bank, to breeding new rice cultivars for India. The IRRI germplasm bank is one of the largest and best organized germplasm banks. The analysis demonstrates the cost-effectiveness of conserving a broad array of rice germplasm, and it exemplifies the importance of properly managed and accessible germplasm for all crops.

MICROORGANISM COLLECTIONS

Microorganisms are an important germplasm resource. Defined here as algae, bacteria, fungi, certain protista, and viruses, they have a wide range of uses that involve the food, chemical, and pharmaceutical industries as well as agriculture. Conservation of microorganisms in situ is largely impractical because of the major effort required to isolate, characterize, evaluate, and improve cultures with particular qualities that make them valuable and worth conserving for future use. For example, plant breeders routinely use collections of different races of pathogens to differentiate disease resistance genes. These pathogen-reference collections are assembled as plant pathologists follow the evolution of new races in the field. They are crucial in defining breeders' goals for disease resistance.

Conservation of microorganisms has all of the problems associated with conserving other kinds of germplasm including maintaining genetic stability in culture, protection of intellectual property rights, data banks, and the establishment of global networks.

PART I

BASIC
SCIENCE
ISSUES

1

Genetic Vulnerability and Crop Diversity

Over the ages the tendency of crop improvement efforts has been to select varieties with traits that give the highest return, largely by concentrating on genetic strains that combine the most desirable traits. The resulting genetic homogeneity and uniformity can offer substantial advantages in both the quantity and quality of crop harvested, but this same genetic homogeneity can also reflect greater susceptibility to pests or pathogens. Thus it appears the more that agricultural selection disturbs the natural balance in favor of varietal uniformity over large areas, the more vulnerable such varieties are to losses from epidemics. The increased risks presented by genetic selection and the increased cultivation of only a few selected cultivars are easily perceived. Chapters 1 and 2 of this report focus on crop vulnerability, because it is a broadly recognized problem. The issue of genetic vulnerability, however, is only one of several important problems affecting the management of global genetic resources.

Significant efforts have been made by national and international institutions to collect and preserve crop genetic resources as an insurance policy against future disasters. The recognition, however, that the products of a few major breeding programs are now planted over entire continents under increasingly intensive conditions raises new concerns that are global in scope. These concerns prompted the committee to reassess where the world agricultural community stands today with respect to the risks of genetic vulnerability and progress in diversifying crop gene pools.

47

The concerns became apparent in 1972 with the report *Genetic Vulnerability of Major Crops* (National Research Council, 1972). It was prepared by scientists who, alerted by the 1970 epidemic of southern corn leaf blight (*Helminthosporium maydis*) in the United States, became concerned about the potential for similar outbreaks in other major crops. They sought to warn the agricultural community about trends in modern plant breeding programs that may have contributed to the crisis and recommended that the genetic foundations of major crops be diversified to reduce the risk of future outbreaks. Nearly 2 decades later, the issue of genetic vulnerability is still fresh, and debate continues on the risk that it poses.

This chapter addresses genetic vulnerability, first by defining it. The chapter discusses when vulnerability becomes a cause for concern and how the risks it presents can be measured. It examines the trends that contribute to genetic vulnerability and the strategies for reducing it. It evaluates the impact of modern plant breeding programs on genetic vulnerability, and the changes that have occurred in recent years that have increased or decreased genetic vulnerability. Finally, it assesses what remains to be done to minimize vulnerability.

WHAT IS GENETIC VULNERABILITY?

The term *genetic vulnerability* has been widely used to invoke fears of disastrous epidemics and threats to global food security, often without clarifying what is meant by the term. This chapter uses genetic vulnerability to indicate the condition that results when a crop is uniformly susceptible to a pest, pathogen, or environmental hazard as a result of its genetic constitution, thereby creating a potential for disaster.

Critical Factors and Assumptions

Two important factors interact to increase the potential for crop failure: (1) the degree of uniformity for the trait controlling susceptibility to the hazardous agent or environmental stress, and (2) the extent of culture (often monoculture) of the susceptible variety. The greater the uniformity for a susceptible trait and the more extensive the area of cultivation, the greater the risk of disaster. In the case of pest or pathogen attack, two additional factors enhance the risks: (1) a highly dispersible disease or insect agent, and (2) favorable environmental conditions for the multiplication of the agent.

Susceptibility to diseases, insects, and environmental stresses is a natural phenomenon characteristic of all plants, no matter how primitive or highly bred. Natural epidemics, droughts, or other adverse conditions may affect any landraces or cultivars grown by farmers. The extent and impact of the disaster on the food supply and regional economy may, however, be lessened in agricultural systems that incorporate a variety of crops and landraces.

There is cause for concern when extensively planted cultivars of major crops are derived from limited gene pools and, hence, are uniform for a high percentage of traits with narrowly based resistances to common pathogens or other agents. These concerns have prompted surveys of plant breeders on their perceptions of the gravity of the problem and a reevaluation of trends in international varietal development and distribution.

Although some breeders and scientists are encouraged by the wider availability of crop gene pools into which exotic plant genes have been introduced, others are worried that genetic uniformity may be increasing on a global scale because of the widespread adoption of modern varieties with similar genetic backgrounds across continents where large numbers and mixtures of landraces were formerly grown. Farmers in developing countries who intensify their cultivation practices and adopt improved varieties may increase their chances of losses from epidemics.

One of the rationales for crop germplasm conservation is that plant genes have utilitarian value. When they are effectively used they can decrease crop susceptibility to natural predators, pathogens, and stresses. Thus, genetic vulnerability can be characterized in terms of genetic resources use. Genetic vulnerability results from the improper or inadequate deployment of these genetic resources, in conjunction with biotic and abiotic stresses. It may also be more inherent in some agricultural systems than in others.

The Role of Uniformity

Genetic uniformity is desirable for many agronomic traits of concern to farmers, processors, and consumers and does not, by itself, make a crop vulnerable. In other words, uniformity and susceptibility need not be synonymous. Crops can be vulnerable even when they are genetically quite dissimilar if they have in common a trait (often governed by a single gene) that renders them susceptible to a pathogen, pest, or environmental stress. The chestnut blight (*Cryphonectria parasitica*) epidemic is a classic case in which a genetically heterogeneous species was virtually destroyed by a fungal pathogen that is highly damaging to the American and several other species of chestnut trees (Anagnostakis, 1982). However, such cases are unusual.

More commonly, genetically similar crop varieties that may have in common the same critical gene for resistance to important diseases or insects are grown over large areas in successive seasons. This dependence on a single source of resistance is, arguably, the crux of the problem. Breeders can strengthen plant resistance against epidemics by broadening the diversity of resistance genes and "pyramiding" multiple genes from different sources and genes controlling other mechanisms of resistance. One key to the uniformity issue is understanding the genetic basis of the plant-pathogen, plant-pest, or plant-stress interaction and identifying preventive measures that breeders can take to build in greater resistance. Another is to reduce contiguous areas of land planted to the same varieties.

The Role of Monoculture

Intensive and continuous cultivation of uniform crop varieties enhances opportunities for pathogen or pest evolution and the natural selection of new strains able to attack their hosts successfully. In a monoculture of a single variety or genetically uniform group of

As Russian wheat aphids feed, the wheat leaves tend to roll around them, making them difficult to kill with conventional pesticides. Credit: U.S. Department of Agriculture, Agricultural Research Service.

varieties, the pest must overcome only one genotype, as opposed to numerous genotypes in mixed cultures. Dense stands can facilitate the spread of the attacking population. The effect is particularly severe in tropical environments, where natural reproductive cycles of the insect or disease often are not curtailed by changing climatic conditions. Failure to rotate crops, or sequential planting of two crops that share a common genotype and the same pathogens or insect pests, produces similar results. Nevertheless, the requirements of mechanization and consumer demands for low-cost, high-quality food make it difficult to alter these patterns.

Extensive monocultures clearly pose a risk, yet the spread of closely related, high-yielding varieties (HYVs) continues, despite the risk of vulnerability. Farmers are generally not concerned about vulnerability until production is affected. They depend on plant breeders to supply replacement varieties when existing ones fail. Replacement varieties are often readily available in industrialized countries, but may not be in many developing countries because breeding or seed production programs may be limited or nonexistent.

The Role of Environment

Environmental conditions play both direct and indirect roles in triggering genetic and physiologic plant responses. Weather conditions may promote crop susceptibility. Warm conditions tend to favor the development of fungal and bacterial pathogens, whereas climates that are only periodically moist may enhance damage from insect-vectored viral diseases and seedborne toxins (for example, aflatoxins or toxins of *Fusarium* kernel rot). Also, the weather greatly influences off-season survival of the attacking population and the amount of off-season inoculum available to initiate a seasonal epidemic. In these situations, networks of collaborating scientists monitoring the effect of environmental conditions on the development and spread of pest problems provide the best prediction of possible vulnerability.

Vulnerability to abiotic stresses may also arise from genetic uniformity for susceptibility. Plants may vary considerably in their tolerance of or susceptibility to abiotic stresses. Drought or excessive salt levels can cause complete crop failure. Cold or hot weather during flowering may induce pollen sterility, and high temperatures may cause flower drop and poor seed set. Although it is sometimes possible to incorporate resistance into crop varieties, the genetic control of stress resistance traits is usually complex. Experimental testing, and thus varietal development, are often difficult.

Measuring Genetic Vulnerability

Three components must be considered when assessing genetic vulnerability. First, the area devoted to major cultivars must be monitored by analyzing volume and distribution of seed sales (which is closely held information in the private sector) through actual farm surveys. Such information was more readily available in the past than it is at present, because many national agricultural agencies (including the U.S. Department of Agriculture [USDA]) have ceased gathering such data in recent years.

Second, trends in pest and pathogen evolution must be monitored to identify those that pose the greatest threats to common cultivars under specified conditions. Given the global patterns of varietal distribution, this should be done on a worldwide basis. For rice, maize, wheat, beans, and the several other commodities, information could be developed through the international agricultural research centers (IARCs). Hundreds of collaborating scientists around the world grow and evaluate breeding lines developed by the IARCs. They collect data on agronomic factors and on susceptibility to pests and pathogens.

Similar public networks (although much more limited in scope) operate for a few crops in the United States through the USDA and state agricultural experiment station collaborators and in Europe. These networks can be excellent sources of information on developing pest or pathogen problems. However, information is not readily available for minor crops and commodities traditionally developed in the private sector (for example, hybrid corn, coffee, bananas, and other plantation crops), although major multinational seed companies often find it worthwhile to develop their own networks. Networking can be a very cost-effective means both of disseminating germplasm and monitoring evolving pest problems.

Third, the extent of genetic uniformity among major cultivars must be determined. In comparison with determination of acreage devoted to major cultivars and pest and pathogen evolution, data on the extent of genetic uniformity among major cultivars are scarce and difficult to assemble. The number of traits involved is enormous, detection methods are laborious, and cultivar pedigrees (when available) usually do not reveal the ecogeographic origins of important resistance genes. Biotechnological methods (see Chapter 7) could be used to develop gene probes and to construct genetic maps, circumventing some of these problems, but the costs are high and the techniques have not yet been broadly applied.

One of the few thorough studies of genetic uniformity was done on rice by scientists based at the International Rice Research Institute

(IRRI). Hargrove et al. (1985) analyzed the diffusion of rice genetic materials among 27 plant breeding programs over a 20-year period and found a disturbing similarity in the genealogies of improved rice varieties across Asia. Some narrowly focused studies have been done on the geographic distribution of specific traits, such as genes for fertility restoration in hybrid rice based on cytoplasmic male sterile germplasm (Li and Zhu, 1989). Nevertheless, there are few such studies, and key pedigree information is not available for varieties created by the private sector.

The product of these analyses is not a quantitative measurement but a pattern formed by correlation of the trends in the amount of land planted to certain crops, pest-pathogen development, and extent of uniformity for major resistance genes. Congruent with the definition used here for genetic vulnerability, the greatest risk results when widely planted cultivars are exposed to increasingly severe pathogen or pest problems for which they lack broadly based resistance.

Measurement of the trends discussed above, as well as others, indicates the degree to which crops are at risk from genetic vulnerability, a risk that is aggravated by intensive cultivation practices. The following is an examination in greater depth of biological interactions that lead to vulnerability and the impact of breeding programs designed to manipulate those interactions.

PLANT, PATHOGEN, PEST, AND ENVIRONMENTAL RELATIONSHIPS

The genetic theories of mutation, selection, population biology, and host-pest genetic interaction apply equally to both plant pathogens (bacteria, viruses, and fungi) and invertebrate pests (aphids, ants, weevils, other arthropods, and nematodes), both of which are used throughout this section as examples. When plant, pest, or pathogen responses to environmental stresses are genetically controlled, the same theories apply.

Evolutionary Nature of the Relationship

A crop variety's response to a pathogen—whether it is resistance or susceptibility—is a result of its evolutionary and genetic relationship with that organism. This accounts for the phenomenon known as the breakdown of resistance, in which a variety previously resistant becomes susceptible to new forms or races of the pathogen population. In fact, the crop's genetic resistance is unchanged and is still effective against the original pest or pathogen population. It is the genetics of the pathogen that has been altered through natural selection so that it can attack the crop variety. The variety is then considered suscep-tible, although no change in its genetics has occurred. The task of the plant breeder becomes to incorporate into the now-susceptible vari-ety additional resistance alleles that are effective against the new pathogen population.

Seldom, if ever, is total genetic protection from disease possible. Resistant varieties are those with significantly decreased disease or pest susceptibility that does not have a negative impact on the growth and biomass production of the plant, and therefore the yield. This decreased susceptibility, however, lasts only until the pathogen or pest population gains new genetic ability to overcome the resistance.

Simmonds (1991) recognizes four broad types of genetically con-trolled resistance (Table 1-1): major gene or vertical resistance, which is pathotype specific; polygenic or horizontal resistance, which is

TABLE 1-1 The Four Main Kinds of Resistance

Kind of Resistance	Specificity	Genetics[a]	Durability
Pathotype-specific or vertical, VR/SR	Very high	Oligogenes	Mobile pathogens, durability usually bad
Pathotype-nonspecific major gene resistance, NR	Nil	Oligogenes	Immobile pathogens, durability may be good
General or horizontal resistance, HR/GR	Nil/low	Polygenes	High
Interaction or mixture resistance, IR/MR	Some	Heterogeneous oligo[b]	Probably good

[a]Oligogenes are single genes that produce a pronounced phenotypic (expressed character) effect, as opposed to polygenes, which have individually small effects.

[b]Some authors (Simmonds, 1979) are inclined to attribute some weight to heteroge-neity for polygenic systems, but this matter appears to be undecided.

SOURCE: Adapted from Simmonds, N. W. 1988. Synthesis: The strategy of rust resistance breeding. Pp. 119–136 in Breeding Strategies for Resistance to the Rusts of Wheat, N. W. Simmonds and S. Rajaram, eds. Mexico, D.F.: Centro Internacional de Mejoramiento de Maíz y Trigo. Reprinted with permission, ©1988 by Centro Internacional de Mejoramiento de Maíz y Trigo.

pathotype nonspecific; major gene resistance, which is pathotype non-specific; and mixture or interaction resistance.

Types of Relationships

Under the genetically controlled resistance patterns described above, the susceptibilities of crop plants to pests or pathogens can develop in many different ways. These host-pathogen relationships fall into three general groups: those characterized by (1) changes in the pathogen or its effect, (2) movement of varieties or pathogens to new biomes, or (3) changes in agronomic practices or pest control strategies. In the first group, major pathogens may evolve, minor pathogens may become more serious, or the environment may enhance the severity of the pathogen's attack. In the second group, a crop may prove to be susceptible to newly encountered pathogens when grown in a new location, or exotic pests introduced into the crop's home region may be successful in attacking it. In the third group, new agronomic traits or practices may render the crop susceptible to new pathogens, or the crop may be vulnerable to previously discounted pests when control strategies or priorities change. Most of these responses are intensified by genetic uniformity, ecologic uniformity of agronomic conditions, or both. Examples illustrating the complexity of interactions that plant breeders must deal with when seeking pest and pathogen resistance in their crops are discussed below.

Evolution of Major Pathogens

Plants with resistance conferred by a single gene or with resistance to a specific pathogen race are said to have vertical resistance. This means they can be damaged by a different virulent race. There are major concerns about varieties that contain vertical resistance genes effective for dealing only with the prevailing races in the breeders' plots. The critical question is: How likely is a new race to develop or to be introduced? Put differently, how soon will there be severe losses because of breaching of the resistance by the pest or pathogen species? The answer depends on how frequently different races arise in the existing pathogen population, how well adapted they are for survival in the off season, and the size and dispersion of the pest or pathogen population.

The record of single major genes in providing lasting or durable resistance is mixed. The *R* genes for potato late blight (*Phytophthora infestans*) resistance and the *Sr* genes for stem rust resistance in wheat are two classic examples. In both cases, at least some resistant varieties collapsed because of the selection and spread of new pathogen races able to attack varieties carrying the resistant genes. In wheat, certain combinations of *Sr* genes in new varieties led to longer-term resistance to stem rust (*Puccinia graminis*). In the United States, no widespread epidemic of wheat stem rust has occurred since the late 1950s. However, outbreaks have appeared in other regions.

The biology of the pest and its interaction with the host plant greatly influence the epidemic potential of any new race or biotype that may be selected (Simmonds, 1979). For example, a soilborne fungus that is not also seedborne may develop a new race in a particular field and remain highly localized for decades. In general, vertical resistance genes are ephemeral and, when introduced in crops, are soon rendered ineffective by new races of pests and pathogens. This is especially true for cases of pathogens that are windborne (for example, the powdery mildews) or of pests that are highly mobile or that have high reproductive rates (such as the brown plant hopper, *Nilaparvata lugens*).

It is important to consider the evolutionary relationship of the resistance genes and the pest or pathogen. If the two have coevolved in the primary center of diversity, it is highly likely that the pest population contains genes effective against each plant gene conferring resistance (Ashri, 1971; Browning, 1972). In developing new varieties, plant breeders must consider whether resistance is sufficiently durable to be economically useful. Several guidelines emerge from past experience. For pests with a high epidemiologic potential that have coevolved with the host plants, major genes for resistance are ephemeral in their utility, if used singly. In some coevolved systems, pyramiding of many genes may work in augmenting the durability of the resistance. Genes for vertical resistance can be used for cases in which experience has shown that selection of new races is slow, the spread of new races is slow, or pathogen survival from season to season is low.

Minor Pathogen Enhancement

In addition to the major pests and pathogens that a breeder addresses in a breeding program, there are many others that have only a minor impact on crop performance. It is assumed that selection for high yield will ensure sufficient resistance against yield reduction by these minor pests.

However, minor pathogens can become important if there is a lack of challenge in breeders' plots and a difference in environmental conditions between the plots and farmers' fields. The inoculum for

the minor pathogen may be so low in the plots that increased susceptibility will not be detected. Changes in the agronomic environment
could result in an increased pathogen population. Thus, without any
shift in the genetics of the pathogen, a minor pest may become a
major problem. A good example is the recent development of gray
leafspot disease (*Cercospora zeae-maydis*) of maize into a disease of
potential importance in the United States.

Environmental Enhancement of Pathogen Effects

The environment affects disease levels by altering, increasing, or
decreasing pest development. Thus, levels of resistance adequate in
one environment may be much too low in another.

If the environment is not conducive to pathogen or pest increase,
it is difficult to select for resistance. In general, selection in an arid
environment not favorable to airborne fungal or bacterial diseases
often results in plant genotypes inadequately protected genetically
against many fungal and bacterial pathogens when the crop is grown
in wetter climates. For viral diseases with insect vectors and for
many insect pests, climates that are only periodically moist may favor heavier crop damage.

In 1985, rains broke the worst
drought in a decade in the Sa-
hel, but they also caused an
outbreak of the Senegalese grass-
hopper. An observer (right)
from the Food and Agriculture
Organization examines the remains of a farmer's crop of millet. Credit: Food and Agriculture Organization of the United
Nations.

Varieties Vulnerable in New Regions

Not all pests are widely distributed. At one time the major continents contained separately evolving host-pathogen systems. The equilibria of these systems have been breached by the movement of crop species and pests into new regions (Buddenhagen, 1977).

When crops are moved between continents or regions, they may encounter pathogen or pest species for the first time, with unpredictable results. Examples of new encounters resulting from the introduction of crops into North America are fire blight (*Erwinia amylovora*) of apples and pears, Pierce's disease of grapes (*Xylella fastidiosa*), and dwarf bunt of wheat (*Tilletia controversa*). Examples of such encounters in the tropics are African cassava mosaic virus, cacao swollen shoot virus, maize streak virus, and Moko disease (*Pseudomonas solanacearum*) of banana (Buddenhagen, 1977).

Lack of pests or pathogens can be a serious impediment to an effective breeding program. The potential impact of the organisms on the crop cannot be assessed when new varieties are developed in the absence of a pathogen or pest.

Varieties Vulnerable to Exotic Pathogens

Many of the great epidemics of the past have been caused by the movement of pests into new areas where a crop species has developed in the absence of the pathogen. The late blight (*Phytophthora infestans*) epidemic of potatoes in Ireland in the 1840s is a well-known example. Dutch elm disease (*Ceratocystis ulmi*), white pine blister rust (*Cronartium ribicola*), and chestnut blight (*Cryphonectria parasitica*) in North America are three more recent examples. Maize, barley, and wheat have all succumbed to rust epidemics (caused by *Puccinia* spp.) where they had previously escaped on continents distant to their origin—maize in Africa in the 1950s, barley in South America in the 1970s, and wheat in Australia in the 1970s (Simmonds and Rajaram, 1988).

Some of the great tropical plantation cultures have developed specifically because the crop escaped potentially devastating coevolved pathogens by being moved to distant continents. Coffee in Latin America escaped coffee rust (*Hemileia vastatrix*) from Asia and Africa until very recently. Bananas in Latin America escaped Sigatoka (*Mycosphaerella musicola*) until 1932 and the more virulent black Sigatoka (*M. fijiensis*) until the 1970s. Rubber in both Africa and Southeast Asia still escapes South American leaf blight (*Dothidella ulei*) of its Amazonian homeland (Buddenhagen, 1977).

Agronomically Related Vulnerability

New cultivars may be more vulnerable to pests because of either new agronomic traits or practices that are changed to use the new cultivars to their fullest yield potential. This is especially true when breeders target whole continents, where there is much variability in agronomic technology.

The semidwarfing genes of both wheat and rice have affected the epidemiology of fungal pathogens by reducing plant height and interleaf distance, consequently reducing air circulation and increasing the humidity around the plant that can contribute to enhanced pathogen densities and rate of development. This has been studied intensively in *Septoria* leaf blight of wheat (Scott et al., 1982, 1985). In another crop, such as beans, changes in plant form for easier mechanical harvesting also reduces the potential of infection by a pathogen. An upright plant shape reduces contact with the soil, promotes improved air circulation and lower humidity, and helps reduce infection and spread of diseases.

Breeders often seek to develop varieties with shorter life cycles. Fewer days to maturity may mean fewer pests or pathogen generations during crop growth, or less chance of exposure to weather conditions favorable to the pathogen. However, in tropical latitudes, shorter life cycles may lead to the raising of two successive crops (double cropping) with increased exposure to pests.

A more subtle, highly unpredictable case is when an agronomic gene actually modifies the intrinsic level of pest susceptibility or resistance in the crop. This may be due to multiple characteristics affected by a single gene (pleiotropy), or an agronomic gene may be on the same chromosome and near a gene for susceptibility (linked).

The corn blight epidemic of 1969–1970 was the result of susceptibility conferred by a pleiotropic gene. A cytoplasmic (mitochondrial) gene was used by breeders to produce maize plants that did not have viable pollen. This simplified the production of commercial F_1 hybrids. However, it was later discovered that this gene, producing what is called Texas (or T) cytoplasmic male sterility, had an additional effect of increasing susceptibility to a race of the fungus *Bipolaris maydis*. When climatic conditions were right, the fungus reproduced explosively on millions of acres and over many varieties of maize.

Change in Vulnerability to Discounted Pests

All new varieties are, to some degree, susceptible to various pathogens and pests, but it is assumed that this will be relatively insignificant.

For some crops, chemical treatment or other practices are used as controls to protect them. In other cases, occurrence of the pest or pathogen is infrequent, sporadic, or absent; or the level of susceptibility is so low that little damage will result. For many horticultural, orchard, or ornamental crops, product quality is so important that resistance or susceptibility to pests and pathogens is often largely ignored in breeding, on the assumption that these high-priced crops will be protected by chemicals.

BREEDING STRATEGIES AND THEIR IMPACT ON GENETIC DIVERSITY

Modern plant breeding is essentially an evolutionary process, characterized by a more rapid rate of change than that which occurs in the slow natural selection of wild species or the landraces of primitive agriculture. Two major differences exist between the processes of early domestication of crop plants and current breeding practices. First, early development of the crop frequently occurred in the center of diversity for coevolving pests and pathogens. Second, during the early evolution of the crop, hybridization often occurred with related wild species. Today, many crops are grown outside their centers of origin and are distant from related wild species or coevolved pathogens. Few breeding programs use wild species or primitive cultivars in the breeding process; instead, they rely on proven elite breeding lines (Marshall, 1989; Wilkes, 1989a).

Genetic Diversity in Primitive and Modern Varieties

In most primitive farming systems, genetic diversity within and among crops was the norm. Genetic heterogeneity, however, did not provide absolute assurance against epidemic pest attacks nor against environmentally caused yield losses. Rather, the rate of loss varied within the population because of both genetic heterogeneity and crop admixtures. Information on the genetics of landraces in primitive agriculture is fragmentary (Janzen, 1973). Little is known about gene flow and interpopulation versus intrapopulation variation, especially when wild relatives are included (Altieri et al., 1987; Ladizinsky, 1989).

Modern varieties, in addition to their high-yield capacity, possess resistances or tolerances to many disease and insect pests and to normally expected environmental stresses. It is often alleged that new hybrids and varieties are weak and prone to disease, and that they must be given plentiful supplies of water, fertilizer, and pesticides to do well (Fowler and Mooney, 1990; Mooney, 1979, 1983).

Most modern varieties, however, show superior yield under the intensive cultural conditions of modern agriculture that increase competition for water, nutrients, and sunlight and favor the development of pest populations. Under these conditions many older varieties exhibit sterility, poor root growth, lodging, and susceptibility to pests and diseases (Duvick, 1987).

Many genetic traits give varieties their higher-yield potential. For example, modern maize hybrids show improved root strength, resistance to fungal species that cause stalk-rots, resistance to heat and drought, ability to withstand poor nitrogen nutrition, resistance to the European corn borer (*Ostrinia nubilalis*), and ability to grow when densely planted. Because these genetic improvements are typically present with the new resistance genes, the net result is a so-called pyramiding of resistance and tolerance. Except in extreme environments, new hybrids yield more whether grown under good or bad conditions (Anderson et al., 1988)

Germplasm Varietal Development and Vulnerability

The breeding system of a crop species—whether self- or cross-pollinated—can have a great effect on its genetic diversity. Those that are cross-pollinated, such as many perennials, are highly heterozygous, and self-pollinated crops are more generally homozygous. However, there are many exceptions. Even though maize has a long history of cultivation as a cross-pollinated, heterozygous crop, modern maize hybrids are genetically highly uniform. Although every plant is usually highly heterozygous because it results from the cross of two homozygous inbred lines, it is genetically identical to all other plants of the hybrid.

Most perennial crop species are obligate outcrossers and exhibit a wide array of genetic variability in subsistence agriculture. But modern perennial cultivars intended for large-scale production such as sugarcane, banana, apple, and citrus are often selected as individual plants and are then multiplied clonally to produce many replicates of one heterozygous individual. Such clones are genetically identical to one another as the single-cross maize hybrids are, they differ only in the way the selected genotype is multiplied.

Self-pollinating annuals, such as wheat, are typically grown as genetically uniform cultivars tracing back to one nearly homozygous selected plant. However, blends of several uniform, but genetically distinct, cultivars are sometimes purposely grown together, thus producing genetically heterogeneous populations. These can impede the spread of disease and prevent an epidemic provided that effective resistance genes are present in some of the cultivars.

Performance superiority in crop species results from the accumulation of favorable alleles. The use of unadapted germplasm in crosses introduces unfavorable genes and increases the likelihood of breaking favorable linkages that may be difficult to recover. Therefore, breeders tend to rely on crosses between adapted lines in which many favorable alleles and linkage groups may be present in both parents. They are reluctant to use unadapted exotic germplasm for crossing purposes (particularly if they continue to make acceptable progress without its use) and therefore do not make heavy use of landraces and wild and weedy relatives (Duvick, 1984a; Peeters and Galwey, 1988). This is especially true for crops for which there have been long-standing breeding programs, such as the cereals. This strategy increases the odds for steady performance improvement, but it also tends to enhance genetic uniformity.

If a new crop-threatening hazard (for example, a disease or an insect) appears for which genetic resistance is unavailable in the breeder's own nurseries, the breeder is forced to look elsewhere for the needed genes. Improved germplasm in the major crop species can be found in national, regional, and international screening nurseries and as a result of trials conducted by international agricultural research centers, the USDA Agricultural Research Service, and a few land-grant university experiment stations. Advanced lines and varieties undergoing evaluation in such trials are heavily used by many breeders as parents in new crosses. Performance data on unfamiliar material, when available, are more useful to breeders than are germplasm bank descriptor data, because they provide information on productivity and adaptation that is particularly sought by breeders in choosing parental materials.

Exotic germplasm is most often used in the pursuit of long-range goals. It reaches large breeding programs via national, regional, and international evaluation networks and by personal contacts. These same programs generally maintain large breeders' collections of germplasm that are heavily used for hybridization with exotic materials (Centro Internacional de Mejoramiento de Maíz y Trigo, 1985a, 1987a).

The greatest reliance on germplasm bank collections is made by breeders of plant species that have little or no history of improvement. Generally, these are pulses (beans or lentils), some root crops, vegetables, and some industrial crops. For these crops, the breeder who has initiated an improvement program often has no alternative but to use exotic collected material for local evaluation and selection.

The most frequently voiced criticism from breeders who use germplasm banks is the lack of reliable and complete descriptor information and passport data. This forces breeders to screen large numbers of accessions more or less blindly. They would prefer to identify, for initial screening, a smaller number of accessions best suited to the production environment for which they are breeding, but performance data are rarely available. Finally, the diversity of accessions in a collection may be more indicative of the history of collecting efforts in selected geographic regions than of the breadth of genetic diversity of the species (Peeters and Williams, 1984).

Genes identified in primitive materials usually need to be introgressed or backcrossed repeatedly to improved genotypes before breeders will use them. This process is known as germplasm enhancement, or prebreeding. However, unless there is need for an all-out search for a particular trait, breeders may be reluctant to use the older, more primitive accessions preserved in banks. Rather, they would prefer the enhanced materials, which may not be in germplasm banks (see also Chapter 6).

Gains in Productivity

Progress in plant breeding of the major grain crops can, in part, be measured as increases in productivity or yield per unit area (Buddenhagen, 1985). Changes in farming practices, such as increased use of nitrogen fertilizer and better weed and pest control, have also been responsible for major gains in productivity in the major U.S. crops (maize, wheat, sorghum, and soybeans). The most important input, however, has been that of improved varieties and hybrids from breeding programs (Duvick, 1987).

Repeated studies for each of four major crops have shown that about 50 percent of yield gains in the United States over the past 30 to 50 years were due to varietal improvement (Fehr, 1984). The rate of yield gain as a result of genetic improvement has been estimated at about 1 percent per year for each crop and has been greater for wheat (2 percent) and maize (3 percent). Parallel trends are documented in other developed countries, for example in the United Kingdom, where yield gains for wheat progressed at the rate of 3 percent annually during the period from 1949 to 1984 (Silvey, 1986).

Vulnerability and Varietal Replacement Over Time

Varieties are adopted, become popular, eventually lose their popularity and are discontinued for many reasons. Yield is an important consideration of variety adoption by farmers. The prevailing markets also affect popularity, the rate at which a variety may be adopted, and the rate at which it eventually will be abandoned.

The area devoted to major U.S. crop varieties in 1969 is shown in Table 1-2. Varietal dominance for major grains ranged from six lines used in hybrids grown on at least 71 percent of the land in corn to nine varieties on 50 percent of the land in wheat. Such data raised serious concerns about greater vulnerability due to narrowing of the genetic base (Duvick, 1984a). Unfortunately, USDA no longer monitors the amount of land devoted to certain crops or the varieties of crops that are planted. Thus, it is no longer possible to assess vulnerability in this way.

A question remains if crops are more or less vulnerable today than in 1970. In 1981, Duvick (1984a) examined data from the seed

TABLE 1-2 Area and Farm Value of Major Crops in the United States and Extent to Which Small Numbers of Varieties Dominate Crop Area, 1969

Crop	Area in Hectares (10^6)	Value in Dollars (10^6)	Number of Total Varieties	Number of Major Varieties	Area Planted to Major Varieties (percent)
Bean, dry	0.6	143	25	2	60
Bean, snap	0.1	99	70	3	76
Cotton	45.3	1,200	50	3	53
Maize[a]	26.8	5,200	197[b]	6	71
Millet	0.8	—	—	3	100
Peanut	0.6	312	15	9	95
Peas	0.2	80	50	2	96
Potato	0.6	616	82	4	72
Rice	0.7	449	14	4	65
Sorghum	6.8	795	—		—
Soybean	17.2	2,500	62	6	56
Sugar beet	0.6	367	16	2	42
Sweet potato	0.04	63	48	1	69
Wheat	18.0	1,800	269	9	50

NOTE: Dashes indicate that the information was not available.

[a]Maize includes seeds, forage, and silage.
[b]Released public inbreds only.

SOURCE: National Research Council. 1972. Genetic Vulnerability of Major Crops. Washington, D.C.: National Academy of Sciences.

TABLE 1-3 Perception of Vulnerability Among Breeders

	Response (percent)		
Crop	Serious	Some Concern	Not Serious
Cotton	7	33	60
Soybean	10	48	42
Wheat	25	50	25
Sorghum	0	62	38
Maize	5	32	64
All crops	10	44	46

NOTES: Breeders' perception of vulnerability was determined by their response to the question: "How serious is the problem of genetic vulnerability in your crop?" Numbers may not add to 100 due to rounding.

SOURCE: Duvick, D. N. 1984. Genetic diversity in major farm crops on the farm and in reserves. Econ. Bot. 38:161–178. Reprinted with permission, ©1984 by Society for Economic Botany.

industry to estimate the concentrations of the six most widely planted cultivars of cotton, soybean, and wheat and the correspondingly important lines of maize (Table 1-3). Duvick argued that, from 1969, diversity increased and that the lead varieties were replaced in most cases. However, Smith (1988a) argued that increased numbers of released varieties do not necessarily reflect an increase in genetic diversity because many varieties share common ancestries.

The average time for wide-scale use of a successful variety from the time of its introduction is about 7 years (Duvick, 1984a). Farmers often continue to grow a variety as long as it continues to perform well for them, and they sometimes persist for years after experts have warned against its use. This happened in India, where the wheat variety Sonalika was removed from the recommended list of wheat because of known susceptibility to leaf rust (*Puccinia recondita*). However, it continues to be planted over vast areas because it can be harvested earlier than many varieties, has desirable grain color, and is suitable for multiple croppings (Dalrymple, 1986a).

The Seed Savers Exchange has documented varietal losses in the North American vegetable seed industry on the basis of U.S. and Canadian mail-order seed catalogs. In 1984, a total of 4,963 nonhybrid varieties were available from 230 companies, but 54 percent of these were offered by one source only. When a second inventory was completed in 1987, there was a net loss of 15 companies, the majority of which were smaller companies that had been rich sources of unique varieties (Whealy, 1988). The last such comprehensive inventory of commercially available varieties by the USDA was in 1901 and 1902.

In general, most annual field crop varieties change considerably over time. A very different picture emerges for long-lived tree crops such as apples, for which turnover is measured in decades. In these crops there is less of a tendency for genetic change and more of a dependence on chemical and cultural practices to overcome pest and pathogen problems. For both tree and annual crops, recent trends away from heavy chemical control combined with less energy-intensive cultural practices require intensive breeding efforts to maintain crop quality and productivity.

Alternative Strategies to Enhance Diversity and Increase Stability

Modern agricultural practices often use large inputs of fertilizers, herbicides, fungicides, and insecticides, in combination with monoculture, to optimize yields. However, it appears that additional applications of inputs have leveled off (and may be dropping) relative to yields (Duvick, 1987).

In recent years, there has been a growing interest in agricultural systems that mimic natural ecosystems. The stability of early farming systems was due not only to the genetic diversity and heterogeneity of their landraces but also to the spatial separation of farms and the temporal separation of crops that reduced vulnerability to pests and promoted more efficient use of water, nitrogen, and light (Gould, 1983). Alternative agricultural strategies use a mixture of management and technological options that try to take advantage of natural cycles and biological interactions (National Research Council, 1989a). These options include crop rotations to reduce biotic stresses and enhance soil fertility, modified tillage practices for soil conservation and weed control, integrated pest management practices for insect and disease control, and the use of cultivars with enhanced genetic resistance to various stresses (National Research Council, 1989a).

Most IARC programs include some multiple-cropping or farming systems research, such as cropping practices with beans for erosion control in Africa (Centro Internacional de Agricultura Tropical, 1989) or rotation systems for vegetables in Asia (Asian Vegetable Research and Development Center, 1988). A range of alternative rice production practices developed at IRRI is tested at over 180 sites through the Asian Farming Systems Network (International Rice Research Institute, 1985).

Alternative genetic strategies include modern multiline mixtures and relay plantings of distinctly different genotypes that enhance agroecosystem stability (Wolfe, 1985, 1988). In a multiline variety, individual lines differ in specific resistance genes. The lines are often segregates from common parents but are not isogenic; varietal mixtures may include lines with different genetic backgrounds and resistance genes. Varietal mixtures appear to be the best option to supply heterogeneity for disease resistance (Browning, 1988; Simmonds and Rajaram, 1988; Wolfe, 1988). Their disease-dampening effect is due to a combination of disease restriction factors, a dilution of inoculum that reduces the success rate of sporulation, and slowing of the spread of disease and postponing the development of a broadly effective pathogen (Marshall and Pryor, 1978). In addition, varietal mixtures make no new demands on plant breeding programs because currently available varieties are adequate, in most cases, for mixture development.

Some reports suggest that a field made up of heterogeneous cultivars provides yields at least as good as those provided by the separate components, whether they are multilines (Borlaug, 1958; Browning and Frey, 1969) or varietal mixtures (Brim and Schutz, 1968; Wolfe, 1985). Such mixtures are usually more stable than their component genotypes with respect to environmental interactions (Schutz and Brim, 1971; Simmonds, 1962). However, multiline or varietal mixtures may present problems for the processing industry, especially when there are industrial restrictions, as there are for wheat.

VULNERABILITY AND CROP DIVERSITY SINCE 1970

The 1970 southern corn leaf blight epidemic focused public awareness on the risks of genetic vulnerability in crops. Trends toward genetic uniformity in crop breeding programs and extensive monoculture were identified by the National Research Council (1972) as major factors contributing to the disaster, but there were no comprehensive data with which to measure changes in crop vulnerability over time. In that report, the data on crop production and the area devoted to major crop varieties served as a benchmark for estimating changes in crop vulnerability from varietal turnover and altered breeding strategies 2 decades later.

Informed measures of vulnerability involve a correlation of estimates of genetic uniformity among major varieties and their land areas with trends in pest and pathogen evolution. An increase in the number of available varieties is not a reliable indicator of increasing genetic diversity, because very similar cultivars may be marketed under different names or may share many genes in common with many other varieties. Modern methods (isoenzyme electrophoresis, high-performance liquid chromatography, DNA restriction fragment length polymorphism analysis, and gene mapping) provide more precise means for measuring genetic similarity between two cultivars, but they cannot be used to predict susceptibility to unknown pests or pathogens. The following are assessments of changes in diversity of dry beans, wheat, maize, rice, and several minor crops.

Dry Beans

Dry beans (*Phaseolus vulgaris* L.) offer a unique example of a crop for which susceptibility to an epidemic was accurately predicted. The National Research Council (1972:225) report recognized that "for a considerable part of the edible dry bean acreage in the United States, annual production rests upon a dangerously small germ plasm base." At that time, two market classes of *P. vulgaris*, the Michigan navy bean and the pinto bean, accounted for 60 percent of 600,000 ha planted in the United States in 1969 and 1970.

Within 5 years of the report, one of the authors had refined a predictive model for dry bean genetic vulnerability, which suggested that pinto beans faced the highest risk of an epidemic of any dry bean commodity class (Adams, 1977). Adams calculated a genetic distance index based on 36 chemical and agronomic traits of each cultivar to assess the homogeneity of beans grown in each production region. By combining these estimates of genetic similarity with the acreage covered by each cultivar, Adams calculated a homogeneity index for each bean-producing state. The assumption was that if the homogeneity of cultivars in a region was great, less time would be required for a pathogen to adapt to them as hosts and the infection rate would be higher.

Within five growing seasons, the pinto bean fields in Colorado and Wyoming—the two states at greatest risk—suffered an epidemic of rust (*Uromyces appendiculatus*) that caused damage of $15 million to $20 million in 1982 alone (Stavely, 1983). Pinto bean yield losses ranged from 25 to 50 percent in this region in 1981 and 1982 (Venette and Jones, 1982), causing the widespread destruction clearly recognized as a potential in dryland regions in the 1972 report. Yet, the majority of pinto bean farmers continued to grow the rust-susceptible cultivars up until the time of the epidemic. Today, susceptible varieties continue to occupy 40 percent of the land planted to pinto beans in Colorado, southwestern Nebraska, and northwestern Kansas, despite another, lighter rust epidemic in this area in 1987 (Stavely, 1988). Nevertheless, the 1981-1982 epidemic prompted a significant continuing trend to develop rust-resistant cultivars.

The severity of rust led to a multi-institutional commitment to resistance breeding that has produced dramatic results. Since 1983, when USDA enabled one of its pathologists to concentrate on bean rust, 70 pathogenic races of *U. appendiculatus* have been identified from field samples, and 1,118 USDA plant introduction bean accessions have been evaluated for rust susceptibility (Stavely, 1988). Two breeding lines of pinto beans have been released that are homozygous for dominant genes for resistance to 33 races of rust and homozygous for a second independent gene for resistance to many of the same races.

There are now both privately and publicly bred pinto bean cultivars in production with resistance to some of the prevalent rust races in the High Plains of the United States. The area planted to pinto beans is shared by several different cultivars and is far more heterogeneous than it was in 1970. Breeders and pathologists point to the exchange of germplasm with the Centro Internacional de Agricultura Tropical (CIAT, International Center of Tropical Agriculture) as one of the major factors that allowed them to diversify the gene pool of pinto beans. Latin American germplasm screened by CIAT has proved particularly valuable, and USDA has sent samples of its rust-resistant accessions to CIAT and other breeding programs overseas as well.

Wheat

Wheat (*Triticum* spp.) illustrates two conflicting trends in genetic diversity. In the United States, there has been a significant increase in the number of cultivars and the range of exotic germplasm from which they are derived (Cox et al., 1988a; Dalrymple, 1988). At the same time, farmers in developing countries are using fewer landraces and cultivars because of the success of high-yielding cultivars. Their replacement of landraces in wheat's centers of origin continues to concern genetic conservationists (Centro Internacional de Mejoramiento de Maíz y Trigo, 1989).

More than 20 years ago, a report of the National Research Council (1972:135) warned that a "highly vulnerable situation exists in the United States, where a relatively small number of varieties dominates the wheat acreage." Duvick's (1984a) survey of plant breeders suggested that the concentration of U.S. land planted to the six leading cultivars of wheat was about as great in 1980 as it was in 1970. A survey covering a longer time span, however, indicated a steady increase in the number of wheat varieties grown each year since 1949; the increase was from 263 varieties in 1969 to 429 varieties in 1984 (Dalrymple, 1988). In addition, the amount of land covered by the leading wheat variety and the other top 10 varieties has declined dramatically since 1969, as more privately bred varieties have come to share the market. Nevertheless, caution must be used in assessing the extent of new genetic diversity, because many varieties may be closely related.

Cox et al. (1986) analyzed the common parentage and genetic similarities of wheat cultivars grown in the United States within the previous 10 years. The pedigrees of all 400 U.S. winter wheat varieties may be traced to 74 ancestors, with the bulk of the germplasm still derived from a small number of pre-1919 varieties. At the same time, the use of exotic germplasm by wheat breeders is increasing. Genes from exotic landraces or other species occurred in a quarter of the soft red winter wheats and 7 percent of the hard red winter wheats grown on at least 16,000 ha in 1984. More than 60 percent of the hard red winter wheat breeding lines in the U.S. Southern Regional Performance Nursery now contain chromosome segments from other species. Of 17 wheat breeders contacted in a recent survey, 14 were making crosses with other species, accounting for about 2 to 4 percent of their crosses, advanced lines, and releases (Cox et al., 1988a).

Exotic germplasm is being used to reduce genetic susceptibility of wheat cultivars to powdery mildew (*Erysiphe graminis*), dwarf bunt (*Tilletia controversa*), stripe rust (*Puccinia striiformis*), Hessian fly (*Mayetiola destructor*), stem rust (*P. graminis*), leaf rust (*P. graminis*), *Septoria* leaf blotch (*Septoria nodorum*), and greenbug (*Schizaphis graminum*) (Cox et al., 1988a). The resistance will sometimes last longer in cultivars in which two or more different genes for it are bred into the same cultivar (Gould, 1986). This strategy, termed *pyramiding*, is particularly promising for cultivars grown in the presence of rapidly evolving Hessian fly biotypes (Gallun and Khush, 1980; Kiyosawa, 1982).

Trends in developing countries provide a dramatically different picture. There, a high percentage of wheat lands are planted to a small number of high-yielding, semidwarf varieties (Figures 1-1 and 1-2). In Bangladesh, for example, HYVs covered about 96 percent of the wheat area in 1984 with 67 percent of the wheat land planted to the variety Sonalika (Dalrymple, 1986a).

In India, about 76 percent of the wheat area was planted to HYVs in 1983. Sonalika, a variety developed in India and released in 1967, alone occupied 30 percent of India's wheat land (Dalrymple, 1986a). India has suffered several severe wheat epidemics since 1969. Shoot fly (*Atherigona* spp.) killed 50 to 100 percent of all tillers on semidwarf and triple-dwarf HYVs in 1974, and Karnal bunt (*Tilletia indica*) affected 67 percent of the Sonalika crop and eight other HYVs in 1975.

The wheat varieties in Mexico appear to be somewhat more di-

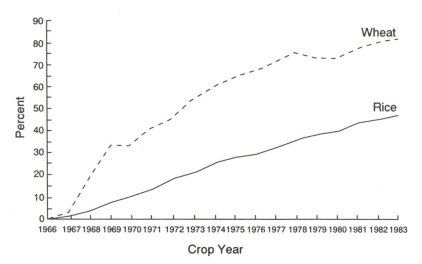

FIGURE 1-1 The estimated proportion of areas planted with high-yielding varieties of wheat and rice in southern and southeastern Asian nations increased between 1965–1966 and 1982–1983. Source: Dalrymple, D. G. 1986. Development and Spread of High-Yielding Rice Varieties in Developing Countries. Washington, D.C.: U.S. Agency for International Development.

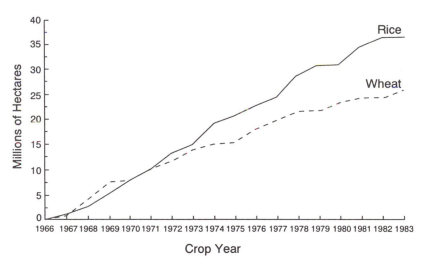

FIGURE 1-2 The estimated area planted to high-yielding varieties of wheat and rice in southern and southeastern Asian nations increased steadily between 1965–1966 and 1982–1983. Source: Dalrymple, D. G. 1986. Development and Spread of High-Yielding Rice Varieties in Developing Countries. Washington, D.C.: U.S. Agency for International Development.

verse than those of other developing countries. In 1983, HYVs occupied 95 percent or more of Mexican wheat fields, with about 32 percent of the area planted to three varieties. From 1976 to 1981, the breakdown of leaf rust resistance in HYVs such as Nacozari caused a rapid turnover in widely planted varieties (Centro Internacional de Mejoramiento de Maíz y Trigo, 1985b).

The past 2 decades have seen positive trends toward the greater use of exotic germplasm in U.S. wheat breeding programs, increased numbers of available varieties, and decreased dominance by a few leading varieties. Similar trends have not been seen in developing countries. Rather there is a worrisome dominance by a small number of HYVs. Multiplication and distribution of a greater number and diversity of varieties are restricted in developing countries by the typically low capacities of national seed agencies and limited private sector involvement.

Maize

The National Research Council (1972) suggested that too little genetic diversity was available in U.S. maize varieties and recommended the restoration of genetic diversity through the development of new and unrelated types of maize resistant to southern corn leaf blight (*Bipolaris maydis*) and other hazards. In 1981, questionnaires were sent to public and private plant breeders in the United States on this and related topics (Duvick, 1984a). The dominance of the six most popular inbred lines of maize decreased from 71 percent of the land planted to hybrids in 1969 to 43 percent in 1980. At the same time, only 5 percent of the maize breeders believed that genetic vulnerability was a major problem for maize, 32 percent felt it was of some concern, and 64 percent doubted that it was a serious concern (Table 1-3). Duvick (1984a) concluded that breeders felt they could produce new resistant cultivars quickly enough to keep farmers from suffering dramatic yield or economic losses.

The major inbred lines used in maize breeding had undergone considerable turnover since 1970, and novel sources of resistance continued to be discovered in elite lines. The message was that the U.S. maize industry had already built in four strategies for reducing the magnitude of its vulnerability: diversity through time (varietal turnover), diversity through anticipation (early-warning alerts), diversity in reserves (numerous advanced lines), and rapid deployment of genetic diversity (the ability of the U.S. seed industry to bring in materials from anywhere in the world).

Other surveys have interpreted changes in the maize industry differently. Zuber (1975) concluded that the U.S. maize germplasm base was nearly static from 1970 to 1975. Only after 1975 did large changes in the frequency of use of major inbred lines begin, but two germplasm sources (Iowa Stiff Stalk Synthetic and Lancaster Sure Crop) continued to dominate (Zuber and Darrah, 1981). Since 1980, the use of privately available inbred lines as opposed to publicly developed lines of maize has increased dramatically (Darrah and Zuber, 1986).

Many of the maize hybrids sold under different names, however, continue to depend on a few closely related inbred lines. Thus, the apparent changes through time may not be as dramatic as some studies suggest. Comparing isozyme data for 138 commercial hybrids used in 1986 with the set of hybrids used in 1981, Smith (1988a) concluded that genetic diversity in U.S. hybrid maize grown in the Corn Belt has remained at an identical level quantitatively. However, the time period for comparison was short in relation to the rate of varietal turnover. Nevertheless, most surveys have shown that there is little immediate prospect for a large-scale increase in genetic diversity of hybrid maize (Cox et al., 1988a; Goodman, 1985; Zuber and Darrah, 1981).

Maize breeders claimed that their programs had broader germplasm bases in 1981 than they did in 1970 (Duvick, 1984a). By 1986, however, very few genes from exotic germplasm and none from related species had been bred into advanced lines or cultivars. Overall, exotic germplasm is rarely used in hybrids grown in the United States. Less than 1 percent of U.S. hybrid maize has exotic germplasm traceable to sources other than North American landraces and lines (Cox et al., 1988a). Lack of adequate facilities to evaluate materials that are sensitive to day length hamper the use of tropical germplasm in the United States (National Research Council, 1991a). Some evaluation of 400 tropical accessions with promising traits for temperate breeding programs has been done in the southern states of Florida and Texas (Castillo-Gonzalez and Goodman, 1989) and breeding materials adapted to the United States are being developed from elite tropical sources (Holley and Goodman, 1988).

Numerous tolerance and resistance traits can be obtained from exotic as well as adapted maize varieties. Exotic germplasm is a source of resistance to maize bushy stunt mycoplasma, rust (*Puccinia* sp.), leaf blight, maize streak virus, maize chlorotic dwarf virus, maize chlorotic mottle virus, rootworms (*Crambus caliginosellus*), stalk borers (*Papaipema nebris*), and earworms (*Heliothis zea*) (Cox et al., 1988a; Nault and Findley, 1981; Prescott-Allen and Prescott-Allen, 1986). Exotic landraces and other species are the only known sources of resistance to certain viruses (Nault and Findley, 1981). Many maize breeders report that they can also find resistance in already adapted elite lines (Duvick, 1984a). They may be picking up resistance factors from the earlier introgression of exotic genes, however. Continuation of introgression of exotic germplasm into adapted materials can be valuable for future breeding maize programs and most likely will fall to public sector and international programs like Centro Internacional de Mejoramiento de Maíz y Trigo (CIMMYT, International Maize and Wheat Improvement Center).

The international maize outlook has changed considerably over the past 20 to 30 years. During the period 1961–1965 to 1983–1985, world maize production increased by 84 percent, representing an annual growth rate of 4 percent (2.8 percent in developing countries) (Centro Internacional de Mejoramiento de Maíz y Trigo, 1987b). By 1985, improved genotypes were grown on 40 million ha or about one-half of the total maize area in developing countries. Thirty million hectares are planted with hybrids, with the People's Republic of China, Brazil, and Argentina accounting for 80 percent of that area (Centro Internacional de Mejoramiento de Maíz y Trigo, 1987b). Ten million ha are planted with improved open-pollinated varieties. In the remaining half of the total maize area in developing countries, however, adoption rates of improved varieties are low. This may be due to seed production systems that restrict the diffusion of improved varieties (Timothy et al., 1988) or because farmers prefer landraces.

The rate of variety turnover to hybrid maize in developing countries can be illustrated by using Kenya as an example. Maize is the major staple in Kenya. Until the 1950s, nearly all varieties grown in Kenya were local landraces traceable to Tuxpeno-Hickory King landraces brought from North America in the nineteenth century (Timothy et al., 1988). In the late 1950s, two exotic varieties from Latin America, Ecuador 573 and Costa Rica 76, were introduced as breeding stocks to hybridize with local landraces. By the mid-1960s, the first inbred lines and hybrids were released in Kenya and were grown on half of the large-scale farms there. The impact of the hybrids' high yields was so great (30 percent higher than earlier synthetic hybrids) that Kenya's hybrid maize area increased from 0 to 600,000 ha between 1963 and 1981. Both large- and small-scale farmers have become dependent on hybrids, so that 61 percent of Kenya's total maize area was devoted to hybrids in 1985, and another 5 percent to improved open-pollinated varieties (Timothy et al., 1988).

CIMMYT has been one of the primary suppliers of germplasm to Kenyan national programs, even though most of the CIMMYT populations are generally susceptible to common rust (*Puccinia* sp.) and turcicum blight (*Helminthosporium turcicum*), both of which are problems in Kenya (Timothy et al., 1988). Streak virus tolerance has been transferred to inbred lines from a maize composite developed by the International Institute of Tropical Agriculture (IITA). Despite these introductions of exotic germplasm, the germplasm base in Kenya remains limited because of the policies of the Kenyan Seed Company. From 1963 to 1981, the Kenyan Seed Company and its government-controlled predecessors released only five open-pollinated varieties, four varietal crosses, three double-cross hybrids, two three-way-cross hybrids, and two top-cross hybrids; some of these used the same CIMMYT parents (Timothy et al., 1988). It is therefore not surprising that a severe outbreak of leaf blight (*Alternaria triticina*) began in 1978 and became widespread by 1979 (Singh et al., 1979). It affected two HYVs and several of the remaining local landraces as well. A major concern is that the Kenyan and other nationalized seed industries in developing countries may be modeling their breeding programs after the IARCs, but are not yet able to respond rapidly to resistance breakdown in released cultivars (Lipton and Longhurst, 1989).

Maize hybrids developed in both the private and public sectors have spread to developing countries where they were previously absent and, for the present, have brought novel germplasm to those countries. Dominance by just a few hybrids or a few sources of germplasm may now be of concern in some of those countries. However, 49 percent of the total maize-growing area in developing countries remains planted to locally adapted landraces and the progeny of purchased, open-pollinated seed (Centro Internacional de Mejoramiento de Maíz y Trigo, 1987b).

Those countries with a high proportion of area sown to the new HYVs face a dilemma. They probably cannot return to cultivating the indigenous maize varieties without reducing yields, yet they cannot continue to plant the same high-yielding varieties indefinitely because new pest races likely will appear, causing disastrous epidemics. The vulnerability is particularly acute in tropical and subtropical areas that lack a cold season (and often little dry season). Lack of support for public plant breeding efforts in many developing countries makes it unlikely that they will be able to mobilize new varieties in sufficient time to prevent disaster.

Rice

Genetic uniformity is a greater concern in rice than it is in wheat because of this crop's limited diversity for three major traits: the single locus for semidwarfism (*sd-1*); a common cytoplasmic ancestry for many HYVs; and a widely used, single source of cytoplasmic male sterility in Chinese hybrid rices (Chang, 1984a; Chang et al., 1985).

Cytoplasmic uniformity, a concern in maize since 1970, is now a concern in rice and other major crops (Hargrove et al., 1980). Resistance or susceptibility to disease is influenced by cytoplasmic inheritance not only in maize and rice but also in strawberries, potatoes, wheat, grain sorghum, cotton, and grapes (Hargrove et al., 1980). In rice, 38 percent of the female parents used worldwide in rice breeding in 1983 and 1984 traced maternally to one Chinese-Indonesian variety named Cina, which is commonly used at IRRI and elsewhere. The use of semidwarf parents in breeding began to increase in 1965. Although the use of locally developed semidwarf lines as females has dominated various national breeding programs since then, nearly all of them can be traced back to *sd-1* semidwarf ancestors. Hargrove et al. (1985:9) concluded that "the genealogies of improved rice varieties across Asia appear disturbingly similar." Although no weakness in Cina cytoplasm or *sd-1* semidwarfs has been reported, strong recommendations were made to diversify the genetic base of future rice varieties.

The area planted to rice HYVs (more than 73 million ha in 1982 and 1983) exceeds those planted to wheat and maize HYVs in developing countries (Dalrymple, 1986b). It is well known that a very small number of varieties accounts for a very large proportion of the area planted to rice HYVs. In Malaysia, for example, three varieties accounted for about 40 percent of all Malaysian rice plantings in 1981 and 1982. In Indonesia, two varieties accounted for 54 percent of the total rice plantings in 1983 and 1984. In the Philippines, it was estimated that two IRRI varieties occupied about 90 percent of the entire rice-growing area during the 1984 dry season (Dalrymple, 1986b).

The widespread planting of a few closely related varieties with intensive cultivation practices and multiple cropping has led to a breakdown in varietal resistance and to outbreaks of tungro virus, grassy stunt virus, and the brown planthopper (*Nilaparvata lugens*) in parts of Bangladesh, southern India, Indonesia, Malaysia, the Philippines, Thailand, and southern Vietnam, although none of these was as destructive as the southern corn leaf blight was in the United States (Chang, 1988). In contrast, no serious epidemics have been reported from parts of Thailand, where traditional varieties are commonly grown in rotation with HYVs (Chang, 1984a). Multiple and continuous cropping of the same HYV and staggered planting dates across vast production areas have contributed to the breakdown of monogenic (or vertical) resistance because of changing biotypes of the brown planthopper in the Philippines, Indonesia, Sri Lanka, and Vietnam. Sequentially released new varieties (for example, IR26 with the gene bph-1, IR36 and IR42 with the gene bph-2, and IR56 and IR60 with the gene bph-3) with related genetic backgrounds were each developed by adding a gene to deal with an emerging insect biotype that had rendered resistance in earlier releases ineffective. Although these varieties provided some relief, new pest biotypes soon evolved (Heinrichs et al., 1985). The new insect populations also transmitted virus more effectively, as in the case of the tungro virus transmitted by the green leafhopper (Nephotettix spp.) (Dahal et al., 1990). Through the misuse of host resistance, rice farmers have suffered severe losses from the resulting "boom-and-bust" cycles of oligogenic resistance (Chang, 1984a).

During the 1970s brown planthopper and the associated grassy stunt virus inflicted heavy losses to the rice crop over a broad region (Dyck and Thomas, 1979). Following a decade of planting, the virus resistance gene (Gsv) obtained from Oryza nivara was rendered ineffective when a second biotype of the virus emerged. Similarly, when semidwarfs were widely grown, the predominant stem borers shifted from striped borers (Chilo zacconius) to yellow borers (Scirpophaga incertulas) (Chang, 1988).

Indiscriminate use of insecticides has also been linked to a resurgence of the brown planthopper in some areas (Chelliah and Heinrichs, 1980). By contrast, Indonesia has made notable progress controlling pest populations by varietal rotation, pest monitoring, communitywide pest control, and synchronization of planting dates (Manwan et al., 1985).

With more intensive cultivation of fewer varieties, rice diseases and insects continue to grow in number, intensity, and geographic distribution (Chang, 1984a, 1988). Examples include bacterial blight (Xanthomonas campestris), which reduced seed yield by 50 percent, largely on an HYV under irrigation in Niger in 1983 (Reckhaus and Adamou, 1986), and rice yellow mottle virus, which spread from irrigated paddy rice in Niger, Burkina Faso, and Mali and has infected HYVs from IRRI in Côte d'Ivoire and neighboring countries (Awoderu et al., 1987).

The Republic of Korea initially reported spectacular increases in grain yield after using the semidwarf gene in Tong-il and other varieties. In 1978, about 76 percent of the rice area was planted to HYVs. However, a cool season and widespread blast epidemic in 1979 affected the HYVs more severely than they did the traditional varieties and led to heavy losses (Rural Development Administration, 1985). The area of HYVs dropped to 20 percent in 1988, and national rice production slumped.

Hybrid rice in the People's Republic of China represents another potentially vulnerable situation. The F_1 hybrids grown in the major rice production areas share a common cytoplasmic male sterility source and the *sd-1* locus. The area planted with hybrids has rapidly expanded from 5 million ha in 1979 to 15 million ha in 1990. So far the Wild Abortive cytoplasm remains the most stable source of male sterility (Yuan and Lin, 1989).

Rice faces special pressures from narrowly based diversity for major traits, continuous use of the same variety or close relatives in multiple cropping, dangerously intensive cultivation practices, reliance on monogenic resistance to important insect pests, and warm and humid environments that sustain the multiplication of pests and rapid shifts within insect populations. The genetic base has also narrowed in temperate regions, such as in Japan (Kimura et al., 1986) and the United States (Dilday, 1990). The inclusion of wheat and rice in continuous cropping systems (Bangladesh, the People's Republic of China, and India) poses new risks of developing new pathogens common to the two hosts (Chang, 1988). The custom of growing traditional varieties in rotation with HYVs in different crop seasons, as is practiced in Thailand, can reduce potential crop losses.

Minor Crops

In 1988, the Agricultural Research Service (ARS) of USDA requested that the Crop Advisory Committees (CACs) assess the status of genetic resources utilization, evaluation, conservation, and vulnerability for their particular crops. The USDA has CACs for all major crops and most minor crops of importance to U.S. agriculture (National Research Council, 1991a).

For apples, two cultivars account for more than 50 percent of the entire U.S. crop. The CAC for apples concluded "commercially cultivated apples are so lacking in genetic diversity that they could easily become victim to a catastrophe" (Apple Crop Advisory Committee, 1987:16). Walnuts are represented by a few cultivars, all of which come from a small gene pool of progenitors. Three diseases and two pests do economic damage to the walnut crop.

Prunus cultivars in the United States—peach, plum, cherry, almond, and apricot—have a limited genetic base. Most commercial peach cultivars can be traced back to fewer than a dozen parents, with the variety Chinese Cling represented in the majority of them. Commercial nectarine cultivars can be traced back to four parents. The entire tart cherry industry is based on a single heirloom cultivar, Montmorency. The sweet cherry industry uses the Bing cultivar on

A crop of sunflowers is grown for the production of sunflower oil. Credit: Food and Agriculture Organization of the United Nations.

75 percent of its acreage in the northwestern United States. Apricots have a narrow gene base, whereas plums have the broadest of any of the *Prunus* crops in the United States. The *Prunus* CAC considered all of these stone fruit tree crops in the United States to be susceptible to insect and disease pests. Because the ecologic risks associated with certain chemical pesticides limit their use in pest and disease control, vulnerability will likely increase.

The *Vitis* CAC said that grape is a prime example of genetic vulnerability because of uniformity. It cites examples of "the narrow genetic base upon which most of the world's commercial grape production rests" (Grape Commodity Advisory Committee, 1987:10). It suggested that a majority of vineyards in the United States contained grapevines susceptible to a wide range of pests. Pests and diseases are being introduced into formerly "clean environments" (Grape Commodity Advisory Committee, 1987:11). Also, new abiotic stresses from herbicides, ozone, and sulfur dioxide contamination are affecting grapes.

Sugar beets also suffer from a narrow genetic base derived from very few parents. They are vulnerable to four diseases and two major pests. Between 1974 and 1976, the sugar beet crop in the central and western parts of the United States suffered from a powdery mildew epidemic (*Erysiphe betae*). Another root crop, the sweet potato, is considered to be in a critical state. One cultivar makes up 75 percent of sweet potato production in the United States, and 79 percent of all cultivars are derived from just three parents.

Sunflower hybrids grown in the United States also have a very narrow genetic base, with most being derived from a few inbred lines produced by USDA. Only one source of cytoplasmic male sterility is used throughout the entire world. Although genes for resistance to some major insect and disease stresses can be found in the remarkably complete assemblage of wild North American sunflowers (Heiser, 1975) now maintained by USDA, the wild sunflower program has been severely curtailed because of a lack of funds.

Although the CAC reports identify other crops for which the genetic bases have been broadened (cowpeas, peanuts, tomatoes, clover), on the whole, genetic diversity in the minor crops appears to be seriously restricted and to lag behind gains that have been made for major crops. The level of resources it has taken to build in early warnings of potential epidemics and alternate protection strategies for major crops is proportionately reduced or absent for minor crops. Yet, the costs in loss of employment, closure of farms, or disappearance of industries and businesses resulting from an epidemic would be severe, both economically and socially.

Genetic Vulnerability Today

For major crops in developed countries, varietal turnover and the number of varieties planted have increased since 1972, indicating that the overall level of genetic vulnerability may have decreased. The genetic basis of elite germplasm, however, was found to be shallow because of extensively shared ancestry and limited use of exotic germplasm. Continued efforts to broaden the genetic diversity of primary breeding pools of major crops would provide enhanced and more stable resistance to biotic and abiotic stresses.

In the United States most minor crops have extremely narrow genetic bases, and progress in diversifying them lags far behind that in major crops. The relatively small industries associated with many of these crops do not have the resources to incorporate rapidly needed genes for resistance into new cultivars. The fruit crops are particularly vulnerable to epidemics in the near future, because many of the chemicals that have buffered them from disaster over the past 3 decades are now being withdrawn from the market. Small farmers and hand laborers may be adversely affected by the lack of genetic protection bred into these crops.

In the developing world, high-yielding hybrids and modern varieties of the major crops (especially rice and wheat) have come into dominance within the past 15 to 25 years. The immediate effect of these introductions is to make available novel genes for resistance to widespread diseases to which landraces may have been susceptible. Simultaneously, rapid losses of native diversity are occurring. As a few releases from the same breeding programs come to dominate a developing country's production, increased vulnerability will almost certainly emerge. It is doubtful whether breeding programs in developing countries can react as rapidly to future epidemics as U.S. maize breeders did between 1970 and 1972.

For the past 20 years, leadership in wheat and rice breeding for developing countries has come from CIMMYT and IRRI, respectively. There are indications that these institutes will devote smaller proportions of their effort to variety development in the future. The national breeding programs in developing countries are not nearly as well funded nor as effective as are IARC breeding programs. There is some question, therefore, whether sufficient breeding efforts worldwide will be devoted to developing and using new wheat and rice varieties at frequent intervals for the developing world. This is of particular significance if the IARCs move away from breeding in favor of other activities, such as biotechnology or development of farming systems for marginal environments before national breeding capabilities are more uniformly strengthened.

At least 5 to 10 years of effort are needed to put strong national breeding programs into place or to improve ineffective ones. Those countries with undersized or ineffective programs that are now depending on the IARCs for new HYVs must begin as soon as possible to build their national breeding programs to the size and efficiency that will serve their countries properly. The pooling of several small programs can, in theory, provide a sufficient base for developing and using improved varieties at frequent intervals. Unfortunately, the histories of planned, government-led, intercountry cooperation show few successes. Remarkable skills in management and diplomacy and stable resources are needed to make such cooperation productive. However, commodity-based networks such as the Programa Regional Cooperativo de Papa (PRECODEPA, Cooperative Regional Potato Program) for potatoes and the Latin American Maize Project for maize promise to be effective in disseminating germplasm and promoting crop improvement.

RECOMMENDATIONS

Assessment of genetic vulnerability involves the synthesis of four components: the area devoted to major cultivars, trends in pest and pathogen evolution, the extent of genetic uniformity for resistance or susceptibility to biotic and abiotic stresses, and the ability for development of replacement cultivars. Such data are essential to monitoring the potential for severe crop losses. Reductions in public sector funding for agricultural research and service agencies have seriously diminished capabilities for monitoring and responding to crop vulnerability in the United States and other nations.

The potential for crop vulnerability must be nationally and globally monitored.

Information on potential threats should be exchanged among cooperating institutions (for example, the Food and Agriculture Organization of the United Nations, IARCs, USDA, and other national agencies). Responsibility for gathering data lies in national authorities that oversee varietal development and use, and in international and multinational organizations that distribute seed widely.

Genetic uniformity per se does not cause crop susceptibility or epidemics; indeed, uniformity for many traits is demanded by farmers, processors, and consumers. When susceptibility to diseases, pests, or environmental stresses exists, then uniformity for that trait (regardless of whether the variety is otherwise homogeneous or heterogeneous) enhances the risk of crop damage.

A wide range of genetic and agronomic strategies should be employed to minimize crop uniformity and consequent susceptibility.

Plant breeders and agronomists should balance the need for uniformity in agronomic traits with the need for diversity in resistance to biotic and abiotic stresses. Extension specialists should work with processor and consumer groups to win acceptance of greater diversity in crop varieties.

Complex interrelationships among plant hosts, pathogens, pests, vectors, and the environment are evolutionary in nature and determine the crop response—whether it will be resistant or susceptible. Pyramiding of resistance genes provides more durable resistance; single-gene resistance is often more easily overcome by evolving pest populations. Varietal mixtures and varietal deployment can be used to supplement genetic manipulation of resistance. Additionally, the genotypes of successive releases must vary significantly to provide genetic diversity over time.

Agricultural systems that mimic natural ecosystems through use of, for example, multiple lines, varietal mixtures, relay cropping, and integrated pest management, are significantly more stable than monocultures of a few genetically close varieties and, in many cases, provide yields at least as good as those of monoculture systems.

Greater emphasis is needed on agronomic management strategies (for example, multiple cropping, crop rotation, tillage practices, crop mixtures, varietal mixtures, multiple lines, integrated pest management, pest trap crops) through extension services and private-sector marketing campaigns as preventative measures to pest and pathogen attack.

2

Crop Diversity: Institutional Responses

The period since 1970 has seen a growth in germplasm collection and conservation. Crop breeding programs have increasingly focused on international needs. Along with this increased interest has come questions about the changing roles of the public and private sectors in collecting, conserving, and using genetic resources. This chapter discusses germplasm collection and conservation efforts worldwide and the global impact of related activities in the United States.

GERMPLASM COLLECTION AND CONSERVATION WORLDWIDE

Several centers of the Consultative Group for International Agricultural Research (CGIAR)—most notably the International Board for Plant Genetic Resources (IBPGR)—have been active in conserving and managing the genetic resources of plants. The establishment of the IBPGR in 1974 by the CGIAR signaled a commitment to conservation. IBPGR works with nearly all countries of the world to promote and coordinate the establishment of genetic resources centers and to further the collection, conservation, documentation, evaluation, and use of plant germplasm. It is a multidisciplinary effort guided, in part, by the needs of national and international germplasm banks (Williams, 1989a). In October 1991, a previously ratified agreement was signed by board members from China, Denmark, Kenya, and Switzerland, which established the International Plant Genetic Resources

Institute (IPGRI). After ratification by the Italian government is obtained, IPGRI will assume the duties of IBPGR and the latter will cease to exist.

Role of IBPGR in Catalyzing Collection and Conservation

Since its establishment, IBPGR has organized over 400 collecting missions in more than 100 countries. Duplicate samples are offered to the country of origin, in accordance with IBPGR policy. Seed handling units have been established in Costa Rica, Singapore, and the United Kingdom to facilitate the distribution of collected samples (van Sloten, 1990a). IBPGR has also published a wealth of literature on the scientific, technical, and organizational aspects of germplasm conservation.

Initially, IBPGR placed a high priority on the collection of major food crop cultivars. However, the importance of wild relatives and primitive landraces has been increasingly recognized in recent years, and these are receiving greater attention (Hoyt, 1988; International Board for Plant Genetic Resources, 1985a). In 1988, IBPGR organized or funded 22 collecting projects that dealt primarily with wild species of roughly 50 commodity crops. Ten projects were organized to collect landraces and primitive cultivars (International Board for Plant Genetic Resources, 1989a). Nevertheless, progress has been hampered by limited knowledge of the distribution of the secondary and tertiary gene pools of many crops (Williams, 1989a).

Although the acquisition efforts of the IBPGR and the other international agricultural research centers (IARCs) are useful for the commodities within their mandates, reviewers have pointed out the need for a strategy dealing with other crops valued by developing countries that do not fall within the CGIAR mandate (Hawkes, 1985).

Germplasm Banks Worldwide

IBPGR has played a role in developing germplasm banks worldwide; this has been done chiefly in partnership with national governments, regional organizations, and the IARCs (Consultative Group on International Agricultural Research, 1985). Germplasm banks have been established in 92 countries and 10 IARCs (Alexander von der Osten, personal communication, CGIAR Secretariat, October 28, 1992). More than 100 countries have some form of genetic resources program or carry out related activities (Williams, 1989a), but it is far from clear how many are investing, or are able to invest, sufficient funds to do the job properly. About one-third of the world's holdings is in the IARCs (van Sloten, 1990a).

There has been dramatic growth in the germplasm bank storage capacity in developing countries, from a handful in 1976 to about 30 today (Chang, 1992). Many germplasm banks have received some technical assistance, financial assistance, or both from IBPGR (Plucknett et al., 1987). Through 1984, IBPGR allocated nearly $2.5 million to 37 countries for conservation purposes (Hawkes, 1985). Several Asian countries have built modern seed storage facilities with aid supplied by the government of Japan (T. T. Chang, International Rice Research Institute, personal communication, October 1990). Some critics question whether so many germplasm banks are necessary and whether the funds might have been better invested in crop improvement pro-

During test crosses of rice plants at the International Rice Research Institute in the Philippines, the flowering spikes of the plants are enclosed in envelopes to prevent further cross-pollination. Credit: Bruce Dale, ©1992 National Geographic Society.

grams (Frankel, 1989a). Furthermore, it is uncertain whether many germplasm banks will ever function without CGIAR support.

Irrespective of germplasm bank facilities, there are about 400 significant crop germplasm collections (Williams, 1989a), but little information on how many of these are financially viable. IBPGR has designated 152 of these at 38 centers as global, continental, or regional base collections (International Board for Plant Genetic Resources, 1989a). They cover the major food crops, vegetables, and forages—roughly 50 commodities. In addition, 23 centers have agreed to conserve global, continental, or regional field collections of 9 vegetatively propagated crops (International Board for Plant Genetic Resources, 1989a). However, how functional many of these are is unclear.

As the number of germplasm banks increases, the need for information exchange and coordination of their responsibilities also grows. The IBPGR originally envisioned a regional framework for germplasm bank coordination. This goal was successful for specific needs, such as the European Cooperative Program for the Conservation and Exchange of Crop Genetic Resources (Williams, 1989a). However, the regional strategy was not effective where significant differences existed in program sophistication and commitment, where there was a lack of historical basis for collaboration, or where permanent financial support was not assured. Latin America, East and West Africa, and particularly, Southwest Asia lagged behind Southeast Asia and Europe in developing regional efforts (Hawkes, 1985). Among the commodity-based IARCs only the International Rice Research Institute (IRRI) and the Centro Internacional de la Papa (CIP, International Potato Center) have been fully effective in working with their crops. Currently, IBPGR is reorganizing its approach to one of crop-oriented networks backed up by crop-specific data bases on the premise that it will facilitate the use of the collections by breeders and other scientists (van Sloten, 1990a).

Various attempts to assess the completeness of existing collections (International Board for Plant Genetic Resources, 1985a; Lyman, 1984) have been criticized for putting too much emphasis on the numbers of accessions rather than patterns of diversity, availability, and security (Chang, 1989; Williams, 1989a). Nevertheless, the continued growth of collections is placing strains on their management and funding. Unfortunately, many collections have never been adequately managed.

Considerable interest and controversy surrounds the concept of core collections or subsets as a strategy for reducing management burdens and facilitating use of the collections (Brown, 1989a) (see Chapter 5). Critics of the scheme argue that large collections (those with a high proportion of the crop's total diversity) offer certain operating efficiencies (Chang, 1989) and that the reserve collection might be neglected.

Other practical issues were recognized early on as a source of serious future problems (Consultative Group on International Agricultural Research, 1985). Below-standard conditions for maintenance, lack of regeneration capability, insufficient passport data, incomplete data bases, insufficient technical staff, and inadequate operating budgets have been identified (Goodman and Castillo-Gonzalez, 1991; Plucknett et al., 1987; Williams, 1989a) as serious constraints to ensuring survival of the germplasm let alone to maximizing benefits from the germplasm bank network.

The potential severity of these problems is illustrated by several findings. In its review of the U.S. National Plant Germplasm System (NPGS) this committee noted that tests conducted between 1979 and 1989 showed that 29 percent of the National Seed Storage Laboratory's 232,210 accessions had seed germination rates that were either unknown (21 percent) or less than 65 percent (8 percent). Of all of the accessions, 45 percent had less than 550 seeds (National Research Council, 1991a). The United States has fewer economic or other constraints than most other countries but, like nearly all of them, it has emphasized storage facilities at the expense of regeneration, evaluation, and utilization.

The inadequacy of international efforts to conserve Latin American maize germplasm has been noted several times (Goodman, 1984; Goodman and Castillo-Gonzalez, 1991; Goodman and Hernandez, 1991; Salhuana et al., 1991). Salhuana et al. (1991), coordinators of the Latin American Maize Project (LAMP), illustrated the fragile status of these accessions. Only about half could be evaluated due to lack of viable seed. Almost one-fourth of the 300 races failed to have even one accession with sufficient viable seed for evaluation. The lack of reliable storage facilities has resulted in the total loss of a large number of accessions and severe genotype deletions (genetic drift) in many more.

Evaluation

Until recently, IBPGR considered only characterization and preliminary evaluation to be within its purview and acknowledged that these tasks lagged behind exploration and collection. However, it is now putting greater effort into collecting botanical and other data on priority crop collections for entry into international crop data bases. In 1988, projects were under way for 54 collections of 30 commodities

(International Board for Plant Genetic Resources, 1989a). Evaluation is costly. In 1989, Williams (1989a) conservatively estimated that immediate needs for evaluation of major crop germplasm would require $30 million over a 5-year period. He compared this figure with total annual expenditures worldwide of $55 million for managing crop genetic resources.

Although in-depth evaluation is the responsibility of interested scientists rather than IBPGR, few germplasm banks of national programs have developed clear links with those breeders and others who evaluate (Williams, 1989b). It is apparent that many national programs do not have the resources or personnel to carry out evaluation or regeneration. An analogous situation exists for enhancement, thus creating a serious gap with a major negative impact for crop improvement in developing countries that falls outside IARC-mandated commodity programs (Hawkes, 1985).

IBPGR Interactions with the IARCs

The working relationships between IBPGR and the IARCs have evolved over the past 15 years as the programs of each have matured. Initially, IBPGR provided assistance to field collections, installing or upgrading some storage facilities, and assistance with documentation. Misunderstanding developed over IBPGR's role and claims, with some IARCs considering it a source of continued funding support for genetic resources work. This was clarified, and the IARCs have substantially increased their allocations for genetic resources work (Consultative Group on International Agricultural Research, 1985). Because IARC genetic resources programs have grown in size and sophistication, several have taken the lead in germplasm activities related to their commodities (for example, IRRI for rice and CIP for potatoes). Most of the IARC collections operate as base collections that are partially backed up by duplicate collections stored at another center. IBPGR now serves primarily in a coordinating role, collaborating when warranted on special projects and copublishing CGIAR publications on plant genetic resources activities.

Interactions with National Programs

IBPGR and other IARCs have had a broad impact in promoting and establishing genetic resources programs at the national level, with impressive progress seen in the growth of germplasm collections in developing countries. However, many problems remain that are beyond the ability of IBPGR, or perhaps anyone else, to resolve. These include continued uneven geographic coverage, operational problems, inadequate budgets and personnel, and neglect because of the low priority of germplasm collections in national development programs (Consultative Group on International Agricultural Research, 1985). With many developing countries hardly able to support minimal agricultural research programs, it is doubtful that their germplasm programs will continue without external assistance.

Several reviews point to the lack of linkages between national germplasm banks and plant breeders as a major weakness obstructing the use of the collections (Chang, 1985a,b; Consultative Group on International Agricultural Research, 1985; Frankel, 1989a). In a few cases, germplasm banks were established in countries that had no plant breeding programs to exploit the resources (Williams, 1989b). In the opinion of one expert, the shortage of plant breeders in small developing countries is a far more urgent problem than is genetic resources programs (Frankel, 1989a).

These critical weaknesses present opportunities where support by bilateral donors would have a major impact on the realization of benefits from national germplasm banks. Linkages between national genetic resources programs and breeding activities should be fostered through the support of germplasm enhancement and through the support of data-base development (Cohen and Bertram, 1989). Such initiatives would help to integrate genetic resources programs into national agricultural development strategies, but only if the support were sufficiently long term.

Roles of Other International Agencies

A number of other international organizations, both governmental and nongovernmental, have made important contributions to conserving plant genetic diversity. Notable examples of organizations pursuing ecosystem-oriented conservation efforts include the International Union for the Conservation of Nature and Natural Resources (now known as the World Conservation Union), the United Nations Environment Program, and the World Wide Fund for Nature (Drake, 1989; Heywood, 1989). The Food and Agriculture Organization (FAO) of the United Nations has expressed interest in the exchange of plant germplasm and genetic resources issues as part of its global program since 1947. It has provided IBPGR with facilities (until 1989) and support (Esquinas-Alcazar, 1989).

Since the late 1970s, FAO has been the forum for debate over the control of genetic resources that intensified after passage of the U.S. Plant Variety Protection Act in 1970 (7 U.S.C. Sections 2321–2583).

Some have asserted that the CGIAR germplasm banks have exploited developing country resources for the benefit of multinational corporations (Brockway, 1988; Fowler and Mooney, 1990; Mooney, 1979, 1983). However, this assertion has been contested by others (Brown, 1988; Plucknett et al., 1987; Witt, 1985).

An FAO resolution was passed in 1983 that established an intergovernmental forum, a financial mechanism, and a legal basis for the coordination of international germplasm responsibilities (Esquinas-Alcazar, 1989). The intent was to give developing countries more control over both the global germplasm bank network to be developed and the use of genetic resources from their countries (McMullen, 1987). Until recently, the United States and many other developed countries declined to participate, in part because of conflicts over plant breeders' rights. After protracted controversy, the various sides appear to be approaching a consensus. Elements of the FAO's proposed program include a global monitoring system, establishment of a network of in situ conservation sites, and periodic reports on the status of world plant genetic resources.

DEVELOPMENT OF INTERNATIONAL CROP BREEDING PROGRAMS

The establishment of IRRI, Centro Internacional de Mejoramiento de Maíz y Trigo (CIMMYT, International Maize and Wheat Improvement Center), and other commodity-based IARCs since the late 1960s has greatly enhanced the use of plant germplasm to develop high-yielding varieties (HYVs) of food crops. The IARCs have become the main sources of externally supplied germplasm used by national agricultural research programs. By 1983, the national programs had released over 1,000 new varieties of cereals, legumes, and root crops developed with IARC-provided germplasm (Anderson et al., 1988) (Table 2-1).

The adoption of the new HYVs has been widespread, and for rice and wheat has contributed to unprecedented yield increases of nearly 2 percent annually in developing countries (Anderson et al., 1988). By 1983, HYVs of wheat and rice had supplanted traditional varieties on approximately half of the lands used for these cereal crops in all developing countries, and in certain countries this occurred to an even greater extent (Dalrymple, 1986a,b). Yield increases attributable to the new cereal varieties alone (excluding increases attributable to fertilizers and other inputs) exceeded 36 million metric tons in 1983 over the yields in 1970, sufficient to meet the annual needs of 500 million people (Anderson et al., 1988). Thus, improved varieties have

TABLE 2-1   Number of Center-Related Varieties Released by
National Authorities in Developing Countries Through 1983[a]

Crop	Sub-Saharan Africa	Asia	Latin America	Middle East and North Africa	Total
Barley	0	2	0	8	10
Beans, field	4	2	90	0	96
Cassava	26	5	32	0	63
Chickpeas	0	1	0	2	3
Cowpeas	14	2	12	1	29
Maize	61	49	126	2	238
Pasture species	0	0	12	0	12
Pearl millet	5	3	0	0	8
Pigeon peas	5	2	0	0	7
Potatoes	31	16	12	2	61
Rice	31	140	129	2	302
Sorghum	8	18	5	0	31
Sweet potatoes	6	0	0	0	6
Triticale	2	2	7	0	11
Wheat, bread	40	44	114	66	264
Wheat, durum	5	3	13	20	41

NOTE: Excludes varieties developed by national programs from sources similar to those used by the international agricultural research centers.

[a]The term *center-related* means that a center of the Consultative Group on International Agricultural Research had direct involvement in developing the plant variety.

SOURCE: Anderson, J. R., R. W. Herdt, and G. M. Scobie. 1988. Science and Food: The CGIAR and Its Partners. Washington, D.C.: World Bank. Reprinted with permission, ©1988 by The World Bank.

had a major impact on reduced food costs and improved nutrition in the developing countries, particularly among the poor. The semidwarfing genes in wheat and rice have also led to increased yields in developed countries (Chang, 1988; Dalrymple, 1980).

Progress with Legumes, Root Crops, and Vegetables

In general, IARCs working on legumes, root crops, and vegetables were established more recently than those that focused on rice, wheat, and maize. Germplasm collections are less complete for these crops than they are for the major cereal crops (particularly for related wild species), with the exception of peanuts, potatoes, and tomatoes (Lyman, 1984). Therefore, the impact of varieties developed from IARC-related germplasm is less dramatic, but the successes are mounting. In

During harvest time at a demonstration potato field in Bolivia, a government extension agent explains ways that farmers can increase their yields. Credit: Food and Agriculture Organization of the United Nations.

terms of numbers of varieties released, the greatest progress for noncereals to date has been achieved in beans, cassava, and potatoes (Table 2-1).

In beans, for example, a network of Latin American researchers coordinated by the Centro Internacional de Agricultura Tropical (CIAT, International Center for Tropical Agriculture) was responsible for the development of varieties resistant to golden mosaic virus. These are now grown in more than 20 different countries, replacing traditional varieties on 40 to 60 percent of the areas planted to beans. Yield increases of 20 to 30 percent have been realized with no other change in production practices (Anderson et al., 1988). Both CIAT and the International Institute for Tropical Agriculture (IITA) maintain and distribute cassava germplasm, from which more than 60 varieties have been developed or released by national programs (Table 2-1). Potato germplasm distributed by CIP is currently under evaluation in 80 countries through 5 networks. By 1984, varieties had been named or released in 23 developing countries (Anderson et al., 1988).

Until very recently, the CGIAR network has not supported work on vegetables. The independently funded Asian Vegetable Research and Development Center has assumed leadership for research on cabbage, mung bean, soybean, sweet potato, pepper, and tomato. Although efforts are oriented toward Asia, germplasm is exchanged with over 100 countries worldwide (Asian Vegetable Research and Development Center, 1988).

Nurseries and the Enhancement of Exotic Germplasm

An extremely important contribution of the IARCs is the role they play in the enhancement of exotic germplasm and in the dissemination of exotic germplasm through international nurseries. No other group of institutions is as well placed for such a job, with respect to the combination of daily access to world germplasm collections, plant breeding expertise, contacts with breeders worldwide, and reasonably stable funding. Because of the difficulty of handling exotic accessions under temperate conditions and the decline in public sector enhancement activities, the developed countries are very nearly as dependent on the IARCs for these services as are the developing countries. In sum, access to a steady stream of freely available enhanced germplasm is arguably a sufficient reason by itself for support of the IARC network by developed countries.

The centers distribute thousands of seed samples annually to nearly every country in the world. The most extensive form of distribution is through international nurseries, which are designed to test varieties for wide adaptability and yield or for resistance to specific pests, diseases, or environmental stresses. Participating countries have the opportunity to test their own materials against the best international ones and to observe the performance of international materials that may be suitable for the conditions in their own countries. Specialized nurseries designed to screen for disease resistance provide early warning of emerging pathogen threats.

Nurseries for testing wide adaptability of major cereals involve hundreds of scientists in 80 or 90 countries. Specialized nurseries usually operate on a smaller scale. Each center may coordinate anywhere from one or two up to a dozen such specialized nurseries for each commodity within its mandate. In 1988, CIMMYT ran five major nurseries and three specialized nurseries entailing the distribution of nearly 3,000 sets of samples (Table 2-2) (Centro Internacional de Mejoramiento de Maíz y Trigo, 1989).

Specialized nurseries are particularly important tools for disease surveillance. Collaborators help monitor prevalent insects and diseases and identify new forms that may pose a future threat. CIMMYT

TABLE 2-2   Number of Wheat Nurseries That Are Part of the
International Programs of Centro Internacional de Mejoramiento de
Maíz y Trigo, 1988

Country	Wheat, Bread	Wheat, Durum	Triticale	Barley	Germplasm Development	Specialized Nurseries
Africa	253	106	59	69	57	71
Asia	289	47	44	69	75	46
Europe	178	115	90	58	48	59
Latin America	307	106	83	42	63	80
Middle East	126	75	24	37	36	40
North America	56	19	20	21	18	24
Oceania	15	10	10	3	3	10
Total nurseries	1,224	478	330	299	300	330
Total countries	83	69	65	56	59	58

SOURCE: Centro Internacional de Mejoramiento de Maíz y Trigo (CIMMYT). 1989.
CIMMYT 1988 Annual Report: Delivering Diversity. Mexico, D.F.: Centro Internacional
de Mejoramiento de Maíz y Trigo. Reprinted with permission, ©1989 by Centro Internacional
de Mejoramiento de Maíz y Trigo.

has used its surveillance nursery to study the genetics of rust resistance in wheat (Centro Internacional de Mejoramiento de Maíz y Trigo,
1989). IRRI organized specialized nurseries for adverse environments,
such as those with deep water, acidic upland soil, and cool temperatures (Seshu et al., 1989).

Analyzing the diffusion of improved rice germplasm through nurseries coordinated by IRRI, Hargrove et al. (1985) urged careful attention to the extent of genetic diversity for important traits. They recommended that IRRI incorporate greater diversity for semidwarfism
and cytoplasm into its trial entries, possibly identifying the sources
of these and other traits in the nursery handbooks. They also suggested that IRRI assemble sets of improved cultivars with different
sources of resistance as a means of facilitating the incorporation of
greater diversity at the national level. Similar analyses are needed
for other commodities.

Networks and Interactions with National Programs

International centers have stimulated the development of crop
improvement programs where none existed previously. For example,
IITA's work on cassava, yams, sweet potatoes, and cocoyam has prompted
the establishment of 23 root and tuber programs in Africa since 1979
(Anderson et al., 1988). Commodity research networks have also

promoted greater investment in improvement programs. However, the growth of new programs has sometimes had the unfortunate effect of pulling scientists away from other programs because of the limited numbers of trained personnel.

National programs view the creation of high-yielding crop varieties and advanced breeding materials as the IARCs' most significant achievement (Anderson et al., 1988). The centers have also made important contributions to improving crop breeding methods and agronomic practices in developing countries through their training programs. Crop-specific techniques developed by the centers for insect and disease screening have also been widely adopted. Additional work is needed on strategies for increasing genetic diversity in national breeding programs through a combination of greater genetic diversity in nurseries, assistance with the evaluation of genetic diversity in national varieties, and assistance in planning breeding strategies, if necessary. IARC guidance with plant breeding technology is helping national programs to upgrade their capabilities to the point that they do virtually all of their own breeding work, when national support is adequate.

The international centers depend on national programs for technology adaptation and transfer. The IARCs can provide ideas, research objectives, methods, experimental germplasm, and training, but these must be adapted to local needs. The IARCs may supply finished materials (elite lines ready for naming), segregating progeny for selection under local conditions, or parental lines for local crossing, depending on the capability of the national program. Most often, the centers make crosses and send them to national programs for evaluation of segregating progenies. More advanced programs, such as the bean breeding program of the Instituto de Ciencia y Tecnología Agrícola (ICTA, Institute of Agricultural Science and Technology) in Guatemala, make their own crosses and exchange materials with CIAT and other countries.

It is important to recognize the growth in capabilities that many programs like ICTA have achieved, so that they are now carrying out their own crosses and developing their own varieties with minimal or no IARC input. Through a survey of Latin American rice breeding programs, CIAT found that half of them (representing 90 percent of the rice area in Latin America) generated their own genetic variability by making an average of nearly 1,900 crosses per year during the period from 1983 to 1987, which was more than CIAT had made (Centro Internacional de Agricultura Tropical, 1989). However, many subsequently experienced delays in completing yield trials and in producing basic seed for large-scale release.

The centers do not foresee an end to their collaboration with advanced national programs. Rather, they project a continuing role in the international testing and distribution of germplasm—tasks more effectively handled by an international organization. Support for international networks is another area of comparative advantage. The centers are also aware that many national improvement programs have serious constraints. To meet their needs, CIMMYT plans to continue development of ample quantities of experimental maize varieties for rapid local adaptation with a minimum of adjustment. CIMMYT is also continuing its commitment to practical training for maize improvement as well as outreach activities (Cantrell, 1989).

Research networks, such as the Asian Farming Systems Network, the Trypanotolerance Network, the International Network for the Improvement of Banana and Plantain, and the International Rice Testing Program, that have focused on specific commodities or problems have proved to be highly cost-effective mechanisms for advancing commodity improvement in national programs, particularly where scientists are isolated or few in number. Networks provide access to new technologies or strategies through, for example, regional monitoring tours and meetings, and collaborative research projects, with minimal cost beyond direct expenses. Many others exist outside the CGIAR system. Support for networks may come from the IARCs, national research institutions, bilateral donors, professional organizations, and so on. They present excellent opportunities to donors for high-impact, low-cost investments.

The role for donors in support of national germplasm systems has already been discussed, particularly for commodities not within center mandates (Cohen and Bertram, 1989). An equal need exists for support to crop improvement programs and public or private seed production systems that will help realize the benefits from germplasm. National programs must have the capacity in their crop improvement programs to take advantage of the germplasm available through the international network.

Concerns

There have been critics of the green revolution and, by association, of the IARC plant breeding programs (for a general overview see Simmonds [1979]). It has been argued that varietal development at the IARCs is done under favorable conditions, that it emphasizes heavy use of inputs that farmers cannot afford, and that disease resistance strategies in the early years were based on vertical resistance rather than horizontal resistance, which is believed to be more stable

(Anderson et al., 1988). These criticisms have been noted by the centers, and their breeding programs have been modified to meet national needs better (Anderson et al., 1988). The collections and plant genetic resources programs have expanded to include more land-races and related wild-type species in an effort to broaden the genetic base of crop improvement.

It is paradoxical that the success achieved by the IARCs, as demonstrated by the widespread adoption of their products, appears to be contributing to increased crop vulnerability on a global scale, although major disasters have so far been averted. Nevertheless, the situation warrants careful attention to increasing the genetic diversity of germplasm distributed through international channels and to other preventative agronomic measures.

International programs—both public and private—share a responsibility for ensuring that adequate genetic diversity is incorporated into the most widely used varieties or parental lines. Enhancement efforts should receive high priority and stable funding at increased levels. This is a particular charge of the public sector. Continued farming systems research is needed to come up with better alternatives to monoculture and promote more stable agronomic practices. Finally, the IARCs must plan carefully for their continued (although changing) responsibilities to national breeding programs. The IARCs have done a good job of tailoring their services and materials to a range of national capabilities, from quite limited to quite advanced. The centers must retain this spectrum of services, balancing the need for breeding finished varieties with the need for assistance with more basic research in new technologies.

GLOBAL IMPACT OF ACTIVITIES IN THE UNITED STATES

This section looks at the global effects of changes and other developments occurring in the United States. Areas of discussion include the U.S. National Plant Germplasm System, proprietary initiatives, and the shifting balance between public and private sector roles in varietal research and development.

Changes in the U.S. National Plant Germplasm System

A reorganization of federal and state efforts dating from the 1940s and 1950s created the NPGS in the United States in the early 1970s. The National Plant Genetic Resources Board was established by the secretary of agriculture in 1975 because of concerns raised by the 1970 corn blight and the 1972 report by the National Research Council (Butler and Marion, 1985). However, this board was abolished in 1992. The NPGS was initiated before many countries recognized the importance of conserving crop genetic resources. It is the world's largest distributor of plant germplasm (National Research Council, 1991a).

The NPGS is a diffuse network under the leadership of the Agricultural Research Service of the U.S. Department of Agriculture (USDA), with important contributions made by the Cooperative State Research Service of USDA and state agricultural experiment stations, as well as other public and private groups (Shands et al., 1989). Its mission statement, drafted in 1981, states in part that "the National Plant Germplasm System (NPGS) provides the genetic diversity necessary to improve crop productivity and to reduce genetic vulnerability in future food and agriculture development, not only in the United States, but for the entire world" (Shands et al., 1989:98). Many recommendations have been made in recent years to strengthen the NPGS, streamline its organization, improve its responsiveness to the user community, enhance its utilization of germplasm collections, and encourage greater collaboration with international efforts (Council for Agricultural Science and Technology, 1984a; General Accounting Office, 1981; National Research Council, 1991a; Office of Technology Assessment, 1987a; U.S. Department of Agriculture, 1981). However, few of these recommendations have been followed (National Research Council, 1991a).

Impact of Plant Varietal Protection and Patents

The debate over proprietary rights for plant and other life forms spans a range of technical, economic, legal, and ethical issues that are examined in depth in Chapter 12. This section discusses the impact of proprietary initiatives on public and private sector plant breeding efforts and genetic diversity.

Genetic Diversity and Germplasm Exchange

The passage of the U.S. Plant Variety Protection Act (PVPA) has clearly stimulated the development of new crop varieties at modest costs, in terms of increased seed prices, market concentration, and constraints on information exchange (Butler and Marion, 1985); but it is not evident that the new varieties are any more or less genetically diverse than they were before. It was thought that PVPA would reward breeders for turning out substantially different cultivars. Unfortunately, it has encouraged some breeders to produce near duplicates of elite cultivars through backcrossing and similar breeding techniques, an acceptable and financially more rewarding strategy (Duvick, 1987). The criteria for utility patents are more stringent than the criteria for certificates under PVPA, limiting the number of near-exact patented copies (Jondle, 1989).

The extent of diversity in new varieties is also obscured by the use of multiple brand names for similar or even identical products. For example, certain maize and sorghum hybrids based on crosses of publicly available inbred lines are sold under many different designations, and wheat and soybean varieties are given multiple names in the private seed trade. Whether or not hybrids and self-pollinated cultivars are correctly designated has little overt effect on their use as sources for future genetic advances. Once on the market, all are available for breeding. But the concentration of breeding and marketing efforts on a narrow circle of favored cultivars greatly restricts the possibility for genetic advance in the future. The extent of the reduction in genetic variance is hidden from public notice, because brand name diversity is not an adequate measure of the diversity of cultivar genotypes. Legal safeguards that are intended to prevent duplication in naming and invisible narrowing of the genetic base are only effective when they are enforced.

The problems of minimal numbers of cultivars and of duplication and copying may solve themselves to some extent. Cultivars containing genes that are close copies of a cultivar with a highly successful genotype usually are not quite as productive as is the original cultivar. There is a reward for breeders who lead the way with superior products with genuinely different genotypes. The rewards for such breeders will be minimized, however, if close copies of their creations, sold at discount prices, persistently turn up in minimally short times. Therefore, PVPA protection, or use of the utility patent with its broader protection, may be necessary to protect and encourage breeders to turn out agronomically and genetically distinct varieties.

International Impact

There is essentially no protection of plant breeders' rights in developing countries, because few developing countries recognize such rights. No developing country is currently a member of the Union for the Protection of New Varieties of Plants within the United Nations World Intellectual Property Organization. Disclosure requirements for patents may even improve access of breeders in developing countries to elite materials—particularly inbreds—that would normally be unavailable. Furthermore, it is possible that the value of germplasm may increase as a consequence of patent protection, which will provide a greater incentive for germplasm collection and preservation (Barton et al., 1989). However, patent protection will not help developing countries where breeding programs are inadequate or poorly developed.

This possibility underscores the critical need for a plant breeding capacity in developing countries so that they can realize the value of their genetic resources and ensure access to advanced breeding materials by their farmers. It is argued that the adoption of proprietary rights by the United States and other developed nations creates a responsibility to assist the developing countries and the IARCs in building the capacity to conserve and make use of these genetic resources (Barton et al., 1989). The immediate need is for sources of high-yielding, stress-resistant germplasm in locally adapted materials for the development of new high-performance varieties. In many places these materials exist in breeder's nurseries and trials, but there are no effective seed production or sales organizations to promote them.

Changing Public and Private Sector Roles

The balance between U.S. public and private sector roles in varietal research and development has shifted in the past 2 decades. Funding for public sector plant programs has not kept pace with the need for basic research and the maintenance of an adequate germplasm supply for both public and private sectors. Meanwhile, the private sector role has expanded considerably and is pulling scientists away from the public sector. It is inevitable that the public sector role, if not bolstered, will result in a lowered capacity to train future plant breeders. Moreover, private sector breeding activities are focused on commodities with major markets, and not necessarily on broader germplasm needs. Increased public sector funding for traditional plant breeding is essential to maintain acceptable yield gains and an adequate supply of enhanced materials for national and international needs.

The Decline of Public Sector Breeding Programs

For more than 50 years, production of new cultivars and inbred lines in the United States has been done primarily by breeders at public institutions with support from land-grant universities and USDA. For most crops, the majority of varieties in use today were developed through publicly funded research. Even for maize, for which 80 percent of production is based on privately developed varieties, most of the privately developed varieties were derived from publicly developed lines, which are inbreds freely available to all (McMullen, 1987). Varieties developed in the public sector contribute significant genetic advances and act as a check on seed prices set by the private sector. In addition, public sector programs are an important source of both finished and unfinished materials for private companies as well as for international exchange (Butler and Marion, 1985).

Funds for public plant breeding programs have been seriously constrained from federal sources and, more recently, from state sources as well. Table 2-3 illustrates a decline in scientific human-years assigned to plant breeding research of 10 percent in agronomic crops and 25 percent in horticultural crops since 1970. USDA no longer officially engages in cultivar development and release, except for crops for which no other organization is willing to engage in these practices. As a result of PVPA, however, public sector efforts have been stable for some commodities (rice, turf grasses, and legumes) and have actually increased for others (wheat and soybeans) through the use of royalty income to support these programs. Thus, it cannot be said that proprietary protection has been entirely responsible for the public sector decline in breeding efforts. In fact, PVPA has contributed to the maintenance of programs for commodities that benefit from protection.

The phenomenon of reduced public funding is not limited to plant breeding—it applies to agricultural research in general. Biotechnology is one of the few areas that has seen growth, and the degree to which it has directly or indirectly diverted funding from traditional plant breeding efforts is hotly debated. Equally serious is an inaccurate perception on the part of some public policymakers that private sector breeding efforts can replace those of the public sector. The closure of public sector plant breeding programs for some crops would have serious negative consequences for the private sector, including foreclosure of companies without breeding programs, increased concentration into fewer and larger companies, reduced exchange of breeding materials, and ultimately, a decline in genetic diversity (Butler and Marion, 1985). Furthermore, continued deterioration in public funding will be responsible for an ongoing loss of scientists to the private sector and failure to train the next generation of plant breeders. A permanent shift of leadership in breeding research away from the universities will result in failure to maintain acceptable levels of competition, efficiency, and yield gains in the seed industry (McMullen, 1987).

TABLE 2-3 Scientific Human-Years in the Public Sector Assigned to Plant Breeding Research of Agronomic and Horticultural Crops from 1969 to 1970, 10 Years Later (1979–1980), and Most Recently (1986–1987)

Commodity	1969–1970	1979–1980	1986–1987
Agronomic			
Maize	65.1	68.0	54.5
Grain sorghum	16.4	17.5	15.8
Rice	7.5	10.5	10.8
Wheat	51.2	58.5	62.8
Barley	20.0	13.9	13.8
Oat	15.2	13.7	10.2
Small grains[a]	23.3	11.1	7.3
Soybean	35.9	42.4	45.0
Cotton	45.8	42.0	34.8
Tobacco	32.3	16.5	14.1
Alfalfa and other legumes	24.3	42.7	31.6
Grasses and other forages	33.0	35.9	29.8
Total	370.0	372.7	330.5
Horticultural			
Potato	20.0	15.6	17.7
Carrot	1.5	2.6	2.8
Tomato	20.8	13.1	14.4
Bean and pea	20.1	19.6	23.5
Sweet corn	3.3	4.8	3.7
Cucurbit	10.5	16.4	10.8
Sweet potato	3.7	3.9	4.1
Crucifer	1.2	2.8	1.6
Onion	1.0	1.4	2.5
Vegetable crops	35.1[b]	26.6	24.5
Total	117.2	106.8	105.6

[a]Scientific human-years were assigned to small grains without specification of crop. This value is in addition to the assigned values for wheat, barley, and oats.

[b]Includes lettuce and other crops not itemized separately; in addition, this value would include commitment to the listed vegetables without specific designation of the programs by crop.

SOURCES: Data are from analyses by the U.S. Department of Agriculture of information drawn from the Cooperative Research Information System.

The Growth of Private Sector Breeding Programs

The expansion of markets and the introduction of PVPA have had a stimulatory effect on private sector plant breeding programs for self-pollinated crops, particularly soybeans (Butler and Marion, 1985; McMullen, 1987). The impact on other crops has been mixed, however. Since 1982, the number of companies involved in plant breeding research increased for maize, soybeans, turf grasses, sugar beets, and canola but decreased for vegetables, grain sorghum, wheat, sunflowers, and the small grains of oats, barley, rye, and triticale (Table 2-4). Both groups contain hybrid as well as self-pollinated crops.

Private sector investment in plant breeding research and development increased fourfold since 1970—much more than the public sector increase. Ranked by numbers of scientific personnel in a survey of private companies by Kalton et al. (1989), pesticide research was first, plant breeding second, food product development third, and biotechnology fourth. Numbers of personnel increased 34 percent for 18 crops or crop groups since 1982, particularly for maize, soybeans, and sugar beets, but declined for cotton and sunflowers (Table 2-5). The total number of private sector plant breeding personnel is probably equal to the number in public institutions and substantially exceeds the number in the public sector for crops such as maize, soybeans, forage legumes, grain sorghum, and sugar beets (Kalton et al., 1989).

TABLE 2-4   Number of Companies Conducting Breeding Programs on Major Crops in the United States, 1989 Versus 1982

Major Crop Category	1982	1989
Maize	66	75
Vegetables	44	37
Soybeans	26	34
Turf grasses	8	16
Alfalfa-forage legumes	14	16
Grain sorghum	21	15
Wheat	21	11
Cotton	13	11
Sugar beets	5	10
Flowers, ornamentals	9	9
Sunflowers	16	9
Forage grasses	5	8
Canola	0	6
Oats, barley, rye, triticale	11	6
Rice	5	4
Safflower	3	2
Fruits	2	2
Peanuts	0	1

SOURCE: Kalton, R. R., P. A. Richardson, and N. M. Frey. 1989. Inputs in private sector plant breeding and biotechnology research programs in the United States. Diversity 5(4):22–25. Reprinted with permission, ©1989 by DIVERSITY.

TABLE 2-5 Number of Scientific Personnel and Trained Technicians Involved in Private-Sector Plant Breeding on Major U.S. Crops, 1989 Versus 1982

| Major Crop Category | Scientific Personnel by Degree Level | | | | | | Technicians |
| | Doctorate | | Master's | | Bachelor's | | |
	1982	1989	1982	1989	1982	1989	1989[a]
Maize	155.10	256.89	100.25	114.10	201.90	269.73	303.35
Vegetables	96.35	108.25	59.50	60.30	92.70	147.20	166.80
Soybeans	35.90	59.69	15.90	25.50	41.80	53.00	94.86
Alfalfa-forage legumes	22.95	28.25	18.00	18.25	16.60	29.30	21.00
Wheat	23.40	25.20	18.20	21.30	27.70	32.60	13.30
Grain sorghum	22.45	22.80	12.05	12.10	32.20	42.40	22.50
Sugar beets	14.30	22.00	2.00	14.30	6.00	18.00	31.00
Rice	7.25	9.30	2.00	4.00	4.00	7.00	14.00
Cotton	17.28	11.11	11.00	6.30	19.00	7.00	37.10
Flowers, ornamentals	4.50	8.35	4.50	9.50	12.50	19.00	24.00
Turf grasses	8.50	8.05	2.20	10.20	6.35	18.70	12.90
Sunflowers	15.00	7.26	13.00	7.20	12.80	15.20	15.50
Barley, oats, rye, triticale	7.05	5.50	2.10	1.55	5.60	6.00	5.25
Canola	0.00	4.27	0.00	0.00	0.00	11.35	2.05
Forage grasses	2.40	1.60	2.00	1.55	3.80	1.70	2.10
Peanuts	0.00	1.00	0.00	0.00	0.00	2.0	2.00
Safflower	1.70	0.50	1.00	1.00	1.20	1.0	0.00
Fruits	0.00	0.35	2.80	0.00	3.00	0.00	0.00
Total	434.13	580.37	266.50	307.15	487.15	680.75	767.71

NOTE: Data are based on full-time scientist per year equivalents.

[a]Technicians were not included in the 1982 survey.

SOURCE: Kalton, R. R., P. A. Richardson, and N. M. Frey. 1989. Inputs in private sector plant breeding and biotechnology research programs in the United States. Diversity 5(4):22–25. Reprinted with permission, ©1989 by DIVERSITY.

TABLE 2-6   Approximate Annual Research Expenditures on Private Plant Breeding in the United States, 1989 Versus 1982

Expenditure Group (US$000)	1982		1989	
	Number of Companies	Projected Total (US$000)	Number of Companies	Projected Total (US$000)
0	9[a]	000	8[a]	000
Under 100	44	2,200	25	1,250
100–500	62	15,500	54	13,500
500–1,000	23	17,250	33	24,750
1,000–5,000	17	42,500	27	67,500
5,000–10,000	5	37,500	5	37,500
10,000–25,000	0		3	52,500
Over 25,000	0		2	75,000
Total	160	114,950	157	272,000

[a]Several of the companies contacted conducted no research themselves, but contributed funds to experiment station research on plant breeding, or sold varieties and hybrids developed by others on a royalty basis.

SOURCE: Kalton, R. R., P. A. Richardson, and N. M. Frey. 1989. Inputs in private sector plant breeding and biotechnology research programs in the United States. Diversity 5(4):22–25. Reprinted with permission, ©1989 by DIVERSITY.

Private sector research expenditures for plant breeding in the United States more than doubled from 1982 to 1989 (Table 2-6). Investment in biotechnology related to plant breeding was close to $100 million (Table 2-7), or about one-third of total plant breeding expenditures, with major emphasis on maize and vegetables (Table 2-8). Private companies generally focus their efforts on commodities with large markets and on those that can be protected under PVPA, patents, or trade secrets. This is understandable, because the cost of breeding a new variety is estimated to be between $2 million and $2.5 million (McMullen, 1987). However, this focus gives rise to the concern that other crops (small grains, forage grasses, sunflowers, and some vegetable crops) receive little or no attention by private sector plant breeding and biotechnology research programs (Kalton et al., 1989) as well as decreasing attention by the public sector.

The effect of proprietary protection on genetic diversity in privately developed varieties is probably minimal. Although private sector breeders felt that PVPA increased the genetic diversity of open-pollinated varieties, public sector breeders detected no effect (Butler and Marion, 1985).

The impact of patent protection on germplasm exchange is ex-

TABLE 2-7   Approximate Annual Research
Expenditures on Biotechnology Related to Plant
Breeding in the United States, 1989

Expenditure Group ($000)	Number of Companies	Projected Total ($000)
Under 100	13	650
100–500	9	2,250
500–1,000	5	3,750
1,000–5,000	10	25,000
5,000–10,000	8	60,000
Total	45	91,650

SOURCE: Kalton, R. R., P. A. Richardson, and N. M. Frey. 1989. Inputs in private sector plant breeding and biotechnology research programs in the United States. Diversity 5(4):22–25. Reprinted with permission, ©1989 by DIVERSITY.

pected to be minor for most crops. Even for crops that are protected, the incentive to enhance profits by extensive cross-licensing will most likely relieve some constraints on exchange (Jondle, 1989). However, biotechnology patents are likely to impose serious constraints on exchange because they will prevent other breeders from using patented genes (Day, 1993).

Restructuring of the Seed Industry

The ability to maintain the inbred parents of hybrid crops as trade secrets attracted private enterprise to the breeding and sale of hybrids early in the developmental stages of plant breeding (as early as the 1920s for maize). Numerous proprietary inbred lines of maize and sorghum have been developed for the production of proprietary hybrids, which also serve as germplasm sources for further breeding. Because of the commercial necessity for controlling these privately developed inbred lines, however, the lines themselves have rarely been made available to the public for breeding or other purposes, even once they are obsolete. However, after hybrids made up of one or more private inbreds are offered for sale, the genes, unless separately patented, are legally available sources of germplasm for public use. Thus, hybrid development does not necessarily constrict the availability of germplasm.

Hybrid breeding methods have been the major incentive for the development of the private seed industry in the United States and

TABLE 2-8   Companies, Scientific Personnel, and Trained
Technicians Involved in Biotechnology Research Related to Plant
Breeding on Major Crops in the United States, 1989

Major Crop Category	Number of Companies	Scientific Personnel by Degree Level			Technicians
		Doctorate	Master's	Bachelor's	
Maize	19	90.1	44.8	96.9	26.5
Vegetables	17	31.4	23.6	42.0	25.0
Soybeans	6	17.3	9.0	18.0	2.0
Cotton	5	7.15	5.8	8.0	3.0
Sugar beets	3	6.5	2.0	7.0	0
Canola	3	9.5	3.0	20.0	4.0
Alfalfa	2	2.1	3.1	2.3	0.5
Sunflowers	2	1.0	2.0	4.0	0.0
Wheat	2	1.1	1.1	1.1	0.1
Other small grains	1	0.5	1.0	0.0	0.0
Rice	1	0.25	0.0	0.0	0.0
Turf grasses	1	0.0	0.9	0.0	0.0
Forage grasses	1	0.0	0.1	0.0	0.0
Undifferentiated by crop[a]	2	85.0	20.0	25.0	10.0
Total	65	251.9	116.4	224.3	71.1

[a]One company conducts biotechnology research on canola, tomato, maize, rice, tobacco, sunflowers, sugar beets, ornamentals, cotton, melons, peppers, soybeans, coffee, cocoa, and oil palms. The second company is researching corn, soybeans, wheat,
and alfalfa.

SOURCE: Kalton, R. R., P. A. Richardson, and N. M. Frey. 1989. Inputs in private
sector plant breeding and biotechnology research programs in the United States. Diversity 5(4):22–25. Reprinted with permission, ©1989 by DIVERSITY.

Europe since the 1930s, because hybrids sell for four to eight times
the cost of open-pollinated varieties (Doyle, 1985; McMullen, 1987).
Until 1970, the U.S. seed industry was made up largely of numerous
family firms with regional or crop specializations, except for maize.
Large private companies tended to dominate hybrid markets, whereas
public sector and small companies controlled open-pollinated markets in the United States as well as in Europe. Most private sector
research was done by a small number of companies.

The restructuring of the world seed industry, following passage
of the PVPA legislation in the United States in 1970, drew much attention as large chemical, pharmaceutical, and food processing companies absorbed many independent seed companies in the United
States and Europe. However, the trend was not solely due to PVPA,

but to a range of factors, including the profitability of the seed trade, the potential of biotechnology, the worldwide impact of genetic research as evidenced by the success of the green revolution, and greater awareness of the potential value of plant genetic resources. Large companies, which were better able to absorb the risks and fund the rising costs of research (particularly for biotechnology), were attracted to what McMullen (1987) called the genetic supply industry.

The mergers provided greater resources for research, a broader research base, an enhanced capacity for testing and marketing, and larger markets. However, mergers may have reduced competition and subjected the seed industry to corporate management techniques and quarterly profit expectations. Nevertheless, the concentration of companies in the market is still far lower than that in many other industries. For example, in the world seed market of 1983, the percentage of sales attributable to the top five companies was less than 20 percent, leaving a niche for smaller companies, especially in local markets and for minor crops (McMullen, 1987). Although conglomerates held nearly half of all plant variety protection certificates by 1982, private market shares did not seriously hamper competitive forces in open-pollinated seed markets, where public varieties still dominate (Butler and Marion, 1985). Even in the hybrid maize industry, conglomerates have not been able to outcompete Pioneer—an old, independent seed company whose market share increased steadily from 1973 to 1983 (McMullen, 1987). Overall, however, these recent trends have probably reduced competition and may have also reduced crop plant genetic diversity through the consolidation of plant breeding activities (Butler and Marion, 1985).

Balancing Public and Private Sector Roles

Through the years, a pragmatic balance has developed between public and private sector plant breeding efforts, which has proved quite effective in producing crop varieties for the industrial countries. The private sector has focused on applied research, seed production, and marketing. It concentrates on finished varieties of major crops with large potential markets, such as maize, sorghum, soybean, sugar beets, alfalfa, and cotton. The public sector has performed basic research and some applied research; developed finished varieties for wheat, soybeans, and most minor crops; developed parental lines and introgressed exotic materials; and put major efforts into training and information dissemination. The relationship between the two sectors is not one of competition but, rather, one of logical complementarity based on the interests of each sector.

Many aspects of this complementarity will continue. In practice, private industry lacks market incentives to breed all crop plants and cannot justify the investment to do so. It is appropriate that public sector breeders sustain breeding programs for species not bred by private industry. However, to fulfill their function of teaching and training future plant breeders, the universities must continue to do at least some plant breeding with major crops. The public sector will continue to have a greater incentive and a clearer mandate for long-term genetic enhancement by using wild and unimproved germplasm than the private sector and should make this one of its major commitments. Recognizing this need, the NPGS has placed increasing priority on strengthening enhancement activities (Elgin and Miller, 1989).

The vigor of the system derives from the variety of institutions within each sector and from the roughly equal balance between the strengths of the public and private sectors, at least until recently. An optimum balance between the two sectors would retain a diversity of seed producers and germplasm programs that are in both the public and private sectors, and that would include private companies, public institutions and foundations, grass-roots nonprofit initiatives, and international centers. This mixture would ensure competition to produce better varieties, would place checks and balances on the cost and supply of seeds between the two sectors, and would serve a wider range of farmer and consumer interests (McMullen, 1987).

The growth of biotechnology undeniably has been a major factor in altering the balance between the public and private sectors in recent years. Heavy research capitalization costs, which are more easily borne by industry, have favored the private sector. Public sector funding has also been attracted to biotechnology, although at lower levels than that in the private sector and often at the expense of plant breeding programs. USDA funds are increasingly taken away from even basic plant breeding programs, unless they are related to biotechnology. State legislatures—the other chief provider of funds for land-grant institutions—are generally not interested in funding conventional plant breeding programs, although they often fund biotechnology research in support of plant breeding when they believe it may expand the economic development of the state. The contributions of molecular biology, however, will supplement, not replace, elite and exotic varieties of plants as major sources for genetic advance.

Teamwork between molecular biologists and breeders is essential to harness technological advances to varietal development (Fehr, 1989). Breeders know the appropriate genetic goals and likely gene sources, what it takes to make a commercially viable cultivar, and what are the expectations and needs of the farmers. Molecular systems are not yet ready to deal with the complex traits of economic importance (for example, yield and drought tolerance) that are handled by breeders. However, biotechnology does have the potential, in the near term, to make significant contributions in elucidating genetic control mechanisms, generating novel genetic variation, developing rapid diagnostic aids for pathogen identification, developing or selecting certain types of superior individuals, and a host of technological aids that will accelerate the progress of plant breeders.

The varietal protection that preceded, and the patent protection legislation that has accompanied, the rise of biotechnology appear to have had little impact on the direction of public sector programs, other than a somewhat greater emphasis on germplasm enhancement and basic research and less emphasis on cultivar development than previously (Butler and Marion, 1985). Nevertheless, many public sector breeders find it difficult to adjust to the concept of limiting free access to their cultivars, and public use of legal protection has been slight relative to that in the private sector (Barton et al., 1989; Butler and Marion, 1985). The public sector has begun to consider the extent to which it might exploit legal protection—either to generate revenues or to keep developments in the public domain through nonexclusive or royalty-free licensing (Barton et al., 1989).

Clearly, adequate funding is the major requirement for sustaining public sector strengths to maintain an optimum balance between the public and private sectors in plant breeding and genetic resources programs. As a means of reaching this goal, funding for biotechnology should be linked to support for traditional plant breeding programs. Land-grant universities need to use greater creativity in soliciting public funds for "packages" or teams of molecular biologists, breeders, pathologists, and other essential scientists to work on commodities important at the state level. Industry could be tapped for matching funds—particularly for training components—whose importance they recognize. The team approach could be successful not only in linking funding for breeding to biotechnology, but in stimulating the interface between disciplines, accelerating biotechnology applications to breeding, and attracting talented students into breeding via biotechnology.

International Implications

In dramatic contrast to circumstances in developed countries, the indigenous private sector role in varietal development and seed distribution is minimal or nonexistent in most developing countries. Multinational seed companies may be reluctant to invest in initiatives in developing countries because of the lack of proprietary protection, the small size of markets, and unfavorable national policies.

The current situation results in part from the establishment of seed organizations in many developing countries during the 1960s and 1970s to provide low-cost seed for farmers (McMullen, 1987). These were often given legal monopolies, with budgets heavily subsidized by the government. The private sector could not compete with subsidized low prices, preferential access to improved varieties from national breeding programs and IARCs, and public control, seed certification, distribution, and farm credit systems. Predictably, the local private seed industry was severely weakened, international seed company operations were seriously hampered (if they were permitted to operate in the country at all), and the semi-state-controlled seed operations often became costly and ineffective seed suppliers.

The lack of an effective seed industry—public or private—remains an important constraint to the distribution of improved varieties in most developing countries (McMullen, 1987). National seed policies have often not been successful, and reform is needed to benefit from new genetic developments. There is a clear need for allocation of functions between the public and private sectors that will be most effective in promoting agricultural development. Private seed companies should be strengthened and encouraged to take over seed operations (production, distribution, and customer service) that they perform best. The public sector clearly has an important role to play in research, seed certification, quality control, regulatory and extension functions, but its performance must be improved.

In the 1990s, the challenge is to promote the adoption of improved seed by farmers in developing countries who fear the risks and are reluctant to pay market prices (McMullen, 1987). The governments of developing countries should recognize that private seed companies can contribute to national agricultural progress and should consider modifying seed policies to encourage their participation. Companies must be willing to adapt to the countries' needs and fit into their development strategies, if they are to realize a share of the potential market.

National governments should formulate seed policies that encourage private seed production, either by local companies or with external collaboration. Many different degrees of engagement are possible with international seed companies, from distributional arrangements, to contractual growing, to joint ventures (Douglas, 1980). Local, small-scale seed production is particularly attractive for crops for which the market is small and external seed company involvement is not feasible or desirable. CIAT has promoted this approach successfully for beans (Centro Internacional de Agricultura Tropical, 1987), pasture grasses, and legumes (Centro Internacional de Agricultura Tropical, 1989) in Latin America. The CIAT Seed Unit works through existing local seed companies, farmer cooperatives, or enterprising individuals by providing training and extension support. The approach stimulates farmer interest in improved varieties and may break the bottleneck on an improved seed supply.

All of the options presented here are predicated on the existence of a strong national or regional plant germplasm and breeding system to supply the raw and finished materials for improved seed. At the Keystone International Dialogue on Plant Genetic Resources, it was recognized that the high costs of national plant germplasm systems put them beyond the means of many countries (Keystone Center, 1990). Regional programs are attractive alternatives that can be fostered through the use of commodity research networks. To be effective, these need modest but stable and long-term funding.

The Keystone group saw an urgent need for global coordination and funding mechanisms to meet worldwide genetic resources conservation needs—conservatively estimated at $300 million to $500 million annually (Keystone Center, 1990, 1991). Discussions are continuing on how to raise the funds from public and private sector contributions and how they should be invested. These worldwide issues call for public sector leadership. Although consensus and implementation may still lie many difficult years ahead, the depth of international concern and the extent of the dialogue are encouraging.

RECOMMENDATIONS

The need to broaden crop genetic diversity continues to be critical in the United States, but it is particularly urgent in developing countries, where the potential for vulnerability has increased significantly over the past 15 years. Case studies of diversity in major crops since 1970 indicate two trends: (1) In the United States, the increase in the numbers of plant varieties and the decrease in the varietal dominance of some major crops suggest that genetic diversity has increased. Concern remains, however, over the degree of similarity in the ancestries of major varieties and the amount of reduction in genetic diversity that may have taken place with the consolidation accompanying the structural changes in the U.S. plant breeding industry. (2) Genetic diversity in rice and wheat has decreased in developing countries. Fewer landraces are grown because of increasing dominance by HYVs. Because private seed companies were not involved in the wheat or rice seed business in developing countries prior to 1980, farmer choice (conditioned by financial pressure to grow HYVs and a limited selection of available varieties) is the major reason for the loss of diversity in farm crops. Genetic diversity in rice is also shrinking rapidly in the United States and Japan (Chang and Li, 1991).

For many areas and in many crops, reducing vulnerability remains a challenge. Concern is growing that breeding programs in developing countries are not equipped to react rapidly if faced with major epidemics. Recent examinations of rice breeding in the United States have shown that some cultivars have more than 70 percent of their genes in common (Dilday, 1990). At present, the wheat varieties in the United States seem to have greater genetic diversity than the wheat and rice varieties in many developing countries. For example, in the United States, six varieties of wheat accounted for 38 percent of the total wheat surface area in 1980. In comparison, in India, only one variety accounted for 30 percent of the wheat plantings in 1983, and in Indonesia, two rice varieties accounted for 54 percent of the cultivated rice area in 1983 to 1984.

Countries must make developing capacities for genetic resources management and use, including human and physical resources, a matter of national agricultural security.

This development is essential in view of the grave potential for increasing global genetic uniformity of major food crops and the associated potential risks of vulnerability. Countries should assess the extent to which their needs for major crops are met by national and international agricultural programs and seed companies and should develop or strengthen programs for commodities not adequately addressed by existing systems. Regional capabilities for monitoring, enhancement, and breeding should be shared where national resources are limited.

Plant germplasm conservation and exchange is carried out on a scale never believed possible even as recently as 20 years ago. This is a result of efforts of many national and international programs in the intervening decades. The burgeoning of germplasm initiatives worldwide, however, is creating crises of management, staffing, communication, equity, and funding.

Global efforts are needed to enable broad and effective conservation and use of genetic resources.

Utilization of germplasm bank resources has lagged far behind conservation efforts because of inadequate linkages among the plant breeders and other germplasm workers. A global effort should include cooperation among the various existing institutional, national, regional, and international germplasm collections. The lack of a global data base providing information and access to the vast collections that continue to accumulate is a critical constraint to the development and management of such a system. Inadequate management and funding of genetic resources conservation risks potentially serious problems of vulnerability in the future. These issues are addressed in subsequent chapters.

3

In Situ Conservation of Genetic Resources

A species or a population sample of a particular part of its genetic variation can be maintained through in situ or ex situ conservation. In situ conservation is the preservation of species and populations of living organisms in a natural state in the habitat where they naturally occur. This method preserves both the population and the evolutionary processes that enable the population to adapt by managing organisms in their natural state or within their normal range. For example, large ecosystems may be left intact as protected reserve areas with minimal intrusion or alteration by humans. Ex situ conservation is the preservation and propagation of species and populations, their germ cell lines, or somatic cell lines outside the natural habitat where they occur. This method maintains the genetic diversity extant in the population in a manner that makes samples of the preserved material readily available. It includes botanical gardens, greenhouses, and the preservation of seeds or other plant materials in germplasm banks under appropriate conditions for long-term storage. This chapter discusses the role of, and barriers to, in situ conservation.

THE IMPORTANCE OF IN SITU CONSERVATION

In situ conservation is an important component of the conservation and management of genetic resources. It supplements the ex situ conservation efforts of local, national, and international collections and provides some important advantages. In situ conservation

117

sites preserve potentially important and useful genes, many of which may be unrecognized today. Their existence enables the selective and adaptive processes that give rise to new genetic traits to continue in response to environmental stresses. These areas can be sources of genetic traits not already captured in ex situ collections. In situ reserves can also provide living laboratories for studying the genetic diversity of the wild species that are the progenitors of modern crops.

Conservation of ecosystem and species diversity has traditionally been dealt with by local or national agencies responsible for wildlife and protected areas. Conservation of genetic diversity, however, has been the concern of those responsible for agriculture (including horticulture) and silviculture. This difference may partially be responsible for the perception that efforts to conserve genetic resources in situ are inadequate (International Board for Plant Genetic Resources, 1985a). More often, however, the lack of an adequate scientific and economic basis for establishing and maintaining in situ conservation efforts is seen as the major difficulty (Hoyt, 1988; International Board for Plant Genetic Resources, 1985a; Noy-Meir et al., 1989; Plucknett et al., 1987).

In situ conservation has been proposed for preserving wild species that are related to domesticated crops and perennials such as forest trees, tropical fruits, or species with short-lived seeds (Ford-Lloyd and Jackson, 1986; Hoyt, 1988; International Board for Plant Genetic Resources, 1985a; Plucknett et al., 1987). For those species, in situ conservation provides the relative stability of species diversity within a coadapted community (Frankel and Soulé, 1981). Some have also suggested that maintaining landraces in traditional farming systems also constitutes a form of in situ conservation (Altieri et al., 1987; Altieri and Merrick, 1987; Oldfield and Alcorn, 1987). In situ conservation may be viewed as a dynamic process that allows the continuance of the evolutionary processes that result in genetic diversity and adaptation.

In situ and ex situ conservation methods are complementary options essential to the maintenance of crop and biological diversity (Office of Technology Assessment, 1987a). It has been argued that although ex situ conservation methods allow more immediate access to genetic resources, in situ conservation methods are essential for the conservation of a broader range of species (Brown et al., 1989; Office of Technology Assessment, 1987a; Oldfield, 1984; Plucknett et al., 1987). In addition, an in situ conservation area can encompass a broad range of species and genetic diversity, much of which may not even be described. In situ conservation is particularly important for trees (Ford-Lloyd and Jackson, 1986; National Research Council, 1991b).

The lengthy generation times make conservation of tree populations in their natural environment essential for any sort of experimental use.

WILD SPECIES AS GENETIC RESOURCES

Wild species are often used to improve established crops and occasionally used to develop new ones. A survey of 18 crops grown in the United States from 1976 to 1980, revealed that from 1 percent (sweet clover) to 90 percent (sunflower and tomato) of the available cultivars had been improved in part using wild germplasm. The combined average annual farmgate value of these cultivars was $4.8 billion (the annual value of the improvements to eight of the crops was estimated by one report to have been $170 million) (C. Prescott-Allen and R. Prescott-Allen, 1986). Use of wild relatives in crop breeding has obvious economic significance and is growing (C. Prescott-Allen and R. Prescott-Allen, 1986; R. Prescott-Allen and C. Prescott-Allen, 1983).

Colorado potato beetles take only a bite or two of this insect-resistant potato plant before they are repelled. The plant has been genetically engineered to contain a gene from wild potatoes that produces a substance distasteful to these insects. Credit: U.S. Department of Agriculture, Agricultural Research Service.

The likelihood that a wild genetic resource will be used in crop breeding is a function of the economic importance of the crop, which determines the existence and size of any improvement program; the rarity of the character being sought and the possibility of locating it in the gene pool of the domesticated crop; and the ease with which the character can be transferred from the wild relative to the domesticated crop. Breeders in search of a particular character may explore quite thoroughly the variability of the crop itself. However, they seldom explore much of the variability contained within its wild relatives. Breeders generally test small numbers of accessions from limited portions of the wild species' range (Rick et al., 1977), but the size of most collections of wild relatives available in ex situ storage are small and often do not reflect the full variability available in situ.

Wild gene pools are an important biological resource for developing new crops, particularly for the timber industry, the livestock industry (forage and fodder crops), and rural development (fuelwood). Like the breeders of established crops, domesticators of new crops differ in the extent to which they explore the genetic variability of the species concerned. In general, the economically most successful new domesticated crops are those that have tapped a diversity of germplasm sources (C. Prescott-Allen and R. Prescott Allen, 1986).

Genetic Conservation Areas

Wild genetic resources may be conserved in situ in a protected area. This is an area of land or water allocated to some form of conservation management. It may be established expressly to maintain the genetic resource, or it may have other objectives as well. Both types of genetic conservation areas are included in the term *genetic reserves* (Jain, 1975a). The principal objective of a genetic reserve is to maintain the individual- and population-level variation of one or more species in their natural range or habitats.

Genetic reserves have the following characteristics:

• An explicit objective to maintain population variation.

• An established protocol for providing information on and access to the protected resources by ex situ collections, breeders, researchers, and other germplasm users, including a procedure for the sustainable collection of reproductive material by authorized agencies and individuals.

• A procedure for monitoring the status of the populations conserved as part of a national genetic resources information system.

The primary role of the genetic reserves is to secure the long-term availability of their wild genetic resources and to preserve the adaptive processes therein.

Consideration of the extent to which intraspecific variation is encompassed in an area is required to preserve the genetic structure of the target species. Multiple sites may be required to capture a reasonable amount of the allelic variation of a species (Food and Agriculture Organization, 1989a). For example, the committee has examined the challenges of preserving the genetic structure of forest trees (National Research Council, 1991b), the in situ conservation of which is particularly important because of their lengthy generation times (Ford-Lloyd and Jackson, 1986).

As discussed in greater detail in Chapter 4, the targets of germplasm sampling strategies are common alleles (population frequency, equal to or greater than 5 percent) that are widespread (found in many populations) or local (found in one or a few populations) (Brown and Moran, 1981; Marshall and Brown, 1975). These common alleles define representative and unique gene pools that could be candidates for in situ conservation.

However, the characterization of gene pools is not necessarily a simple task. The variation among alleles can be inferred individually through studies of intraspecific variation in biochemical characters (for example, enzymes, other proteins, or terpenes), morphology, phenology, growth rate, environmental adaptations, and many other characters. Direct measures of DNA variation are also possible using restriction fragment length polymorphisms or randomly amplified polymorphic DNA markers (see Chapter 7). Genetic variation in characters strongly influenced by environment (phenology, growth rate, morphology of vegetative parts) must be measured through tests that compare responses with different environments. There are, however, few studies of the degree to which biochemical variability reflects morphologic or other phenotypic variation (National Research Council, 1991b).

THE STATUS OF IN SITU CONSERVATION OF WILD TYPES

The need for in situ conservation of wild genetic resources has been widely acknowledged (Food and Agriculture Organization, 1989a; Hoyt, 1988; International Board for Plant Genetic Resources, 1985a; International Union for the Conservation of Nature and Natural Resources et al., 1980; Jain, 1975a,b; Noy-Meir et al., 1989; Office of Technology Assessment, 1987a; Oldfield, 1984). However, real efforts at in situ conservation have been slow to emerge (International Board for Plant Genetic Resources, 1985a; Noy-Meir et al., 1989). At least

In a test field at the Forage and Range Research Laboratory of the U.S. Department of Agriculture, a research geneticist examines hybrids made by crossing native American wild ryegrasses with a wild species from the former Soviet Union. These new, taller ryegrasses stay green later, reducing fire hazards, and may enable animals to graze 2 months longer than usual each year. Their leaves cure well and protrude above the snow so cattle and sheep can graze them well into the winter. Credit: U.S. Department of Agriculture, Agricultural Research Service.

three comprehensive national initiatives have been reported. In Brazil the Centro Nacional de Recursos Geneticos (CENARGEN, National Genetic Resources Center) has established 10 genetic reserves to maintain timber, fruit, nut, forage, and palm species and wild relatives of crops such as cassava and peanut (Giacometti and Goedert, 1989). Germany is using its system of nature reserves as the basis of in situ conservation of the wild progenitors of apples and pears and other wild genetic resources (Schlosser, 1985). The Commonwealth of Independent States is reported to have established 127 reserves for the protection of wild relatives of crops (Korovina, 1980).

A few reserves have been designated for the protection of particular wild relatives or for timber genetic resources. They include the Sierra de Manantlán Biosphere Reserve in Mexico for teosinte (*Zea diploperennis*) (Hoyt, 1988; Russell, 1989) and two reserves each for Zambesi teak (*Baikiaea plurijuga*) in Zambia and *Pinus merkusii* in Thailand (Food and Agriculture Organization, 1989a).

Few reserves have been established or are managed as genetic resources conservation areas. A European survey of wild crop relatives of apples, plums, cherries, peaches, almonds, and *Allium* species found that although reserves existed within the range of these wild species, lists of the plant species they contained were available for few of those areas. No information on genetic diversity was available (Hoyt, 1988).

The goals of genetic conservation can be combined with those of natural or biosphere reserves. A notable example of in situ conservation is teosinte, a wild relative of maize, which is found in Mexico and Guatemala (Centro Internacional de Mejoramiento de Maíz y Trigo, 1986; Plucknett et al., 1987; Wilkes, 1977). In an effort to preserve the genetic resources of teosinte, the Centro Internacional de Mejoramiento de Maíz y Trigo (CIMMYT, International Maize and Wheat Improvement Center) in Mexico and others are using a combination of methods that include in situ monitoring of teosinte populations and preservation in reserves (Centro Internacional de Mejoramiento de Maíz y Trigo, 1986).

Teosinte is generally found in the untilled soil bordering maize fields and, to a lesser degree, throughout some maize fields. There are eight geographically isolated population clusters of annual teosinte. Six of these are found in Mexico and two are found in Guatemala. There are also two perennial populations in Mexico. Teosinte populations range in size from 1 to 1,000 square kilometers.

It is estimated that the current distribution of teosinte is half of what it was in 1900 (Centro Internacional de Mejoramiento de Maíz y Trigo, 1986). Three of the annual populations are considered rare, occurring at single locations. Most of the populations are considered vulnerable and are declining at a rate such that they could become endangered.

There are three principal threats to teosinte populations (Centro Internacional de Mejoramiento de Maíz y Trigo, 1986). As land use is intensified, teosinte is squeezed out of the margins bordering maize fields. The replacement of maize with cash crops, such as short-stature sorghum, leaves teosinte as a visible weed, making it easier to remove. Finally, because of outcrossing to maize, small and isolated stands of teosinte can lose their ability to disperse.

CIMMYT staff make yearly visits to identified annual teosinte population sites to assess the status of each population. If a population is found to be endangered, CIMMYT and national program staff can cooperate to preserve it. This allows monitoring of teosinte without incurring the costs of establishing and maintaining an in situ preserve.

There are other examples of genetic conservation goals being combined with those of natural or biosphere reserves. In 1988, Mexico established the 139,000-ha Sierra de Manantlán Biosphere Reserve (Hoyt, 1988). A small portion of this reserve (about 10 ha) contains the only known stands of the primitive wild relative of maize, *Zea diploperennis*. Earlier, in 1959, the 23,868-ha Sary-Chelck Reserve was established in the Commonwealth of Independent States as part of the Chatkal Mountains Biosphere Reserve, near western China. The area contains wild species related to walnuts, apples, pears, *Prunus* species, and other temperate fruit and nut crops (Hoyt, 1988).

Obstacles to In Situ Conservation of Wild Genetic Resources

Two main obstacles to in situ conservation of wild genetic resources are sectoralism and lack of knowledge (R. Prescott-Allen, resource policy analyst, personal communication, June 1990). The conservation focus of protected areas is typically on the level of ecosystems and species, not on the maintenance of crop genetic resources. To the extent that agencies responsible for protected areas are aware of the need for conserving genetic resources, they tend to regard it as an additional responsibility for which additional resources are generally not forthcoming. Ministries of agriculture or their equivalents have a direct interest in conserving wild relatives of crops, but they may be ambivalent about the importance of in situ conservation. In part, this may be because they often lack authority over the appropriate lands. Thus, difficulties in establishing a protected area may quickly outweigh the benefits of doing so, especially if the goal is protection for only one or two wild relatives of a single crop.

Lack of knowledge of the degree and distribution of interpopulation genetic variation in the wild relatives of crops is another obstacle (International Board for Plant Genetic Resources, 1985a; Noy-Meir et al., 1989). This information is needed for answering questions such as where in situ conservation areas should be established, how large should they be, and what ways should they be managed. Ecogeographical surveys (International Board for Plant Genetic Resources, 1985a) that assess the genetic variation of a species across its entire geographical and ecological range are needed (Hoyt, 1988). It can take years to obtain a complete ecogeographical survey. Although

TABLE 3-1   Crops for Which In Situ Conservation and
Ecogeographic Surveys Are High Priorities

Crop	Wild-Type Relative	Location
Groundnut	Perennial *Arachis* spp.	Latin America
Oil palm	*Elaeis* spp.	Africa, Latin America
Banana	Wild-type diploid *Musa* spp.	Asia
Rubber	*Hevea* spp.	Amazonia
Coffee	Coffee, *Arabica* spp.	Africa
Cocoa	*Theobroma* spp.	Latin America
Onion family	Selected wild-type *Allium* spp.	Worldwide
Citrus	Wild-type *Citrus* spp.	Asia
Mango	Wild-type *Mangifera* spp.	Southeast Asia
Cherries	Wild-type *Prunus* spp.	Europe, Asia
Apples	Wild-type *Malus* spp.	Europe, Asia
Pears	Wild-type *Pyrus* spp.	Europe, Asia
Forages	Hundreds of species	Worldwide

SOURCES: International Board for Plant Genetic Resources. 1985. Ecogeographical
Surveying and In Situ Conservation of Crop Relatives. Rome: International Board for
Plant Genetic Resources; Hoyt, E. 1988. Conserving the Wild Relatives of Crops.
Rome: International Board for Plant Genetic Resources.

such information is of potential value to all crops, related wild species with high priority have been identified (Table 3-1).

Population dynamics studies of a wild emmer wheat (*Triticum dicoccoides*) in Israel promise to provide important answers to basic questions about in situ conservation (Noy-Meir et al., 1989). The study is aimed at addressing questions about the genetic structure of an in situ population and how it changes over many years. The effort combines ecogeographic survey information with topographic studies and population dynamics. Morphologic, phenologic, yield, phytopathologic, and biochemical (allozymes, seed storage proteins) data are all being gathered (Noy-Meir et al., 1989). Although they are expected to yield a wealth of information useful to the management of wild species in situ, the data are also likely to provide results relevant to the sampling of populations for ex situ conservation (see Chapter 4).

The Potential of Using Existing Protected Areas as Germplasm Banks

The International Union for the Conservation of Nature and Natural Resources (1978) classifies protected areas into eight categories according to broad management objectives. All could provide for conserving wild genetic resources, but five are particularly suitable: nature reserves, national parks, natural monuments, managed nature reserves, and managed resource areas.

The first three are suitable for maintaining climax populations because they do not permit artificial maintenance of seral or subclimax stages. Managed nature reserves and managed resource areas usually permit such intervention, and so are suitable for maintenance of pioneer and subclimax as well as climax populations.

Protected areas in the first four categories above cover 4.5×10^6 km^2, an area almost half the size of the United States (R. Prescott-Allen, resource policy analyst, personal communication, June 1990). Many of them contain major wild genetic resources species. For example, Kora National Reserve in Kenya contains seven such species; they are *Vernonia galamensis, Cenchrus ciliaris, Panicum maximum, Sorghum arundinaceum, Acacia senegal, Gossypium somalense,* and *Populus ilicifolia* (Kabuye et al., 1986). St. Lawrence Islands National Park in Canada contains 27 such species (R. Prescott-Allen and C. Prescott-Allen, 1984).

While acknowledging that many wild genetic resources species may be found in existing protected areas, the International Board for Plant Genetic Resources (1985a) has noted several deficiencies that could impair their utility for conserving genetic resources. (1) Most protected areas lack adequate inventories of species or genotypes. (2) Many populations of wild relatives of crops in existing protected areas are too small for maintenance of allelic diversity or even for survival of the species. (3) Minimum conservation requirements for maintaining intraspecific diversity are not considered. (4) The limited monitoring that may be carried out is often insufficient to guarantee conservation.

Most of these deficiencies can be corrected when conservation of wild genetic resources becomes an objective of the protected area. However, the problem of population size casts doubt on the adequacy of existing protected areas for conserving genetic resources. The complexity of the concept of a viable population (Soulé, 1987) and the dearth of information on viable populations means that the extent to which protected areas maintain adequate populations of genetic resources species is not known. Although most protected areas were not designed to maintain allelic diversity of the wild relatives of crops, and although they are not managed to conserve intraspecific diversity, it does not follow that they do not do so, albeit fortuitously.

Those responsible for existing protected areas could add in situ genetic conservation to their objectives for the use of those areas. This would require a listing of genetic resources species for in situ conservation, a check of the inventories of existing protected areas for the presence of populations that warrant conservation, and modification of the management plan to include monitoring of the status of the populations and to facilitate access to information on the resources.

These steps can be taken now, without waiting for information on the degree and distribution of genetic variation within priority species. As such information is obtained through genetic and ecogeographical surveys, it can be used to amplify and refine conservation efforts. In this way, proposed conservation can complement existing protected areas and close serious gaps.

THE STATUS OF IN SITU CONSERVATION
OF DOMESTICATED TYPES

It has been argued that maintaining traditional landraces of domesticated crops in the peasant agroecosystems in which they developed is a form of in situ conservation (Altieri and Merrick, 1987; Altieri et al., 1987; Oldfield and Alcorn, 1987; Wilkes, 1989b). In situ conservation in agroecosystems requires maintenance of the particular socioeconomic conditions (or their substitution with equally favorable incentives) as well as appropriate ecologic conditions (Altieri et al., 1987; Oldfield and Alcorn, 1987).

The introduction of modern crop varieties has led to the decline and loss of many traditional landraces and the agricultural systems that produced them (Altieri and Merrick, 1987; Frankel and Hawkes, 1975a). There are few studies of the factors that promote loyal use of traditional landraces and local varieties (one example is Brush [1977]). There may be circumstances in which the special qualities of a landrace outweigh the modern cultivar's advantages of high yield and high return per unit of labor. An understanding of the cultural and economic factors that may promote the loss of some landraces and the persistence of others and that may lead some individuals and communities to adhere to their local cultivars while others discard them is essential for long-term in situ conservation of these resources.

In situ conservation of landraces has been proposed to preserve not only crop resources but also to perpetuate the adaptive evolutionary processes that produced them (Brush, 1977; Nabhan, 1985, 1989; National Research Council, 1978; Oldfield, 1984; Oldfield and Alcorn, 1987; Wilkes and Wilkes, 1972). Traditional or peasant agricultural systems are frequently polycultures that include minor crops and other potentially useful plants (Alcorn, 1981; Oldfield, 1984; Oldfield and Alcorn, 1987; S. Brush, University of California, personal communication, May 1989). In addition, the in situ maintenance of landraces and their agroecosystems preserves the complex relationship and competition between wild crop relatives and the weeds that may be associated with them (Oldfield and Alcorn, 1987).

There are, however, several constraints to conserving crop landraces in traditional agroecosystems (Ford-Lloyd and Jackson, 1986; International Board for Plant Genetic Resources, 1985a; Oldfield and Alcorn, 1987; Plucknett et al., 1987). It is unlikely that any mechanism could be developed for more than a small portion of the large number of landraces of even the major crops. Such programs will require a substantial degree of monitoring to ensure that farmers do not abandon cultivation of landraces or traditional varieties in favor of new varieties. Ultimately, incentives may be necessary to encourage farmers to retain and continue traditional practices (Altieri et al., 1987; Altieri and Merrick, 1987; Oldfield and Alcorn, 1987; Plucknett and Smith, 1987). This will necessitate a permanent commitment to a long-term program similar to that which is necessary to maintain an ex situ collection.

The activity of Native Seeds/SEARCH is a model of how in situ conservation can be accomplished if the local community is given appropriate incentive. This group combines in situ and ex situ methods in the conservation of traditional cultivars of the southwestern United States and northwest Mexico. The organization provides encouragement and assistance to Native American and other farmers in these areas to grow their traditional cultivars; it also conserves their traditional varieties and landraces in medium-term, ex situ collections (Nabhan, 1989).

In situ conservation of domesticated genetic resources is usually more difficult than that of wild genetic resources, and little is known about the factors that would favor it. It has been argued that in situ conservation has a potentially valuable role to play in an integrated system for maintaining genetic resources. It may be particularly valuable for conserving landraces in regions with crop diversity, thus allowing continued adaptation and evolution.

RECOMMENDATIONS

Genetic resources must be an integral part of the objectives of existing conservation efforts. In situ conservation provides the capacity to protect a wide range of genetic and species diversity and the adaptive processes that shape them. The scientific understanding necessary to achieve the most effective in situ conservation is, however, only just beginning to emerge. Cultural and economic factors likely to impede or promote long-term, in situ conservation of domesticated genetic resources often still need to be identified.

Research is needed to elucidate the components for establishing viable and genetically diverse populations of wild species.

At present, no integrated predictive capability exists for determining the potential longevity of populations in relation to their size and structure (Shaffer, 1987). Protected populations should be large enough to be self-regenerating and to minimize loss of rare alleles. In the absence of information on the species concerned, a minimum population size of 500 individuals has been suggested (Frankel and Soulé, 1981). However, it has since been shown that such numbers fall far short of encompassing the range of demographic, life history, genetic, and environmental factors that confront in situ conservation (Soulé, 1987).

Unique, endangered wild populations that have present or potential value as crop genetic resources should be conserved in situ.

In general, in situ conservation will focus on wild species. When establishing reserves for crop genetic resources, consideration should be given to wild relatives of crops that are the subject of improvement programs. Attention should also be given to wild species for which there is information on both crossing ability with the crop and on potentially useful genetic traits. Finally, wild species that are in the early phases of domestication are also important. They include many timber and forage species for which the interpopulational genetic variation of the wild species is being explored and used (see National Research Council, 1991b).

Existing protected areas should form the nucleus of a system for in situ conservation of genetic resources.

The protected areas would benefit from serving as wild genetic resources areas since it would increase their value to society. The alternative—a separate network of genetic resources conservation areas, developed more or less independently from conventional protected areas—is unlikely, since competition for land is so intense that it is increasingly difficult to establish new protected areas, particularly if their purpose is narrow, such as the protection of a single species.

A redirection of existing activities, although simpler than establishing new reserves, will still require investment. In many developing countries, none of this will happen without permanent financial assistance.

4

The Science of Collecting Genetic Resources

This chapter examines the scientific base for collecting plant genetic resources. It examines the questions of assuring that collections capture a maximum amount of the genetic diversity in a plant population, given limited resources. How many samples must be collected? How many seeds should a sample contain? What is the likelihood of capturing a very rare gene in a collection?

TYPES OF COLLECTIONS

Several kinds of germplasm collections have evolved over the years in response to particular needs: base collections, back-up collections, active collections, and breeders' or working collections. To a certain extent these divisions may be somewhat artificial, because some collections may fulfill more than one purpose. A number of active collections were formerly breeders' collections, for example. The following discussion is intended to describe the variety of purposes that collections must serve.

Base Collections

Base collections provide for the long-term preservation of genetic variability through storage under optimal conditions. Base collection materials are not intended for distribution except to replace materials that have been lost from back-up or active collections. Base collections include the most comprehensive sample of the entire genetic variability of a species group. They are static in that they attempt to preserve the genetic variability that exists in accessions in its original, unchanged state. However, they are also dynamic in that additional accessions, such as newly collected materials, new cultivars produced by plant breeding, and populations of genetically enhanced materials, are added as they become available. Storage lives of many decades now appear to be possible, so that the losses of genetic variability that occur during storage and regeneration can be held within acceptable limits.

Base collections are typically held under conditions of low relative humidity and at subfreezing ($-10°$ to $-20°C$) or cryogenic ($-150°$ to $-195°C$) temperatures. Difficulties are encountered with seeds of some species that cannot withstand chilling or drying. Alternative methods of long-term storage, such as cryopreservation of in vitro cultures, are needed (see Chapter 7).

A global network of germplasm collections of various crops has been initiated with the guidance and help of the Food and Agriculture Organization of the United Nations and the International Board for Plant Genetic Resources, by designating different agencies that serve the base and back-up collections for principal crop species. However, different agencies vary markedly in their capabilities to fulfill their designated responsibilities.

Back-Up Collections

Back-up collections supplement base collections at another location. For example, the U.S. National Seed Storage Laboratory at Fort Collins, Colorado, holds duplicate back-up samples of portions of the maize collection from the Centro Internacional de Mejoramiento de Maíz y Trigo (CIMMYT; International Maize and Wheat Improvement Center) in Mexico and much of the rice collection from the International Rice Research Institute (IRRI) in the Philippines. These back-up holdings are insurance against loss in the primary CIMMYT and IRRI collections.

Active Collections

Active and base collections often include the same materials. Active collections provide seeds or other propagules for distribution to plant breeders or other users. Consequently, attempts are made to maintain sufficient quantities of materials of each accession in active collections, particularly those in heavy demand, so that requests can be met quickly. Materials in active collections are usually managed

Seeds are counted, packaged, and sealed in foil-laminated moisture-proof bags for storage at the National Seed Storage Laboratory. Credit: U.S. Department of Agriculture, Agricultural Research Service.

under shorter-term and more variable storage standards than those in base collections. Growouts or multiplications to replenish seed supplies in active collections are therefore usually necessary at shorter intervals than they are for base collections, thus putting the genetic integrity of accessions more at risk.

Breeders' Collections

The breeders' or working collections include materials of frequent use in breeding programs and are usually short term in nature. Breeders know from experience that superior performance in their local region of adaptation has resulted from the long-term assembly of favorable combinations of alleles at many different genetic loci and that attempts to introgress alleles from exotic sources into such adapted materials are almost always detrimental to performance, at least in the short term. Hence, breeders' collections are made up almost exclusively of advanced lines developed in their own programs, together with elite cultivars, advanced breeding lines, and genetically enhanced populations obtained from breeders in regions with similar ecological conditions. It is widely recognized by breeders that reliance on adapted stocks may limit the potential for continued advances in performance. Modern breeders, however, infrequently turn directly to exotic materials in active collections for potentially useful variability. Rather, they usually obtain their exotic variability indirectly from genetically enhanced populations, breeding stocks, or both into which potentially useful alleles have been introgressed. Breeders' collections have, consequently, come to include sharply increasing proportions of genetically enhanced stocks that carry potentially useful alleles in adapted genetic backgrounds.

SAMPLING STRATEGIES

Sampling plays a critical role in the conservation process. However, consideration of optimum sampling strategies requires identification of appropriate measures of genetic diversity, especially measures of potentially useful genetic diversity. Marshall and Brown (1975:53) emphasized, "There are definite limits to the numbers of samples which can be handled effectively [in conservation programs that are] imposed by the financial and personnel resources available to carry out each stage in the process." For some species the major limiting factor is an incapacity to collect endangered materials before they are lost; for others, it is the difficulty of preserving materials after they have been placed in collections. For most species, however, the major limiting factors are an incapacity to regenerate and evaluate the materials in collections. Limiting factors highlight the need for judicious allocation of resources with the goal of enhancing efficiency within each of the main elements of the global genetic resource system.

PARAMETERS OF GENETIC DIVERSITY

The most basic parameters of genetic diversity are the numbers and the frequencies of alleles at Mendelian loci in the sampling unit under consideration (species, ecogeographical region, or population). Each Mendelian locus, however, is capable of mutating to many allelic states, and the potential number of different alleles, summed over all loci, is so large that it is impractical to think in terms of identifying, collecting, or conserving all of them. Appropriate sampling procedures must be used to ensure adequate representation of potentially useful alleles in the target species in samples of specified sizes taken from an appropriate number of collection sites.

Neutral Allele Model

The neutral theory of molecular evolution suggests that evolutionary change at the molecular level is caused by random drift of selectively equivalent mutant genes. These mutants are not subject to selection relative to one another because they do not affect the fitness of the carrier nor do they modify their morphological, physiological, or behavioral properties. Evolution consists of the gradual, random replacement of one neutral allele by another that is functionally equivalent to the first. The theory assumes that favorable mutations are so rare that they have little effect on the overall evolutionary rate of nucleotide and amino acid substitutions (Ayala and Kiger, 1980; Kimura, 1968).

The neutral allele model is the simplest model (null hypothesis) for determining the expected number and frequency of alleles at a locus (Kimura and Crow, 1964). In this model, the expected numbers and frequencies of alleles are a function of the effective population size and the rate at which mutation produces novel neutral alleles at any locus. When either or both of these is small, it is expected that most neutral alleles will be either very common or very rare within populations but that, as population size or mutation rates increase, a higher proportion of neutral alleles will occur at intermediate frequencies. When effective population size is very large, numerous neutral alleles are expected to be present, but most of them will be present at a very low frequency ($p < .001$). For biologically plausible values of population size and mutation rates, it is unlikely that a population will carry more than a single neutral allele per locus at a high frequency ($p > .5$), more than two neutral alleles at intermediate frequencies ($.5 > p > .1$), more than two or three neutral alleles at low frequencies ($.1 > p > .01$), or more than three neutral alleles at very low frequencies.

If evolution is driven by natural selection, however, the frequencies of alleles will differ from expectations under the neutral model. Thus, if selection strongly favors one allele, the expected equilibrium situation within populations is that the highly favored allele is likely to be more frequent than $p = .5$, that fewer alleles will be present at intermediate frequencies, and that there will be more rare alleles than the number under the neutral model. Balancing selection, in contrast, is expected to lead to greater numbers of alleles present at intermediate frequencies, and if it is strong, it is expected to lead to maintenance in populations of large numbers of alleles, each at a low frequency (that is, allelic distribution profiles are expected to be more flat, platykurtic, than they are under the neutral model).

EMPIRICAL STUDIES OF GENETIC DIVERSITY: NUMBERS OF ALLELES AND ALLELIC PROFILES

The procedures used to determine the extent of genetic diversity within populations and species fall into two general classes, depending on the nature of the data: (1) those based on measurements made on quantitative traits (traits for which phenotypes are distributed in a continuous metrical series) and (2) those based on counts of numbers of alleles governing qualitative traits (traits for which phenotypes can be classified into discrete categories and the numbers and frequencies of alleles in each category can be enumerated by simple counting rules).

Quantitative Characters

Many studies of genetic diversity during the past half century were based on means, variances, and covariances estimated from measurements of polygenically controlled quantitative traits. However, the effects of any single locus on a quantitative trait cannot be disentangled from the joint effects of the several to many loci that affect these traits. Consequently, such studies provide, at best, only indirect and imprecise estimates of those measures of genetic diversity that are most informative in the context of sampling for purposes of conservation (for example, numbers of loci, numbers and frequencies of alleles per locus, the proportion of polymorphic loci, and the degree of association among alleles of different loci). Thus, for the purposes of estimating amounts of genetic diversity, population geneticists and conservationists have increasingly turned to loci for which the effect of each allelic substitution is unique and unambiguously distinguishable from allelic substitutions at all other loci.

Qualitative or Discretely Inherited Characters

The earliest studies of discretely inherited characters were based on the loci governing morphology, pigmentation, physiology, and other discretely inherited variants of single Mendelian loci. Perhaps the most thoroughly studied system is the series of eye-color alleles in natural populations of *Drosophila melanogaster* and other *Drosophila* species. It was shown in detailed experiments involving the extraction and measurement of eye pigment that allelic variants, isoalleles, exist in flies with various eye-color classes. The normal wild phenotype (red eyes) includes a range of variants based on the amount of eye pigment; these variants are designated normal isoalleles. Those phenotypes deficient in eye pigment (for example, white or apricot eyes) are mutant classes and designated mutant isoalleles (see Strickberger, 1976). The allelic profile of the red-eye/white-eye locus thus features a cluster of normal wild-type isoalleles that can be distinguished from one another by special pigment measurement tests and several additional clusters of mutant isoalleles. In segregating populations derived from hybrids between wild-type and mutant individuals, it was typically found that there was an excess of wild-type individuals and a deficiency of mutant ones. It was also found that individuals with mutant eye-color phenotypes were unable to compete with individuals with wild-type phenotypes under laboratory conditions.

The conclusion was that normal isoalleles produce gene products that are necessary for normal functioning of the organism, whereas mutant isoalleles produce only partly functional gene products. Furthermore, mutant isoalleles are usually infrequent or rare (frequency of .001 or less) under conditions of competition in wild-type and experimental populations. This is consistent with the expectation that the mutants are inferior adaptively. Although the majority of loci that govern traits that can be classified into discrete categories, such as color or morphology, appear to be nearly monomorphic for a wild-type allele or a cluster of wild-type isoalleles, some loci appear to have allelic profiles featuring one or two frequent alleles plus a number of infrequent or rare alleles. Also, a few loci are known that feature large numbers of nearly equally frequent alleles; among the best authenticated of such loci are those that govern incompatibilities between haploid pollen and diploid styles in plants. Rare self-incompatibility alleles, especially novel, newly mutated incompatibility alleles, have a selective advantage over frequent alleles; consequently, it is not surprising that the equilibrium status for such systems is one that features many alleles per locus, with each allele being present at low and more or less equal frequencies.

In summary, studies of loci whose variants can be classified into discrete categories indicate that allelic profiles for most such loci in *Drosophila* and other species tend toward near monomorphism ($p >$ .9) for a single wild-type allele, or a cluster of nearly indistinguishable wild-type isoalleles, accompanied by a number of infrequent to rare mutant alleles or clusters of mutant isoalleles. Furthermore, it is widely accepted that such allelic profiles represent selectional-mutational equilibrium situations that develop because one of the normal wild-type isoalleles has a greater or lesser selective advantage over all other alleles at a locus.

Enzyme Variants

In recent years studies of enzyme variants (isozymes) have added greatly to the amount of information on allelic frequency profiles for single loci. The use of isozymes for this purpose has several advantages. (1) Enzyme specificity allows alleles to be identified unambiguously with single loci. (2) Allelic expressions are usually codominant; that is, all alleles at a locus are expressed in single individuals. (3) Several to many isozyme loci can be assayed in single individuals. Thus, even though only one-third to one-quarter of mutations alter enzymes in ways that are detectable by protein electrophoresis, isozyme techniques allow a much higher proportion of genetic variability to be detected at the level of the individual locus than previously available methods do.

The statistics most often reported in surveys of enzyme variability are the numbers and frequencies of the alleles observed at various loci. However, both of these measures are heavily dependent on sample size. Samples larger than $N = 20$ gametes are required to obtain reasonably precise measures of the numbers of alleles at a locus; this is because infrequent or even moderately frequent alleles may often go undetected, even when sample sizes per population are as large as $N = 80$ gametes. It is appropriate to emphasize that several alleles (usually about 3 to 10) occur at many isozyme loci and that estimates of the frequencies of each of these alleles at each locus are very much more sensitive to sample size than are estimates of the total numbers of alleles per locus. Therefore, caution should be used when making comparisons of the allelic diversity reported to be present in different species and even in different populations of the same species. This is because the kinds and numbers of enzyme systems examined, as well as sample sizes, vary widely among investigators and each of these factors can strongly influence the numbers and frequencies of alleles reported.

FORMULATION OF SAMPLING STRATEGIES

Marshall and Brown (1975) defined four conceptual classes of alleles as follows: (1) common, widespread, (2) rare, widespread, (3) common, localized, and (4) rare, localized.

They postulated that the first and fourth classes are likely to be of little concern in formulating sampling strategies. Alleles of the first class (common, widespread) are almost certainly included even in very small samples collected from only a few populations; for example, the probability is greater than .999 that an allele present at a frequency of $p > .5$ will be included in a sample of only 10 gametes from a single population. In contrast, for alleles of the fourth class (rare, localized), the inclusion of a rare allele will be unusual and serendipitous, even in very large samples taken from a large number of populations. (For example, to detect a rare allele present at a frequency of .001 in only 1 among 100 populations of a species, a sample of 2,994 gametes from each of the 100 populations [about 300,000 gametes] is required to obtain [at a probability of .95] a single copy of this rare allele.) It follows that formulation of sampling strategies for the first and fourth classes of alleles does not lead to general guiding principles in sampling, other than the trivial principle of ignoring both classes or of collecting impractically large samples to detect the rare alleles. However, Marshall and Brown (1975) considered the two remaining conceptual classes of alleles, rare, widespread and the common, localized, to be relevant to the formulation of sampling strategies.

Alleles of the second class (rare, widespread) behave as if the species (or population or target area) to be collected is a single, large, unstructured population. It follows that sampling of this class depends only on the total number of gametes in the combined sample from all of the populations in which collections are made. The appropriate strategy is therefore to draw a total sample large enough to ensure that the desired number of rare widespread alleles is included. Table 4-1 gives the sample sizes required, at various probability lev-

TABLE 4-1   Numbers of Alleles That Must Be Sampled To Be Certain of Including, at Various Probability Levels, at Least One Copy of an Allele Present at Various Frequencies in the Area To Be Sampled

Probability	Number of Alleles Sampled for the Following Allele Frequency in Target Area				
	.1	.05	.02	.01	.001
.99	43	88	228	458	4,603
.95	27	59	148	298	2,994
.9	22	45	114	229	2,301
.8	15	31	79	160	1,608
.5	7	13	34	69	691

els, to ensure the inclusion of an allele that is present, at various frequencies, within the target area. For alleles that are present at a frequency as low as $p = .001$, a total sample of 4,603 gametes (approximately 100 gametes from each of 50 populations) will ensure ($p = .99$) that at least 1 copy of any such rare allele is included. Thus, the detection of widespread rare alleles seems attainable for many species.

Alleles of the third conceptual class (common, localized) are alleles that occur in only one or a few habitats, where they reach a high frequency. They may be biologically specialized alleles that enhance adaptation only in certain habitats. These are often the class of alleles of most interest to breeders, because breeders are concerned with improving performance in the specialized habitats of their own ecogeographical regions. Widespread, common alleles have almost certainly been introduced into all habitats, whereas introduction of the localized common alleles of special habitats is likely to have been sporadic. Such considerations led Marshall and Brown (1975) to suggest that locally common alleles are, at least conceptually, the key class of alleles in formulating sampling strategies, whether for capturing the maximal amount of useful variation within populations (within practical limits of sample size) or determining the distribution of environmentally influenced genetic variation within species.

Very large numbers of accessions have been accumulated in germplasm collections for many different species and species groups, and these large numbers in themselves have often been taken to indicate that collections include all or nearly all potentially useful genetic variability. However, existing germplasm collections suffer to a greater or lesser extent from four important deficiencies. First, records made at collection sites have often been inadequate in various respects (for example, a description of ecological features of sites, the sizes of samples, and the manner in which samples were taken). Also, clerical and other errors made subsequent to collection have led to inaccuracies in records, with the result that the quality of the sample cannot be evaluated. Second, the numbers of individuals taken from individual sites have often been very small, so that potentially useful alleles may have been missed in the original sampling of many sites. Third, few or no individuals have been taken from some ecogeographical areas, especially the less accessible sites within areas; hence, species with rare but locally frequent alleles may have been missed in particular. Fourth, the management of accessions within germplasm collections has often been inadequate, with the result that there has been a decay of genetic variability within accessions.

EMPIRICAL STUDIES OF ALLELIC DIVERSITY AND ENVIRONMENTALLY INFLUENCED GENETIC DIFFERENTIATION

Two aspects of genetic variation are of major importance in formulating sampling strategies: (1) the numbers and the frequencies of alleles at a number of representative loci in single populations (which is important in determining the numbers of individuals to sample within single populations) and (2) on patterns of genetic and evolutionary differentiation among populations (which is important in determining the number and distribution of populations to be sampled). The allelic and genotypic variabilities within and among populations of *Avena barbata*, a heavily self-pollinating wild relative of cultivated oats, and allelic variability within and among indigenous cultivated races of maize (*Zea mays*) are used as examples.

A Wild, Predominantly Self-Pollinating Species

Although samples of *A. barbata* taken from single sites (populations) have usually been small, and the sampling variances have consequently been large, there have been a few studies of wild populations in which large numbers of individuals from numerous single populations representing a diversity of ecological situations have been examined. *A. barbata*, the slender wild oat, ranks among the most thoroughly studied of wild species. Large samples have been examined from specific collection sites from throughout the range of the species, including southwestern Asia and the eastern Mediterranean Basin, where *A. barbata* is endemic (Kahler et al., 1980); Spain, where it is naturalized (Garcia et al., 1989), and California, where the species rapidly became a major component of the annual flora and a highly useful forage species subsequent to its introduction from Spain about two centuries ago (Clegg and Allard, 1972). In the study of Garcia et al. (1989), 4,011 individuals (approximately 100 from each of 42 sites in Spain) were assayed for 14 isozyme loci representing nine enzyme systems. Among the 14 loci, 2 were entirely invariant (only one allele was observed in the total sample of 8,022 gametes) and 3 other loci were entirely invariant in 41 of the 42 populations sampled (each of these 3 loci was weakly polymorphic for a single infrequent allele in the same single population). All of these 5 loci were entirely invariant in California but weakly polymorphic for 1 to 3 or more infrequent ($p < .01$) or rare ($p < .001$) alleles in southwestern Asia. Allelic profiles were closely similar for 4 additional isozyme loci; these 4 loci differed from the 5 invariant or nearly invariant loci discussed above primarily because many of the Spanish populations were weakly polymorphic for 1 to 3 or more infrequent or rare alleles at each locus and that both the southwestern Asian and Californian populations also tended to be slightly more variable for rare or infrequent alleles. Thus, the gene pools of southwestern Asia, Spain, and California can be characterized as nearly identical in allelic compositions for these 9 invariant or weakly polymorphic loci.

Nearly all of the 42 Spanish populations were moderately to highly variable for the 5 remaining isozyme loci among the 14 studied. In the majority of Spanish populations, allelic profiles for these 5 variable loci typically featured a single predominant allele ($p > .7$) plus 1 allele present at a low frequency ($p = .01$ to $.10$) and several infrequent ($p < .01$) or rare ($p < .001$) alleles. Occasionally, allelic profiles within populations featured a single predominant allele at a very high frequency ($p > .98$) plus 1 or 2 infrequent or rare alleles. Alternatively, allelic profiles occasionally featured 2 or 3 alleles present at intermediate frequencies ($p = .2$ to $.6$) together with several infrequent or rare alleles. As was the case with the 9 nearly invariant loci, the southwestern Asian populations were more variable than the Spanish populations for these 5 moderately to highly variable loci.

The empirical results presented above indicate that the alleles of *A. barbata* fall into three categories relevant to sampling. Alleles that are highly frequent ($p > .6$) in any single population or region are also nearly always highly frequent in all population and all regions. These predominant alleles appear to be the wild-type alleles of classical genetics and the class one (common, widespread) alleles postulated by Marshall and Brown (1975).

Alleles that are rare ($p < .001$) in any population or region are also nearly always rare in all populations and regions. These alleles apparently contribute significantly to high levels of adaptation in very few, if any, environments; they evidently remained rare in populations because selection removed them at the same rate at which they were produced by recurrent mutational events. They appeared to be the deleterious mutants of classical genetics, and thus, they fit into either the class two (rare, widespread) or the class four (rare, localized) allele of Marshall and Brown (1975). Because of practicalities of sampling, however, it is probably not worthwhile in practice to attempt to distinguish between these two conceptual classes.

Finally, there are alleles whose frequencies vary widely from population to population. The pattern of distribution most commonly observed for these alleles was one in which a given allele was infrequent ($p < .01$) or rare ($p < .001$) in the majority of populations but present at low to moderate frequencies ($p = .01$ to $.05$) in an occasional population. However, in a few cases a given allele was observed in many different populations but at frequencies that varied from $p < .01$ to $p = .5$. Alleles whose frequencies vary from population to population thus appear to fit into Marshall and Brown's common, localized category (class three), although the fit appears to require somewhat flexible definitions of what constitutes common and localized.

Among the 137 alleles that have been recorded for these 14 loci of *A. barbata*, about 10 percent (1 allele for each locus) were consistently highly frequent and omnipresent. This class of alleles (Marshall and Brown's class one alleles) therefore contributed relatively little to either allelic variability within or environmentally influenced genetic differentiation among populations. The majority of the 137 alleles (about 65 percent) were consistently rare ($p < .001$) and were usually found in only 1 or few populations. Because of their infrequency, these rare alleles (classes two and four of Marshall and Brown) contributed little to allelic variability within or among ecogeographic populations of *A. barbata*. About 35 (26 percent) of the 137 alleles were found in most but not all populations and regions, often reaching frequencies of between .01 and .1 and occasionally as high as .5. Such variations in allelic frequencies were often correlated with readily observable and measurable environmental factors (rainfall, temperature, slope, exposure, soil type). Alleles whose frequencies vary widely from population to population (class three of Marshall and Brown) therefore appear to be responsible for most of the allelic variation that exists within and among populations of *A. barbata*.

The gene pools of different populations of *A. barbata* within a region and those of different regions are closely similar in allelic content when they are compared on a locus-by-locus basis. In contrast, the arrays of multilocus genotypes found in different regions and different populations, including closely neighboring populations occupying different habitats, often differ strikingly. Similarly, arrays of multilocus genotypes found in colonial gene pools are often very different from those found in ancestral gene pools; for example, multilocus allelic complexes adapted to contrasting extremely hot and arid versus cool and moist habitats in California have not been found in Spain or southwestern Asia (Allard, 1988; Garcia et al., 1989).

Isozyme Variation Within and Among Cultivated Races of Maize, An Outbreeding Species

Mexico is the homeland of teosinte, the only close relative of maize (Wilkes, 1967); Mexico is also the cradle of maize domestication (Mangelsdorf, 1974). Patterns of isozyme variation have been studied in races of maize from all parts of Mexico, as well as from South and North America and the West Indies. Consequently, Mexican maize provides a convenient standard against which to compare patterns of isozyme variability within and among races of maize from various geographical areas and within and among other cultivated outcrossing species. Doebley et al. (1985) examined 12 plants (24 gametes) from each of 93 collections of maize from Mexico and 1 collection from Guatemala. Their total sample size was 1,128 individuals (2,256 gametes). The 94 collections represented 34 races (Hernandez and Alanís, 1970; Wellhausen et al., 1952). Each plant was analyzed for 13 enzyme systems; 163 alleles representing 23 distinguishable loci were observed (an average of 7.09 alleles per locus).

Table 4-2 gives the distribution of the 163 alleles in frequency classes based on the total sample and also the number of races in which alleles of each frequency class were observed. A single highly frequent allele ($p > .6$) was observed for 20 among the 23 loci ($p > .9$ for 10 loci); the frequencies of the most common alleles at the 3 remaining loci were .432, .402, and .255. All alleles that were present at a frequency of $p > .6$ were ubiquitous, occurring in all 34 races (Table 4-2). Among the 163 alleles, 20 (12 percent) fell into this category and they can reasonably be placed in the common, widespread class (class one) of Marshall and Brown (1975).

In contrast, 93 alleles (57 percent), which were present in the total sample at a frequency of $p < .01$, were nearly always confined to a single race. Thus, nearly all rare alleles appear to fit into class two or class four of Marshall and Brown (1975). The 50 remaining alleles (31 percent), which were present in the total sample at frequencies of .01 to .5, occurred in many but not all populations (Table 4-2). Other studies (Bretting et al., 1987, 1990; Goodman and Stuber, 1983) show

TABLE 4-2 Numbers of Alleles Observed in Various Frequency Classes and the Numbers of Maize Races in Which Individual Alleles Were Observed

Frequency Class (p)	Number of Alleles in Frequency Class	Number of Allele-Race Combinations Observed	
		Mean	Range
>.6	20	34	34–34
.3–.5	6	33	31–34
.2–.3	2	29	29–30
.1–.2	8	25	9–31
.05–.1	9	20	10–23
.01–.05	25	9	2–19
.001–.01	77	2	2–7
<.001	16	1	1–1

NOTE: A sample of 2,256 gametes (1,128 individuals) from 34 Mexican races of maize was used.

SOURCE: Data derived from Doebley, J. F., M. M. Goodman, and C. W. Stuber. 1985. Isozyme variation in races of maize from Mexico. Am. J. Bot. 72(5):629–639.

that the frequencies of such alleles tend to vary over a wide range from race to race. Thus, alleles that vary in frequency from race to race appear to fit into Marshall and Brown's class two.

Doebley et al. (1985) reported that 88 percent of the total allelic variability resides within races and 12 percent resides among the 34 Mexican races of maize; they attributed a large part of this variability to differences in the frequencies of alleles that are neither ubiquitous nor rare. Principal component and cluster analyses showed that variation among races was continuous and that there were no well-defined race complexes; weakly defined differentiated groups were apparent (high-elevation races, northern and northwestern races, southern and western low-elevation dent and flour maize races), but in general, races of maize were not sharply differentiated. Overall, the observed allelic variability within races of maize (a cultivated outbreeding species) is thus remarkably similar to that observed within populations of *A. barbata*, a wild-type inbreeding species.

SAMPLE SIZES FOR EACH COLLECTION SITE

The first step in determining the optimum partition of resources within and among sites in the region(s) from which samples are to be collected is to determine, at some acceptable level of probability, the optimum number of seeds or other propagules needed to capture potentially useful alleles at single collection sites. It is wasteful of resources if the sample sizes taken from single sites are too small to capture potentially useful alleles present at the site. The time and resources required to travel to a site, record ecological and other relevant information about the site, send the collected materials through customs, and so on are also wasted. However, it is also wasteful of resources if the collector takes samples from single collection sites larger than those needed to capture the alleles and provide for long-term storage in germplasm banks or the distribution of accessions to users.

Marshall and Brown (1975:62–63) argued that

> the great majority of common alleles or allelic combinations (whether widespread or local) presumably represent adaptive variants maintained in populations by some form of balancing selection (Dobzhansky, 1970). . . . Consequently, common variants are likely to be of far greater interest to plant breeders than rare variants. . . . [Hence] the aim of plant exploration can be defined as *the collection of at least one copy of each variant occurring in the target populations with frequency greater than 05.*

If only 2 alleles, a_1 and a_2, are present at a locus at frequency p, $(a_1) = .95$ and p_2 $(a_2) = .05$ at a collection site, a sample of 59 gametes will provide at least 1 copy of each of the 2 alleles with a 95 percent certainty (Equation 4-1). If more than 2 alleles per locus are present at a frequency of $q > .05$, the sample sizes required to achieve this goal are not drastically larger; for example, for a locus with an allelic profile of .8, .05, .05, .05, a random sample of 80 gametes will achieve the objective of capturing at least 1 copy of each of the 4 alleles with a probability of $p = .98$. Considerations such as these led Marshall and Brown (1975:73) to the following conclusion: "In most circumstances, sample size should not exceed 50 plants per population and in no circumstance is it desirable to collect more than 100 plants per population."

There are, however, two important additional practical considerations that the collector must take into account: the need to capture lower-frequency alleles in sampling, and how large a sample should be.

Capturing Lower-Frequency Alleles

Experimental results indicate that many localized alleles are present at local frequencies of no greater than $p = .05$, and these variants may be of interest to collectors. Capturing an allele ($p = .95$) present at a frequency of p = .01 at a single site requires a sample size of N = 298 gametes, which is 5 times as large as that necessary for an allele present at a frequency of p = .05. Samples of 100 individuals (200 gametes), the maximum recommended by Marshall and Brown (1975), may thus often fail to capture some of the potentially useful alleles at a single site.

The usual procedure, however, is to collect more than a single sample per location, which greatly increases the probability of capturing rare, but locally important alleles. A single gametic sample of size 160 has only an 80 percent probability of capturing an allele with a frequency of .01, but 2 such samples have a 96 percent probability of capturing the allele. There are strong advantages to collecting more samples per location rather than collecting more individuals per sample. Perhaps the most important of these is that any individual sample may be chosen from a population which, by accident such as local drift (the "bottleneck" effect), may not even possess the allele(s) of interest. It is also important that the alleles sampled should represent the population, hence one should choose from distinct individuals rather than choose two peas from the same pod or 300 sorghum seeds from a single head.

Alternatively, an increase in sample sizes to 200 or more individuals (400 or more gametes) substantially increases the probability of capturing potentially useful alleles if present, usually without a significant increase in the resources expended at the site. This is because a large proportion of the total time, effort, and expense involved in collecting usually goes for traveling from location to location, inspecting and choosing the sites to be sampled at each location, and recording the ecological features of the sites rather than for actually collecting seeds or other propagules from the location. However, given that 400 gametes are to be sampled at a small village, four samples of 100 gametes each are more likely to capture alleles of local importance than a single sample of 400, unless the alleles are distributed very uniformly throughout the village.

Size of the Sample

Additional practical considerations include the size of the sample that (1) can be collected and transported from plants that produce very large seeds or from plants that must be propagated vegetatively, or (2) may be limited by problems of transporting bulky materials or survival during transport. Ideally, the sample size from each population unit should be large enough to provide not only for conservation but also for distribution to users without the need for immediate multiplication, if field seed quality is acceptable. Ex situ multiplication is difficult for many species, nearly always costly in terms of resources and time, and nearly always hazardous to genetic integrity. Consequently, if feasible, sufficient seeds or other propagules should be collected from each site to provide for both long-term storage and immediate use. If this is not feasible, the sample from each site should, if at all possible, be large enough so that the first multiplication (regeneration) will meet both of these goals. The sample should always, except under severe conditions, be adequate to provide sufficient regenerated seeds for long-term storage.

The identification of appropriate populations to sample is usually relatively straightforward for annual, cultivated crop species such as cereal grains, pulses, and oil-seed crops. Farmers harvest seeds in bulk, usually by field, and set aside part to be sown as the next crop. Seed numbers are very large, and the mixing during harvesting and cleaning virtually guarantees that all large samples from a given seed source will include all potentially useful genetic variability. A field or a farm seeded from a common source is the unit of sampling. It is

---

CAPTURING LOW-FREQUENCY ALLELES

The relationship between the size and number of collection samples and the probability of recovering low-frequency alleles is illustrated by the following.

Suppose at one diploid locus there are 2 alleles, a_1 and a_2, that occur in a population at frequencies of p_1 = .90 and p_2 = .10, respectively, and that a sample of 10 zygotes or seeds (N = 20 gametes) is assayed. Even with a sample this small it is virtually certain ($p > .9999$) that at least 1 copy of allele a_1 will be included, that is, allele a_1 can be detected in every sample of N = 20 gametes. The probability is also high ($p = .88$) that the less frequent allele, a_2, will also be detected; that is, in a group of 10 different samples there will be a copy of allele a_2 found in 8 or 9 of those samples (where each sample is N = 20 gametes). If, however, alleles a_1 and a_2 are present in the population at frequencies $p_1(a_1) = .95$ and $p_2(a_2) = .05$, the probability of detecting allele a_1 increases slightly, but the probability of detecting allele a_2 decreases to .64; that is, only about two-thirds of samples of size N = 20 gametes are expected to include a copy of a_2. If p_1 and p_2 are .99 and .01, or .999 and .001, respectively, the probabilities that samples of size N = 20 gametes will include at least 1 copy of allele a_2 fall to .18 and .02, respectively, that is, only about 2 of 10 and 2 of 100 samples of size N = 20 gametes are expected to contain allele a_2. The table below gives the probabilities

often possible to purchase large enough quantities of seeds from such a field to satisfy all needs associated with conservation or short-term distribution to users if collection can be timed with harvest. Market samples should be avoided because their origins may be obscure, they may be mixtures from different regions, or in rare cases, they may represent imported seeds.

How many seeds or other propagules should collectors take under such ideal conditions? Assume that a sample from each collection site is to be distributed to a base collection, a back-up collection, five active collections, and 50 samples for immediate distribution to users. Assume further that the size of each sample must be 50 seeds, the minimal number suggested by Marshall and Brown (1975) on the basis of sampling considerations alone. This requires that the original sample must be no smaller than 2,850 seeds. However, some will be lost during transport or passage through quarantine and perhaps half of all individuals distributed for long-term storage will produce few or no progeny. Hence, samples distributed for these various purposes ideally should be at least twice as large as indicated by

---

that the less frequent allele, a_2, will be detected in sample sizes of 20, 40, 80, or 160 gametes when it is present in a population at frequencies of .10, .05, .01, and .001.

The use of multiple or larger samples greatly increases the probability of capturing an allele. The formula for N samples is $1 - (1 - p)^N$. For example, an allele with the frequency of .01 has a probability of .18 of being found in 1 sample size of 20. Finding the same allele in a collection of 8 samples of size 20 is $1 - (1 - .18)^8 = .80$ probability, which is the same probability as finding it in 1 sample size of 160. To reach a probability of .96 in capturing an allele with a frequency of .1 would require 16 samples of size 20 or 2 samples of size 160.

Probabilities That the Less Frequent Allele (a_2) Present at a Diploid Locus at Various Frequencies Will Be Detected in Samples of N = 20, 40, 80, or 160 Gametes per Population or Site

Frequency	Probability for a Sample Size (N) of			
	20	40	80	160
.10	.88	.99	>.99	>.99
.05	.64	.87	.98	>.99
.01	.18	.33	.55	.80
.001	.02	.04	.08	.15

sampling considerations alone. At harvest, sample sizes of several thousands of seeds or other propagules are not difficult or impracticable to obtain from a single field. Sufficiently large samples may, however, be out of reach when there are limited numbers of individuals at the collection site, when seeds or other propagules are bulky and difficult to transport, or when the reproductive unit is actively growing somatic tissue. In such cases, the collector should attempt to collect the equivalent of no fewer than 60 gametes (30 diploid individuals) to satisfy minimal sampling requirements and should arrange for the first multiplication to be large, stress-free, and carried out promptly.

Natural populations of wild relatives are often distributed in more or less disjunct patches over a much wider range of habitats than their domesticated descendants. For both wild relatives and cultivated plants, several ecological and population factors must be taken into account in deciding where samples should be collected. Ecological factors include slope, soil type and drainage, and differences in the flora and fauna in the various habitats. The most important population factors are the mating system and the mobility of the species (especially the mobilities of pollen and seeds in plants); these factors often have very large effects on within-population organization as well as on genetic differentiation. The goal of capturing useful alleles is most likely to be met by taking stratified samples from diverse sites. Stratified sampling is recommended even if the target area appears to be ecologically homogeneous.

It may be troublesome to obtain adequately large samples, especially from wild-type populations, when (a) few individuals exist, (b) many individuals produce very few seeds or other propagules, or (c) the collecting season is short (for example, because of seed shattering) and varies in time from year to year. Limited reproductive capacity is most likely encountered in more severe, harsher habitats or during years with limited rainfall. Under such conditions, repeated visits will be necessary to obtain adequate samples.

NUMBER OF SITES TO SAMPLE

Allelic and genotypic diversity is not distributed evenly over the range of most species. The collection sites should represent as many distinctive environments as possible within the collector's resources. The total number of sites from which samples can be obtained depends on factors such as the size of the region to be sampled, the quality of transportation, the terrain, the length of the collecting season within different habitats in the region, the time required for collection at each site, and the amount of cooperation available from local collaborators. Careful advanced planning is important to ensure optimal coverage at the proper time.

The mating system of the target species can also be a guide to the optimum number of stratified sites from which samples should be obtained. Allelic and genotypic variability tends to be more uniformly distributed for outbreeders than it is for inbreeders. In either case, the number of distinct sites sampled should be the maximum possible, provided that the samples taken at individual sites are large enough to capture potentially useful local genetic variability.

RECOMMENDATIONS

The scientific bases for the efficient collection, preservation, and distribution of plant genetic resources are all well understood. Particular attention is directed to the following recommendations concerning the size of samples to be collected, strategies for sampling genetic diversity from different ecogeographical regions, and the need for back-up storage of collections.

All major germplasm collections must be protected against catastrophic loss by back-up storage at other institutions.

Duplication of accessions at one or more institutions provides security against loss. Cooperating institutions do not need to test or regenerate seeds in back-up collections. Rather, primary collection managers should ensure that the samples sent to back-up collections are viable. More efforts are needed to back up collections.

Strategies for sampling genetic diversity of crop species should be based on an understanding of the ecogenetic structure of species and populations.

Materials as they now exist in germplasm collections, including the best managed collections, are rarely adequate for study of population genetics of species. Information on the distribution of the allelic and genotypic variations within and among populations is not available for the majority of domesticated plants or animals or their wild relatives. Therefore, in formulating sampling strategies at present, there is no alternative other than to take advantage of information that is available from those species that have been studied most thoroughly and to extrapolate from that information to species about which little is known.

The size of collected samples ideally should always be sufficiently large to minimize or eliminate the need for regeneration prior to storage.

Repeated regenerations result in loss of alleles and sometimes of entire samples and should be avoided whenever possible. However, field-collected seeds may not meet minimum standards of viability for base collections. In such cases it is essential to minimize the genetic shifts or losses that can accompany regeneration (see Chapter 5).

The size of a collected sample should, where practical, be adequate for deposition in both base and back-up collections. Duplicate samples should be supplied to host country institutions, base collections, and back-up collections. Clear responsibility for back-up collections needs to be provided, along with adequate passport data to enable the recovery of the entire collection in case of natural or civil disaster at the base site.

When collecting seed, samples should be chosen from as wide an ecogeographic and ethnological base as is possible.

Given a fixed number of seeds to be collected, it is better to collect from more sites than to collect more seeds at fewer sites. Prior to collecting, ecogeographic and sociological or ethnological planning should identify the range of locations most desirable to sample.

5

The Science of Managing Genetic Resources

During the past 25 years there has been a dramatic increase in the number of germplasm collections in the world as well as in the number of samples stored in these germplasm collections. This growth has occurred principally at the national level as more and more countries seek to conserve those genetic resources of historic or strategic importance to their agricultural systems. Much of this growth in number of collections has been by the partial or complete duplication of existing collections.

Although this growth reflects a growing commitment to genetic resources conservation, it has also raised concerns. In addition, base collections of many of the major crops have now grown so large and diffuse that they may inhibit, rather than promote, effective management of genetic resources and their use by plant breeders. Because of increased emphasis on the collection of wild and weedy relatives of crop plants and the continuing development of new elite gene combinations by plant breeders, collections that are already large may quickly grow progressively larger. The problems faced by curators charged with characterizing, monitoring, and periodically regenerating these collections are also increasing. The various operations described in this chapter must all function well if germplasm banks are to be able to encourage germplasm use rather than merely acquire and maintain specimens.

Germplasm banks may contain seeds or plant materials that are vegetatively propagated as clones. This chapter is concerned largely with the special problems of managing genetic resources in the form of seeds. Chapter 7 discusses some new developments in propagating and maintaining clonal materials.

SEED REGENERATION

Regeneration practices must be related not only to the life cycles (including cultural needs) and the reproductive biology of individual crops and species but also to the cost of obtaining sufficient quantities of good-quality seed. The problems (and costs) vary greatly with the breeding system; they also depend on other features, such as the reproductive rates of individual genotypes. There are special problems for wild species, including seed dormancy, seed shattering, and high variability in flowering time and seed production.

Seed regeneration is thus costly in terms of resources and time, while the risks of genetic drift (change in allelic frequencies due to sampling accidents in small populations) and genetic shift (change in

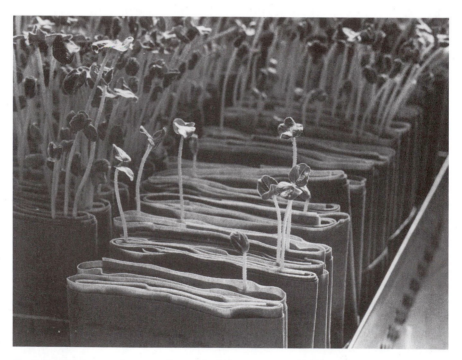

Germination tests ensure the viability of okra seeds kept in storage at the National Seed Storage Laboratory of the U.S. Department of Agriculture. Credit: U.S. Department of Agriculture, Agricultural Research Service.

allelic frequencies due to selection during regenerations) are compounded over each regeneration event. Consequently, the most cost-effective way of minimizing the loss of genetic integrity is to keep the frequency of regeneration to a minimum. Earlier, a distinction was made between regeneration needs for the rejuvenation of stored accessions and multiplication for distribution and use. If sufficient numbers of seeds are stored in long-term base collections, samples may be drawn off periodically as required for multiplication in active collections to satisfy user demands. Such seeds would not be used for conservation purposes thereafter, thus making it possible to carry out their regeneration under less rigid requirements.

By judicious planning, the number of generations required for both rejuvenation and multiplication could be limited to as few as two in the foreseeable future for many species with orthodox seed. Thus, for cost-effectiveness, as many seeds as possible should be collected during sampling or the first regeneration should be maximized to provide adequate seed for immediate use and long-term storage. A better integration of long-term (base) and medium-term (active) storage facilities may be an advantage, and several schemes have been proposed (for example, Linnington and Smith, 1987).

Breeding Systems

The nature of the breeding system is of paramount importance in establishing the tactics of regeneration. Seed bank managers must therefore be thoroughly familiar with the breeding systems of their material and must pay particular attention to the extent that the breeding system varies within the species (Widrlechner, 1987). This section describes the general characteristics of the most commonly encountered breeding systems.

Breeding systems determine to a considerable extent the patterns of ecotypic differentiation and, consequently, dictate subsequent regeneration procedures.

Complete self-pollination is the most extreme form of inbreeding. It results in the rapid fixation of allelic combinations into homozygous, true-breeding genotypes (those in which the offspring and parents are genetically alike). This promotes the buildup and maintenance of multilocus complexes (genotypes) that are preserved intact over several generations and which may be adapted to particular habitats.

Sexual reproduction in outbreeders (those that cross-pollinate) involves crossing, segregation, and recombination. Outbreeder populations thus retain high levels of potential heterozygous and homozygous variability. Because of gene flow, neighboring populations are often less sharply differentiated ecogeographically. Outbreeding populations also show less distinct multilocus structural organization within their populations than inbreeding populations do (Allard, 1988). Important adaptive allelic combinations are, however, preserved over sexual generations through chromosomal linkage, whereas phenotypic correspondence is often further secured by dominance and epistasis (where one gene masks the effect of another) of favored alleles. In preserving the genetic integrity of populations during regeneration, it is important not to disrupt this multilocus genetic organization of the population. The break-up of genetic organization through segregation nearly always leads to loss in vigor and reduced reproductive fitness.

Apomixis (asexual reproduction) often allows highly heterozygous individuals to produce seeds or other propagules that breed true. However, apomixis is frequently facultative with the result that all progeny do not necessarily have genotypes identical to that of the parental individual.

The breeding system, chromosomal organization, and reproductive mode (asexual versus sexual) all interact in the release of genetic variability. Together they constitute the genetic system that allows species and populations to adjust to the twin evolutionary demands of preserving and propagating the successful genotype while retaining the genetic flexibility to meet environmental changes or colonize new areas, that is, the opposing demands of short-term fitness and long-term adaptability. Consequently, they are themselves under genetic selection to meet the life-style and life-cycle needs of different organisms.

Principles Involved in Maintaining Genetic Integrity During Regeneration

The initial supply of seeds generally is inadequate for distribution and must itself be multiplied. The viabilities of seeds or other propagules held in storage eventually fall below acceptable levels. Distributions of seeds ultimately deplete stocks to levels at which regeneration becomes necessary. The genetic structures of accessions can be altered during the regeneration process by contamination because of accidental migration or hybridization between accessions, differential survival of alleles or genotypes within accessions, or to random drift within accessions. This section considers methods for protecting the genetic integrity of accessions and minimizing losses in allelic variability for these reasons.

Contamination from Outcrossing

One of the most serious threats to the genetic integrity of accessions arising during the regeneration process is contamination resulting from hybridization between different accessions. Regeneration of self-pollinated species is usually carried out in field nurseries in which different accessions are planted in closely adjacent rows. No species is completely self-pollinated, however. Under close planting conditions, outcrosses occasionally occur even between accessions of species that are considered to be very heavily self-pollinated. Visibly recognizable outcrosses can sometimes be discarded before maturity during the next regeneration cycle. However, the technical expertise to recognize outcrosses is often lacking, and variants may not be eliminated. This results in a genetic shift that has caused accessions in many collections of inbreeders to lose much of their original distinctiveness. The problems of genetic shifts for outcrossed species are even more severe. In many cases, crossed accessions are mixed to the point that the collections are complex hybrid swarms.

Isolation of accessions from one another during regeneration prevents genetic mixing (see Chapter 4). Fortunately, with most inbreeders, minimal isolation, such as that provided by planting accessions in rows sufficiently separated to prevent physical contact between different accessions, virtually eliminates interaccession outcrossing. In other cases, for example, when bees serve as pollen vectors, techniques such as caging or larger separation distances may be necessary.

Interaccession outcrossing is usually more difficult to eliminate with cross-pollinated species than with self-pollinated ones. The preferred method of regenerating outcrossers is in open-pollinated field plots sufficiently distant from one another to provide isolation adequate to protect against pollen migration. This method is relatively less expensive and more convenient than constructing pollen-proof isolation cages for each accession. For large collections, however, it may not be possible to provide a sufficient number of well-isolated plots or pollen-proof cages to prevent crossing of accessions. In such cases, controlled hand-pollination may be necessary. To compensate for increased costs and greater labor requirements, the number of individuals in each accession that is regenerated may be reduced to an extent that the potential for genetic drift may arise. With fewer plants there also may be insufficient seed for distribution.

Differential Survival

The differential survival or selection of gametes or genotypes that occurs during regeneration is also a threat to the genetic integrity of accessions. Selection that occurs during regeneration can substantially alter the genotypes and hence the phenotypes of accessions. Counterselection for desired individuals is at least partially successful in preserving the original phenotype. Monitoring of isozymes and restriction fragment loci, however, has shown that allele frequencies often shift dramatically or may be lost or fixed in a very few generations in standard, densely planted regeneration nurseries where competition among individuals is high. Changes in allelic or genotypic frequencies are, however, usually much smaller when regeneration is carried out under husbandry conditions that minimize the effects of selection. If individuals are widely spaced so that a high percentage of them germinate, survive to reproductive age, and contribute approximately equal numbers of gametes in the reproductive process and approximately equal numbers of seeds or other propagules to the next generation, only small changes are likely to occur in allelic frequencies.

Isolation

Isolation methods depend on the reproductive biology of the species and on the pollination mechanisms. Studies have shown that pollen dispersal by wind- and insect-pollinated species can result in cross-contamination of up to 5 percent over large distances (Breese, 1989). As a consequence, isolation distances in excess of 200 m are often recommended during the commercial multiplication of allogamous (cross-fertilizing) cultivars. Thus, isolation by distance alone is not normally a feasible method, particularly if large numbers of accessions are to be regenerated on limited amounts of land. However, effective pollen dispersal decreases rapidly over the first 10 or 20 m. Pollen dispersal can be further interrupted and reduced by intervening barriers. This has led to the practice of growing isolation plots interspersed among tall-growing crops, where relatively safe isolation distances of 30 m or less may be achieved for some crops. Other practices help to dilute the amount of unwanted pollen, particularly the planting of square plots rather than rows or using as large a plot as possible and discarding border rows. Plot arrangement is also important in securing effective intra-accession random pollination. Alternative methods use insect-proof cages, pollen-proof glasshouse chambers for wind-pollinated species, or hand-pollination.

Population Size in Relation to Inbreeding Depression and Drift

Genetic integrity during seed regeneration is also threatened by genetic drift. In infinitely large populations the frequencies of neutral alleles do not change from generation to generation. If only limited numbers of individuals participate in producing the next generation, however, random fluctuations in allelic frequencies occur because of sampling. The theoretical consequences of this phenomenon, known as random genetic drift, are well known (Breese, 1989).

Maintaining Effective Population Sizes

In regeneration, the total number of individuals, the nominal population size (N), is less important than the average number of actively breeding individuals, the effective population size (N_e). The practical aim in regeneration is to maximize N_e by ensuring, as far as practicable, complete random mating and equal contribution of as many individuals as possible to the next generation. Hand-pollination and open-pollination by natural vectors are examples of methods used to accomplish this goal.

For amenable species, such as maize and sunflower, pollination control by hand crossing and storage of equal amounts of maternal seed can maintain effective population size at, or above, the nominal population number and, thus, minimize the effects of drift and selective shift. For many crop species, hand-pollination is impractical, and an effective population size can only be maintained by ensuring that random mating is achieved as efficiently as possible by natural pollen vectors (wind or insect) and by equalizing maternal seed sources. The use of natural vectors is critical in these regeneration procedures. Nevertheless, their use will not overcome environmental and genetic variation in pollen production between the different genotypes and its profound effect on decreasing the effective population size.

The effect of numerical differences between the two sexes is very pronounced in dioecious species (those in which male and female flowers grow on separate plants) and the sex ratio of breeding plants. For an extreme case of 2 males and 20 females, the effective population size is 3.6. The calculations are slightly different for monoecious plants (those that have both sexes on the same plant), but the results are much the same.

The Effect of Genetic Drift and Natural Selection

Random loss of alleles due to genetic drift is directly related to the effective population size. Its effect, therefore, can be predicted within statistical limits. The effect of natural selection varies from environment to environment (including different years at a given location) and hence its effects can be determined only through experimentation. In general, random drift causes rare alleles, whether adaptive or not, to drift out of small populations, whereas selection tends to eliminate the less adaptive alleles and to increase the frequency of alleles that are better adapted in the regeneration environment.

Evidence of marked phenotypic changes is now well documented in outbred populations, especially when multiplication is performed in populations growing in regions that differ from their origins (Breese, 1989). Selection operates on individual phenotypes through differential survival or fecundities. Populations are maintained in their natural habitats through the pressures of natural (and farmer) selection. In ex situ situations, the selection pressures may change, and there is almost certainly a change in individual phenotypes as a result of genotype-environment interactions. Such interactions may lead to an increase in the phenotypic heterogeneity of the population. The aim is to minimize genotype-environment interactions and reduce selection pressures. Therefore, important operational considerations are the location of regeneration, growing conditions, effective cross-pollination, and harvesting of equal quantities of seed (Breese, 1989).

Regeneration of Predominantly Inbreeding Species

Completely self-pollinated and genetically homogeneous, accessions present no genetic drift or shift problems, and population numbers are determined entirely by the yield capacity of individual genotypes.

Maintenance of Landraces or Wild Species

Landraces and wild populations are usually genetically heterogeneous and have complex genetic structures, even when the degree of self-pollination is virtually complete. The most effective way of preserving the gene and genotype constitution of highly self-pollinated populations is to maintain the accessions as N subsets of homozygous inbred lines. When $N > 100$, chances are small that no allele or genotype present in the initial sample number will be lost during regeneration. The initial sample number needed to capture a copy of an allele for an inbred population with a similar frequency and probability is double that required for completely outbred populations (on the order of 100 in inbred populations compared with 50 in outbred populations) (Marshall and Brown, 1975). As Gale and Lawrence (1984) show, however, because allele loss (drift) continues in outbreeders, the same efficiency of conservation is achieved over four or five regeneration events for similar constant population sizes of outbreeders and inbreeders when the inbreeders are effectively maintained as inbred lines.

This method of maintaining separate subsets of inbred lines is, however, laborious and costly in administrative terms. Therefore, when large numbers of accessions are involved, it may be preferable to raise the population as a bulk. For highly self-pollinated populations no special isolation requirements are needed. At the same time, it should be remembered that few populations are completely self-pollinated in all environments, so that minimal isolation may require that accessions are sufficiently separated to avoid physical contact. When maintained as bulks, a doubling of the effective population size (N_e) is technically required for inbreeders compared with outbreeders for conserving the same allele frequencies with similar levels of probability (Bray, 1983). Because no crossing is involved, maintenance of effective population numbers (and avoidance of genetic drift) in a nominal population size (that is, the N_e/N ratio) is largely a question of equalizing maternal reproductive outputs. Problems with equalizing maternal reproductive outputs and minimizing genetic shift through natural selection are essentially the same as discussed above for outbreeders.

The mating systems of some populations feature intermediate levels of selfing versus outcrossing. Above particular threshold values of outcrossing (about 5 percent), problems of pollen contamination, crossing requirements, and potential genetic drift and shift require that the populations be handled essentially as described above for outbreeders. Normally, natural open-pollination methods would be used so that the population is maintained as much as possible at its natural level of outcrossing. Considerations for partial inbreeders thus include adequate isolation methods and the provision of the necessary effective pollen vectors.

CHARACTERIZATION

Characterization is a systematic recording of selected morphologic and agronomic traits of an accession. It is typically restricted to those genetically controlled traits that are highly heritable and do not vary with environmental conditions (Frankel, 1989b; Williams, 1989b). Characterization data should be linked to passport and evaluation data. Passport data comprise the information about the site and environment from which an accession originated. Evaluation data for more variable traits, such as productivity, quality of product, and disease resistance are of greatest interest to users of germplasm. It has been argued that (1) when traits such as these are included in data-base systems, such systems become a more useful tool for users who wish to identify accessions for specific purposes, and (2) it is consequently worthwhile for conservation units to assess the potential usefulness of accessions in their collections. For a given germplasm collection, the characterization data form a permanent set of biologic descriptors.

Characterization activities can be carried out concurrently with the regeneration process and therefore typically do not require an extra planting. When the process is divided between two alternating crop seasons, as it is implemented at the International Rice Research Institute, quantitative traits of agronomic importance are recorded in the main (wet) season, whereas qualitative traits can be scored in the off (dry) season.

Wide applicability of characterization data can be ensured by standardized descriptors (traits) and descriptor states (absolute or coded values). The challenges in setting up the descriptors and descriptor states are to choose between an exhaustively large set and a minimal set so as to render the task workable and the data useful. Rice workers have arrived at a consensus, and the standardized descriptors (International Board for Plant Genetic Resources–International Rice Research Institute Rice Advisory Committee, 1980) are widely used. Characterization data are also useful in reidentifying an accession during seed preparation for planting or distribution and when accessions are regenerated. Such checks must be made routinely to ensure the maintenance of varietal identity.

With the advent of powerful electrophoretic and molecular diagnostic techniques, the scope of characterization can be vastly expanded to provide much greater resolution to the genetic differentiation among accessions (see Chapter 7).

EVALUATION

Evaluation is a prerequisite for the use of conserved germplasm. Accessions that are not evaluated mostly remain curiosities. Evaluation has been described as including such fields as crop cytogenetics and evolution, physiology, and agronomy (Frankel, 1989b). It is essential to utilizing the genetic diversity of germplasm collections.

Large-scale systematic evaluation generally falls outside the domain of germplasm workers; rather, it is undertaken by a variety of biologists, including plant breeders, who specialize in crop improvement or production research. However, the initial phase of preliminary evaluation is usually performed by the germplasm workers during the first cycle of multiplication-regeneration and observation. The observations are limited to a small number of morphological features, adaptability at the site of planting, and obvious economic worth. Nevertheless, such observations are crucial in enabling the germplasm worker to forward the materials to the appropriate researchers or breeders for further testing, leading to more systematic evaluation. Plant quarantine precautions should be observed at this stage.

The prerequisites for effective and efficient evaluation are described below.

Increase of Seed or Plant Parts

Sufficient materials in the form of either seed or plant parts are imperative to evaluation experiments. For materials of a heterogeneous and heterozygous nature, a sample size larger than that for the pure-line cultivars will be needed. Wild species generally produce little seed and are difficult to handle. For users of wild species, guidelines should be provided to break strong dormancy, to raise slow-growing seedlings, to provide appropriate solar radiation levels and photoperiod requirements, to follow a taxonomic key for re-identification, and to collect the precious seeds (Chang, 1976).

Germplasm workers are likely to be asked for a second supply of seed or materials by the same or other researchers. Therefore, maintenance of the genetic integrity and the correct nomenclature and accession numbers are essential. It is preferable to maintain a large seed stock rather than making repeated rejuvenations, because the latter leads to genetic drift, outcrossing, and errors in labeling and handling.

Multidisciplinary Approach to Evaluation

Most evaluation operations now require expertise beyond that of one or two related disciplines. A multidisciplinary approach is essential in planning, execution, and assessment of results. The Genetic Evaluation and Utilization Program of the International Rice Research Institute is an example of such multidisciplinary collaboration and interdisciplinary interaction. Effective evaluation experiments often lead to more refined and rewarding research.

Planning and Implementing Evaluation Tests

Every evaluation test should be treated as a scientific investigation. It should embody well-defined objectives, the experimental design should be efficient, and there should be judicious choice of treatments and control varieties. The environmental factors should be controlled, if necessary, and complete statistical analysis is essential. Data should be entered in computerized data files for easy access and monitoring. Results from repeated tests should be compared with previous data before entering the information into the files for updating purposes. The usual progression in the tests is from simple, large-scale screening to verification tests and then to more critical studies.

Choice of Representative Environments or Sites

It is imperative that evaluation experiments be carried out in environments that represent the major target production area(s). For many adverse environments, off-station test sites in stressed environments are necessary. To have a heavy epidemic, disease and insect nurseries need to be planted at "hot spots" where the pest is endemic; artificial infestations should be arranged when possible. The control of temperature, relative humidity, and light intensity in greenhouses or growth chambers may be necessary. Efficient experimental designs are needed to make maximum use of available resources.

Preventing Exotic Germplasm from Becoming a Serious Pest

The extreme seed shattering and strong rhizomatous characteristics of some wild or primitive germplasm pose a threat if such vigorously growing plants or seeds were allowed to escape into water canals or farmers' fields. Measures should be taken to prevent such accidents. At the International Rice Research Institute, wild rices are grown in large pots to control their spread, to ensure their identity, and to obtain harvests from ratooned plants.

Verification of Evaluation Data

Disease- and insect-susceptible accessions may be recorded as such without retesting, whereas populations identified as resistant or tolerant need to be retested under more precisely controlled conditions to establish the validity of the preliminary evaluation. Appropriate control varieties or treatments should be included in each experiment. This step should also apply to the evaluation of other traits, especially those being subject to the effect of variable environments or planting densities in field trials. Sampling of environmental variance and the genotype-environment interaction component adds scope to the evaluation experiments.

Communication Among Plant Germplasm Users

Open and continuous communications among conservationists, scientists in related research areas, such as entomologists and physiologists, and plant breeders is essential to sustain evaluation efforts and subsequent use. Crop advisory committees that bring together a variety of interests from public and private sectors facilitate cooperation and communication. They are important sources of crop-specific expertise for the U.S. National Plant Germplasm System, for example (National Research Council, 1991a).

When evaluation efforts are coupled with in-depth research by members of allied disciplines, further refinements in testing techniques are accelerated. Scientific advances in the discipline(s) concerned are also accelerated as an outcome of the multidisciplinary interaction. Breeding programs benefit from both areas of endeavor. There are many examples of rice breeding programs that illustrate the rewarding results from such collaborations and interactions (Chang, 1985a,c).

Documentation

Despite the accelerated efforts on different fronts, inadequate documentation still constrains the use of exotic or unimproved germplasm. A lack of general adaptation to commercial production areas and poor agronomic and quality characteristics are common complaints aired by breeders as the main drawbacks of exotic germplasm. Low crossability and sterility barriers are additional drawbacks voiced by users. Lack of adequate classification systems, poor passport data sets and sometimes uncertain fertility relationships add to the reluctance in using exotics.

Efficient documentation should extend across all phases of conservation. The availability of evaluation data is crucial for arousing the interest of potential users (for example, breeders, entomologists, plant pathologists) of the germplasm. However, passport data and characterization information are the most important to have. Free and widespread exchange of evaluation results also leads to enhanced exchange of germplasm. The use of standardized descriptors and descriptor states (International Rice Research Institute, 1975) has proved its usefulness in international evaluation programs. A huge backlog of data accumulated by germplasm workers needs to be located, interpreted, transcribed, and entered into the central files. With the constraint of time and space and the death or retirement of earlier workers, a substantial portion of the invaluable data may never be made available to other interested workers.

MONITORING OF SEED VIABILITY AND GENETIC INTEGRITY DURING STORAGE

An important aspect of the viability of stored seeds is the vigor of the germinated seeds. Seed banks should also monitor the appearance of abnormal seedlings, some of which could be mutations that have arisen during storage. Seed production is costly and the viability test destroys stored seeds, however. Other measures to prolong the longevity of stored seeds and reduce the quantity of seeds used in germination tests should be fully exploited to reduce the amount of seeds lost in the process and to decrease workloads.

Nondestructive tests to monitor seed viability should therefore be explored and developed. Processing operations and storage conditions that would prolong viability of stored seeds should also be developed.

The International Board for Plant Genetic Resources has recommended that 85 percent be the threshold viability level for rejuvenation of stored seeds (National Research Council, 1991a). Germplasm banks of the People's Republic of China and the International Rice Research Institute use a 50 percent threshold level, whereas the U.S. National Seed Storage Laboratory has, until recently, used a 60 percent threshold (National Research Council, 1991a). In practice, it may not be possible to obtain high germination rates for some species or accessions. Thus, the viability standards for collections must be a percentage of the optimum attainable germination with carefully produced fresh seeds.

REDUNDANCY AMONG COLLECTIONS

Many national collections (and the collections maintained by international centers) have become very large, and on a worldwide basis, collections of most major crops and some minor crops have become enormous (Holden, 1984). For wheat, for example, there are a total of more than 400,000 accessions in 37 collections worldwide (Plucknett et al., 1987). How many unique accessions exist is not known for any major crop. Large size in itself is not necessarily a problem if the germplasm bank has sufficient storage facilities and management expertise as well as resources—factors that lead to low operation costs (see, for example, Chang, 1989, 1992). Size can, however, limit important activities such as regeneration and evaluation. Management strategies, such as the identification of core subsets and increased efforts to capture passport and characterization data, have been proposed to address the problem of managing large collections (International Rice Research Institute, 1991).

If all duplicate accessions in individual collections were eliminated, the gains in efficiency from a reduction in entry numbers would likely be modest for most collections. The costs of eliminating duplicates may exceed the cost of keeping them, particularly if passport data are not adequate to help in their identification. Isozyme or restriction fragment length polymorphism analyses are far too costly for routine germplasm bank use. Greater gains could perhaps be made by reducing redundancy among collections to the minimum necessary for insurance against catastrophic loss. This is far more difficult to achieve, however, because it requires a high level of cooperation and coordination among national as well as international germplasm banks. The size of collections is seen to be of importance in terms of prestige and funding. Furthermore, many countries have strict quarantine laws, and accessions are retained to avoid the inconvenience and cost of repeated quarantines. When countries heavily depend on particular crops, germplasm collections of those species acquire strategic importance, and larger national collections are seen as insurance against an uncertain future, especially in the light of the growing controversy surrounding the ownership and control of crop genetic resources.

The major problem in redundancy is not within collections, but between them. This redundancy has arisen from the repeated distribution of the same sets of accessions. However, for breeders and curators in countries isolated by distance or quarantine barriers, some duplication between collections offers the advantage of ready access and the opportunity of becoming familiar with a wide range of crop species. Unfortunately, recipient collections do not always retain the original accession identifiers and, thus, it may be difficult to subsequently identify duplicates between collections.

Core Subsets of Collections

The limited scope for eliminating redundancy within collections led to a proposal for developing core subsets in germplasm banks (Frankel and Brown, 1984). Under this concept, the accessions in large germplasm banks would be sampled to form a core subset that contains the genetic diversity in the crop species and its relatives in a base collection, with minimum redundancy. The base collection would include those materials in the core subset plus all other materials and would continue to be maintained (Brown, 1988; Frankel and Brown, 1984; National Research Council, 1991a).

Advantages of Core Subsets

The aim of identifying a core subset is to facilitate use and, in particular, to provide efficient access to the whole collection (Brown, 1989a; National Research Council, 1991a). Consider a breeder faced with a new virulent race of a pest or pathogen. The breeder initially searches for resistance in readily available material, usually a limited working collection. If no resistance is found in the working collection, the next step by the breeder is to screen material from an appropriate germplasm bank collection. If that germplasm bank has a well-defined core, this would be screened first. If no effective resistance is located in the core, then the breeder knows that resistance is relatively rare; the breeder is then faced with a substantial problem and the prospect of screening the other 90 percent of the collection. If resistance is found in the core collection, then the breeder can use that resistance immediately in the breeding program as well as screen additional accessions from the now identified geographic areas held in the base collection.

The core subset concept also has advantages for curators. It is envisaged that more seed of the core subset would be kept on hand, packaged, and ready for distribution to meet general seed requests. It is also envisaged that accessions in the core subset would receive priority in evaluation and characterization, so that in time many more characteristics would be evaluated for all of the core samples than would be evaluated for the remainder reserve samples. In this way, curators could better use a limited budget to promote the distribution of information and material and, hence, facilitate use of the entire collection. With good communication, core subsets may also effectively limit redundancy between collections. The typical accessions of the Latin American races of maize serve as a useful example of a core subset.

Disadvantages of the Core Concept

One disadvantage of the core concept is the possibility, or indeed, probability, that the remainder of the base collection, held in reserve, would erode away and disappear from neglect or, alternatively, would be seen by some administrators as being of less value and therefore dispensable in the interests of economy (National Research Council, 1991a). To minimize this disadvantage, Frankel and Brown (1984) and Brown (1989a) have stressed that this reserve must remain an important and integral part of the base collection. Under the core concept, the reserve serves at least two important functions: (a) as an alternative to be screened if needed variation is not found in the core subset, and (b) as a source of additional diversity when many different genes or alternative alleles are required for the same trait. Nevertheless, the reserve may be more vulnerable to neglect and dismemberment, regardless of how strong a case is presented for its maintenance.

Practical Problems with Core Subsets

There are several practical problems that must be overcome before the core concept can be entirely implemented. One is establishing the appropriate size of a core subset and defining its composition so that it contains a representative sample within the defined size. These problems have recently been examined by Brown (1989a,b). Using the sampling theory of selectively neutral alleles, Brown (1989a) argued that a core should consist of about 10 percent of the collection, up to a maximum of about 3,000 accessions for each species. At this level of sampling, the core subset will generally include over 70 percent of the alleles in the whole collection. Brown (1989a) also argued that, as a general rule, a fixed proportion of 10 percent is more useful than a fixed-number upper limit.

Brown (1989b) examined alternative procedures for choosing core entries. He showed that stratified sampling, in which the collection is first divided into nonoverlapping groups and a sample is taken from each group, is more efficient in establishing a core than is random sampling, in which accessions are chosen from the whole collection at random. Brown also examined options for deciding the number of samples from each nonoverlapping group to include in the core. These included a constant strategy (an equal number of accessions from each defined group) and two variable strategies (with sample sizes correlated with sizes of the groups). Of these, the variable strategies gave broadly similar results and were better than the constant strategy. This leaves the problem of how to stratify the collection into nonoverlapping groups (Brown, 1989a). This can be done rationally only when reasonable knowledge about the materials in question is available (International Rice Research Institute, 1991). Even a knowledgeable curator would need access to the following:

• The origin of the accession (collection site or area of adaptation);

• Characterization data, including taxonomic and agronomic data and information from the analysis of genetic marker loci; and

• Evaluation data for economically important quantitative or qualitative traits (for example, yield, disease resistance, and cold tolerance).

It would be inappropriate for an inexperienced curator to attempt to define a set of core accessions or for anyone to define a core collection for a crop with inadequate passport or characterization data. For large collections, computer-based multivariate methods such as numerical cluster analysis, principal component analysis, or network analysis might be helpful to order the collection into related groups.

RECOMMENDATIONS

Neither the reduction of collection size by elimination of accessions nor the use of core subsets is universally accepted (see, for example, Brown, 1989a,b; Chang, 1989; Marshall, 1989; National Research Council, 1991a). The best strategy for addressing increases in collection size depends on many factors, including the crop species being maintained, the accession data, management practices in use, the needs of users, and the resources available to conserve and manage the collection.

The capacity to perform regeneration of seed accessions when needed should be a major determinant for limiting the size of germplasm collections.

Regeneration of seeds in a germplasm bank collection becomes necessary for two purposes: for rejuvenation, when the viability of seeds in stored accessions falls below acceptable levels, or for multiplication, when high levels of distribution to other breeders or institutions lead to the depletion of seed inventories. Performing regeneration exposes accessions to a variety of risks that could lead to genetic depletion, genetic shift, or mixing with other accessions. With large collections, there may be insufficient resources to grow adequately large populations to preserve genetic integrity. Alternately, insufficient resources may necessitate deciding which of many accessions should be regenerated. In either case, the genetic integrity of a collection can be placed at risk because of the inadequacy to perform this fundamental management practice.

Collections should employ storage and management practices that minimize the need for regeneration.

The risk of allele loss is repeated during each cycle of regeneration. Consequently, the interval between regeneration events should be made as long as possible. Improvements in long-term storage have aided this by reducing the frequency of regeneration needed for orthodox seed. Three other considerations emphasize the need to minimize regeneration: (1) The regeneration process is costly in terms of resources; (2) each regeneration event places accessions at risk from clerical and other human errors, and in danger of genetic shift due to selection or genetic drift; and (3) effective population size is bound to vary with each regeneration event, and over several generations will be much closer to the smallest single value. Even a single regeneration event in which the effective population size is small (less than 50) can result in serious genetic loss within the accession.

Core subsets should be identified for large collections to aid in managing, evaluating, and using accessions.

Core subsets should not be used as a justification for neglecting the maintenance of the other accessions in a collection. Further, identification of core subsets is not to be viewed as a prelude to elimination of other accessions. A core subset should "include, within an acceptable level of probability and with minimum redundancy, most or much of the range of genetic diversity in the crop species and its relatives (typically no more than 10 percent of the whole collection)" (National Research Council, 1991a:125).

Although they should be selected to represent the diversity of the main collection, core subsets are unlikely to capture all of the genetic diversity within a collection. It is probable that if 10 percent of the accessions in a collection are assigned to a subset, no more than 70 percent of all the available alleles will be included (National Research Council, 1991a). Thus, the rest of the collection will continue to serve as a source of diversity. This larger portion may also serve as a resource for genetic traits not captured in the core subset.

Core subsets are management tools that can be used to develop priorities for evaluation and collection. They do not require the construction or establishment of separate collection facilities. Identifying cores does, however, require a basic amount of information about accessions. Minimally this should be the passport data regarding the origins and environment of the accession. Such information is essential to linking core subsets to the potentially wider genetic diversity of the entire collection. An initial step in identifying cores, therefore, is the recording of basic passport information about each accession in a collection.

Redundancies within global collections should be minimized.

The problem of redundancy has arisen because it is easier to add an accession than to discard one. Elimination of redundancy in existing collections is not cost-effective. The costs for biochemical or molecular methods needed to identify duplicates are not justified by the relatively small number of accessions that would be eliminated. A more rational approach is to prevent exacerbation of the problem by a more cautious approach to the addition of new accessions. To accomplish this, basic information (passport data) for existing accessions and new ones is needed. Much of this may exist, but is not often found in the databases of germplasm collections. This information, coupled with the more detailed information that might come from study of selected core subsets would prevent or reduce the addition of duplicative accessions to germplasm collections.

Research is needed to develop long-term storage methods for short-lived, desiccation-sensitive, or cold-intolerant seeds.

At present, many important collections must be maintained as field-grown plants. Many crops are maintained as clones, including important ones such as yams, cassava, sweet potatoes, potatoes, bananas, sugarcane, rubber, cacao, coconuts, and oil palms. Understanding the mechanisms of dormancy in seeds and the physiology of these difficult-to-store seeds may lead to methods of improved storage. It also may be possible to develop methods of storing excised plant parts or embryos as in vitro tissue cultures. Storage of excised embryos at cryogenic temperatures ($-150°$ to $-195°C$) may also prove feasible for some species (see Chapter 7). However, long-lived field collections are necessary for preserving clones and other plants that have short-lived seeds. For example, the seeds of rubber plants survive a few weeks, and attempts at in vitro storage and propagation have been unsuccessful.

Increased efforts should be instituted to train germplasm workers.

Germplasm activities require long-term commitment and continuity in personnel. Being a new and composite technology, the subject is not taught in most universities. The number of trained workers is small and inadequate for the heavy workload. The requisite expertise derives in part from direct experience beyond a training course or university degree. The development of trained professionals and germplasm workers is an imperative for sustaining the future of genetic resources and making advances in agriculture.

6

Using Genetic Resources

The allelic variability within the various kinds of gene pools of all long-domesticated species has become very large. Nearly all domesticated species are now so different from their wild-type progenitors that they can no longer survive in the wild. While there are great stores of genetic variation in wild-crop relatives, experience in breeding indicates that much of this variation is irrelevant to plant breeders. Plant breeders, the ultimate users of germplasm, generally agree that the need for genetic diversity within the economically important plant species has never been greater. The challenge is to detect and transfer those genes that will improve the cultivated species.

AN EXAMPLE OF GERMPLASM USE

The discovery of the Americas by Europeans set the stage for major developments in the recognition and use of crop germplasm. Explorers returned to Europe with New World species that greatly influenced European agriculture. The potato became one of the most important energy foods in Europe, and tomato and corn ultimately achieved status as major crops. Similarly, colonists to the Americas introduced the best germplasm available from their native lands. Introductions of wheat from England, The Netherlands, and Sweden were grown along the Atlantic coast of North America from 1621 to 1638 (Ball, 1930), and numerous additional introductions from a variety of places were made throughout the seventeenth, eighteenth, and nineteenth centuries. These wheats provided the germplasm from which the soft red winter wheat of the eastern United States was developed. Cox et al. (1986, 1988b) have documented the ways in which twentieth century soft red winter wheat breeders took the great genetic diversity available in the large number of landraces that were grown in 1919, augmented this germplasm through hybridizations with other gene pools, and concentrated the best germplasm into approximately 100 different cultivars now grown commercially in the United States.

Turkey wheat, a highly heterogeneous landrace brought to Kansas in 1883 by a group of settlers from the Crimean region of southern Russia, rapidly became the most widely grown variety in the vast hard red winter wheat area of the central Great Plains, where growing conditions are highly variable within seasons, from year to year, and from place to place. Turkey wheat reigned supreme throughout this region for a half century, remaining virtually unchanged.

In the hard red winter wheat area, breeders firmly maintained the core of Turkey wheat germplasm over the years, thus preserving the basic adaptation, yield stability, and product quality of Turkey wheat while improving specific agronomic and quality characteristics. By 1984 the number of commercially grown varieties had increased to 164, but this increase was accompanied by higher than desirable genetic relatedness of the primary cultivars (cultivars that occupied 1 percent or more of the land planted to wheat) in the region. Cox et al. (1988b) evaluated the yields and various agronomic and quality traits of 38 of the major hard red winter cultivars released from 1874 to 1987 to estimate the genetic improvement achieved during that period (Table 6-1). Linear regression analyses of cultivar performance on the year of release showed steady increases in yields (about 1 percent per year) as well as steady improvement in various agronomic and quality traits. They found no indications that a yield plateau had developed.

BREEDERS' PERCEPTIONS AND PRACTICES

The most important scientific questions concerning the use of genetic variability involve (1) ways to identify useful variability and (2) methods to maximize availability to breeders while minimizing the danger of loss.

There is general agreement among experienced breeders that the superior performance of modern cultivars has resulted from the accumulation of favorable alleles and the gradual assembly of these alleles into favorably interacting multilocus combinations. Breeders are consequently reluctant to introduce unadapted materials into their breeding stocks, because use of unadapted germplasm introduces unfavorable alleles and increases the likelihood that favorable combinations of alleles at different loci will be broken up by segregation and recombination. Breeders with established programs thus consider advanced materials in their own breeding nurseries, including commercially grown cultivars, advanced lines under evaluation for release, and sibling lines of such materials, to be the most useful germplasm available to them. They consequently make heavy use of such materials as parents in crosses so that past gains are less likely to be lost as a result of segregation and recombination. Breeders also regard advanced materials from breeders located in ecologically similar areas with favor, and they frequently use such materials as parents in crosses.

Monitoring Advanced Materials

Many breeders (including farmers who practice selection) regularly monitor advanced materials, especially commercially grown cultivars, for potentially useful variants; many useful alleles with major effects on characteristics such as early maturity, height, determinant versus indeterminant growth habit, cold or heat tolerance, and resistance to diseases and insects have been uncovered in this manner. The case of *Periconia* root rot of sorghum, which first appeared in Kansas in 1926, provides a dramatic example. The effects of this disease were catastrophic; the frequency of surviving plants in infected fields was on the order of only one in several hundreds or thousands. The disease soon appeared throughout the southern plains area of the United States and westward into California. Plant breeders throughout the affected area found the same low incidence of surviving plants in local varieties. The progeny of some of these plants were free of disease, whereas others produced both healthy and diseased offspring. Breeders increased the numbers of seeds of resistant plants and were soon able to release resistant strains for commercial production that were indistinguishable from the parent variety but that could be grown in *Periconia*-infested soil without any evidence of injury.

Performance Advances

It has been postulated that although reliance on advanced locally adapted materials improves prospects for continued modest improvements in performance (see above), it reduces the opportunity for ma-

TABLE 6-1 Year of Release and Means Over All Environments for Seven Traits of 38 Hard Red Winter Wheat Cultivars Released from 1874 to 1987

Lodging Cultivator	Year Released	Mean Grain Yield (kg/ha^{-1})	Biomass Yield (kg/ha^{-1})	Volume Weight (kg/m^{-3})	Kernel Weight[a] (g)	Heading Date (May)	Height (cm)	Score[b]
Turkey	1874	1,609	8,533	708	24.5	16	117	2.0
Kharkof	1900	1,426	8,284	699	22.4	15	115	2.0
Blackhull	1917	2,031[c]	9,006[c]	760[c]	26.7[c]	13[c]	115[c]	2.0[c]
Tenmarq	1932	1,589[c]	7,706[c]	722[c]	25.3[c]	14[c]	111[c]	2.0[c]
Cheyenne	1933	1,547	8,054	693	22.9	15	110	1.8
Redchief	1940	1,854	8,742	758	26.9	13	120	1.7
Comanche	1942	1,823	7,766	717	25.2	12	112	2.0
Pawnee	1943	1,712	6,718	726	25.3	11	106	1.8
Wichita	1944	2,174	7,879	779	31.3	8	108	1.7
Ponca	1951	1,911	7,834	722	26.2	13	109	2.0
Bison	1956	1,944	7,715	728	27.2	13	111	1.8
Tascosa	1959	1,880	8,076	740	24.7	10	105	1.5
Warrior	1960	1,915	8,432	711	23.8	15	108	1.5
Kaw 61	1961	2,225	7,856	798	29.0	9	109	1.8
Lancer	1963	1,857	8,293	730	23.9	13	110	1.7
Triumph 64	1964	2,760	8,591	800	32.3	6	101	1.5
Scout 66	1966	2,286	8,662	758	28.9	10	109	2.0
Sturdy	1967	2,239	6,736	752	27.9	7	77	0.0
Shawnee	1967	1,996	8,089	738	23.8	13	109	1.7
Eagle	1970	2,194[c]	8,131[c]	732[c]	28.0[c]	10[c]	103[c]	1.7[c]

Larned	1976	2,396	8,354	758	28.6	11	105	1.3
Vona	1976	2,604	8,162	738	24.4	7	84	0.0
Newton	1977	2,134	7,811	720	24.4	10	90	0.5
Centurk 78	1978	2,367	8,227	749	24.2	10	99	1.5
Arkan	1982	2,494	7,645	740	27.2	7	88	0.7
Brule	1982	2,438	8,683	714	25.0	11	95	0.5
Hawk	1982	2,349	7,692	733	27.8	10	86	0.3
Chisholm	1983	2,761	7,632	779	30.4	6	84	0.0
Mustang	1984	2,389	7,094	759	29.7	8	83	0.0
Siouxland	1984	3,011	9,852	769	30.4	9	104	0.3
Stallion	1985	2,668	7,566	774	25.9	7	80	0.0
TAM 107	1985	2,727	8,034	735	28.6	6	83	0.0
TAM 108	1985	2,520	7,930	705	26.4	11	86	0.3
Victory	1985	2,733	8,524	745	29.4	9	89	0.3
Norkan	1986	2,404	8,482	749	26.3	10	90	0.0
Dodge	1986	2,531	8,015	754	28.8	9	88	0.0
Century	1986	2,982	8,668	765	27.4	9	87	0.3
TAM 200	1987	2,658	7,806	791	24.3	9	76	0.0
LSD (.05)		403	—	42	2.6	2	5	0.05
Change per year[d]		16.2**	0.5	0.4*	0.04*	-0.1**	-0.5**	-0.03**

*, ** Significantly different from zero at the .05 and .01 levels of significance, respectively.

[a]From each plot, 200 kernels were weighed, and kernel weight was expressed in grams per thousand kernels.

[b]Mean over 2 years at Manhattan, with 0 = no lodging, 2 = completely lodged.

[c]Estimated 2-year mean for cultivars tested only in 1986.

[d]Coefficients of regression of trait means on year of release.

SOURCE: Cox, T. S., J. P. Shroyer, L. Ben-Hui, R. G. Sears, and T. J. Martin. 1988. Genetic improvement in agronomic traits of hard red winter wheat cultivars from 1919 to 1987. Crop Sci. 28:756–760. Reprinted with permission, ©1988 by Crop Science Society of America.

jor advances in performance. Another more prevalent view is that modern-day pools of advanced, highly adapted germplasm are, in fact, vastly complex genetic systems containing great reserves of adapted and maximally exploitable genetic variability. There appears to be no solid basis for a choice between these hypotheses at present. Measured rates of progress appear to have been about the same for those gene pools that have been heavily introgressed with exotic germplasm as it has for those gene pools into which there has been only light introgression. Yield plateaus have not been detected in many major crops, even where there has been little or no introgression of exotic germplasm into the breeding populations in recent times. The gene pools in the U.S. Corn Belt from which high-yielding single crosses are produced (Duvick, 1984b) are good examples.

When alleles are needed for specific improvements (for example, resistance to new races of diseases, various aspects of quality, and improved yield), and the needed alleles have not been found in locally adapted materials, the breeder and coworkers are forced to screen exotic germplasm in the search for the needed genetic variability. In this event, advanced exotic cultivars and advanced breeding lines, genetically enhanced germplasm pools, obsolete cultivars and landraces, and wild progenitors are screened, usually in this order. Numerous sought-after alleles have been identified in accessions from active collections.

Recently, materials from national and regional trials, and trials conducted by international centers, have become increasingly important sources of exotic germplasm in recent years. Varieties and advanced lines undergoing evaluation in such trials are often used as sources of desirable alleles, because the substantial performance and evaluation data available from trials conducted under a range of environmental conditions help breeders to identify germplasm with the greatest potential value in their own area. Data from such trials are much more useful than descriptor data, because information about adaptation and productivity are especially helpful in choosing parental materials. Germplasm pools into which alleles for specific attributes have been introgressed into well-adapted genetic backgrounds are increasingly being used as sources of exotic genes. Breeders turn more and more frequently to such populations, because they often provide many different alleles for specifically sought-after traits concentrated from numerous sources into genetic backgrounds that provide adaptedness to specific ecological regions.

Among the heaviest users of accessions from active collections are the breeders of those plant species that have little or no history of breeding improvement. Species that fall into this category include many forage species (grasses, legumes, and forbs) and other plants used as root and industrial crops and medicinal plants. Breeders of the crops of marginal areas, which may have little history of breeding improvement, sometimes also find that primitive materials (including wild relatives) in active collections contribute useful alleles. Breeders who initiate plant improvement programs for such species or areas often have no alternative to evaluating primitive materials and selecting as parents either the more promising accessions or superior individuals within promising accessions.

USERS' PERCEPTIONS OF THE GERMPLASM SYSTEM

Duvick (1984b) has suggested that little more gain in yield in developed countries can be achieved from additional nonbreeding inputs. Among the reasons he cited are that it will be difficult to improve techniques of planting, weeding, and harvesting very much more and that gains from use of insecticides, fungicides, herbicides, and fertilizers appear to have plateaued. There is, on the other hand,

To ensure that only pollen from the desired male parent is applied, the flowers of a female pecan clone are enclosed in a bag at the W. R. Poage Pecan Field Station. A syringe is used to blow pollen into the bag to ensure fertilization. Credit: U.S. Department of Agriculture, Agricultural Research Service.

no indication of reduction in rate of increase that can be attributed to breeding. Germplasm exists that carries any needed traits, and present-day breeding methods are quite capable of transferring these into elite material from which improved varieties can be selected (Duvick, 1984b).

In developing countries opportunities still exist for improvement in crop productivity and stability from nonbreeding inputs, and at the same time, opportunities for inputs from breeding are also higher in developing countries than they are in developed countries. Thus, most countries, developed or not, accept the proposition that if the food, feed, fiber, and other agricultural needs of an ever-increasing world population are to be met, a significant part of the increase must come from genetic improvement of plants and that adequate supplies of useful germplasm are essential for genetic improvement.

Nevertheless, despite the continuing need for genetic diversity, it has often been stated that only limited use is being made of the germplasm resources maintained in national, regional, and international collections. Thus, for example, according to the Five-Year Plan for Action, 1985–1989, of the Regional Committee for Southeast Asia under the auspices of the International Board for Plant Genetic Resources (1984a), very little of the sizable germplasm collections now available in that region has been used in breeding programs or research studies. Limited use is not confined to developing countries: in developed countries, the majority of breeders of principal crops mainly resort to their own working collections for breeding materials (Duvick, 1984b). Frankel (1985a) attributed the limited use of germplasm collections to management problems, such as lack of definition of objectives, excessive size of collections, inadequate evaluation, and lack of breeder participation. Chang (1985a) pointed out the poor communication between germplasm workers and users as the main handicap in using exotic germplasm. These topics are also discussed by Brown et al. (1989).

There has been growing dissatisfaction among breeders with the international germplasm programs and some of the large national collections as they have expanded and grown more complex in recent decades. Some of the most frequently voiced criticisms of these systems are examined below, and an attempt is made to establish scientific criteria for structuring and managing collections to serve their intended purposes.

BREEDERS' PERCEPTIONS OF ACTIVE COLLECTIONS

It is a common perception among breeders that a disproportionate number of accessions in active and base collections are obsolete and that the most modern and useful materials are too frequently not present in such collections. An important step in organizing the structure and management of collections is to identify the unit of utilization. From the breeder's standpoint, the primary dichotomy is whether accessions from collections are (1) used directly as sources of cultivars (individual genotypes, populations, ecotypes, species) or (2) whether they are sources of specific genetic elements (usually alleles of specific loci) to be transferred by breeding into locally adapted genotypes.

The most common type of active collection, a breeding collection, is intended to serve as a reservoir of alleles governing specific traits to be transferred, by appropriate breeding methods, into locally adapted genotypes. Breeders' perceptions of active collections as they are now structured and maintained, together with some suggestions from breeders for improving their usefulness, are given below.

Passport and Descriptor Information

Accessions must be well documented to allow users the means for rationally examining collections. Without relevant data, users must screen accessions more or less blindly in their search for needed alleles. Unfortunately, accurate passport and descriptor information is often not available for many accessions.

For initial screening, breeders prefer to identify those accessions best suited to their own production environment and to screen the remainder of the collection only if they do not find the needed alleles in the initial screening. Efforts have been made to aid plant explorers, germplasm bank managers, and curators by identifying those passport and descriptor data useful to users. The International Board for Plant Genetic Resources has published standard descriptor lists for many crops. However, this crucial information is often missing from the accession record.

Traditional collections are often inadequate for population, ecologic, and evolutionary genetic studies of the structure of populations, ecotypes, and species. Sampling strategies, sample sizes, and documentation have often been inadequate for such purposes. Scientists who undertake such population genetic studies consequently often find it necessary to organize their own collecting expeditions independent of the germplasm system.

Maintenance, Rejuvenation, and Sample Size

One of the most vehement criticisms of germplasm banks concerns failures in the maintenance and renewal (or regeneration) of accessions, which, in turn, have adverse effects on the quality and quantity of materials available to users. In some cases accessions known to be genetically uniform have become hybrid mixtures or swarms, and conversely, accessions known to be genetically diverse have suffered severe genetic erosion. The problems and challenges of adequate regeneration procedures are addressed in Chapter 5.

Redundancy

Many breeders believe that unacceptable levels of redundancy exist in collections. A survey of germplasm conservators and users on the world collections of barley, which contain more than 250,000 accessions, suggests that only about 50,000 are unique accessions (Lyman, 1984). Unnecessary duplication is often obvious and, hence, correctable; for example, adequate passport data often identify redundancy. In the case of barley, a large group of several thousand accessions from the National Small Grains Collection of the United States is duplicated in many collections around the world (Anishetty et al., 1982).

It can be argued that such redundancy provides insurance against loss of important collections. However, it also leads to inflated and inaccurate estimates of the true extent of diversity contained in world collections. Further, if passport data are unavailable, it is not possible to reliably determine which accessions are widely duplicated. New molecular technology (see Chapter 7), coupled with increased efforts to obtain basic passport and characterization information, may in part address the problem of redundancy. Where collections are large, management strategies, such as the designation of core subsets, may also be needed (see Chapter 5).

Obsolescence

A disproportionate number of accessions in active collections are obsolete and are unlikely to serve as useful germplasm in modern breeding programs or in modern basic genetic, cytogenetic, or ecogenetic studies. This problem is discussed further below.

Evaluation

Germplasm systems have done virtually nothing to address the problem of evaluation, even though evaluation of performance (yield, components of quality, disease reaction) in specific habitats, and especially evaluation of prepotency (the capacity of a parent to produce superior offspring) which provides the most useful information for identifying accessions with superior breeding potential. Unfortunately, the substantial information on performance and prepotency that was accumulated over the years has rarely been added to accessions records. Results of even the most simple evaluations are useful, and it is important that managers of active collections add such information to the record of accessions as it becomes available. Even though evaluation is more appropriately carried out by breeders and other researchers than by germplasm personnel, assembly of evaluation data should be a responsibility of the germplasm system. Germplasm collections should devote a portion of their resources to solicitation of evaluation by qualified breeders located in strategic environments, and to assurances that the information gathered is incorporated into the accessions records. Moreover, inadequate communication between germplasm workers and users in the past has impaired the use of unimproved germplasm (Chang, 1992). Core subsets, as described in Chapter 5, would allow collection managers to set priorities for their evaluation resources.

MODERNIZATION OF ACTIVE COLLECTIONS

Often, outstandingly useful modern materials do not find their way into germplasm collections. There are at least five main sources of new accessions:

• New elite lines and cultivars developed by breeders;
• Obsolete but historically important cultivars and stocks (including privately developed stocks) that should become generally available once they are obsolete;
• Genetically enhanced populations and germplasm stocks (discussed in the next section);
• Genetic, cytologic, cytoplasmic, and other stocks developed in basic genetic investigations; and
• Exotic germplasm, including landraces and wild relatives, obtained on collection expeditions.

In North America a system has evolved that attempts to ensure that outstanding products of U.S. breeding programs and other categories of useful germplasm are made known both to curators of collections and to users of germplasm. In 1919 a plan for voluntary registration of field-crop cultivars was developed by the American Society of Agronomy and the Bureau of Plant Industry, U.S. Department of Agriculture (Committee on Varietal Standardization, 1920). This program was subsequently expanded to include the registration of parental lines, elite germplasm, and genetic stocks and the incorporation of the registered materials into the U.S. National Plant Germplasm System (Crop Science Society of America, 1968; White et al., 1988). The materials registered are assigned plant introduction numbers, and the U.S. National Plant Germplasm System assumes responsibility, through the appropriate working collection, for maintaining and increasing the materials when they are no longer available from the developer or commercial sources.

It is the policy of the U.S. National Plant Germplasm System that all plant materials in the system be freely distributed in small quantities for research purposes (there may be an embargo on distributing selected elite germplasm up to 7 years). Very large numbers of cultivars, parental lines, elite germplasm, and genetic stocks have been registered and distributed on a worldwide basis through this program since its inception, and the program has served a useful role in the modernization of many germplasm collections.

The development of the registration programs for plant materials was important for gene resource conservation and utilization. In the United States, registration and publication of descriptive articles in the journal *Crop Science* provide a widely available and permanent record of the most useful modern elite materials.

Registration also fosters greater utilization of the most modern materials and provides appropriate recognition for the developers of such materials and their institutions. Incorporation of the materials into the national germplasm systems attempts to preserve the materials for use in the future. Enhancement programs provide a cost-effective means of using large numbers of alleles previously scattered in hundreds of obscure accessions. Many additional enhancement programs are called for to concentrate and recombine useful genes into adaptive combinations that promote high performance and stability.

EVOLUTIONARY PROCESSES AND GERMPLASM USE

Evolutionary processes can be effective in concentrating and enhancing genetic variability from a small sample of selected materials. Two studies, one with barley and one with maize, indicated how these processes can be effective. These examples illustrate an essential step in using germplasm, which is the transfer of desired genetic traits into breeding lines that have agronomic, physiologic, and morphologic traits compatible with modern production systems. Other examples of such efforts include the Latin American Maize Project and the government, industry, and university cooperative effort in the United States to adapt important tropical sorghum germplasm to more temperate daylength conditions (Maunder, 1992).

Barley

Harlan and Martini (1929) appear to have been the first to recognize the problems associated with obtaining useful alleles from large populations. They noted that most of the 5,000 items then available in the barley collection of the U.S. Department of Agriculture had been found to be of little value. However, they said that some barleys had promising qualities and that hybridization would allow the recombination of characters to produce superior offspring. They selected 28 outstanding accessions from the major barley-growing areas of the world, thus including materials that encompassed a wide range of ecological conditions. They crossed these 28 accessions in all possible pair combinations (378), mixed equal numbers of F_2 hybrid seeds from each cross, and distributed the mixed hybrid population (Composite Cross II [CC II]) to barley breeders in various countries. They predicted that natural selection would eliminate many of the weaker combinations and that, at the end of 5 years, selections could be made with the reasonable hope of isolating superior types. This prediction was fulfilled. Within a few years many outstanding lines had been selected from CC II, including more than 50 lines that became named varieties, as well as numerous other lines that were used as parents in networks of subsequent successful crosses. Furthermore, long-term genetic studies have shown that the 28 parents of CC II contributed much of the total agriculturally relevant variability of the entire barley species to the population and that very useful evolutionary changes continued in CC II for more than 50 generations when it was grown under standard agricultural conditions in each of a number of ecogeographic regions (Allard, 1988).

Maize

The Iowa Stiff Stalk Synthetic population of maize was synthesized in 1933 and 1934 by Sprague and Jenkins (1943) from 16 inbred lines. They noted that many inbred lines of maize possess important desirable characteristics but, because of some specific fault, are unsuited for use as parents in commercial hybrids. They also noted that such lines might be of use as sources of desirable gene combinations in reservoirs synthesized from a number of selected lines. A large number of inbred lines that have been extracted from Iowa Stiff Stalk Synthetic were outstanding in their performance in hybrid combinations, both in yield and in resistance to lodging. Lines selected from it have, over the years, been used extensively in hybrids in the Corn Belt and elsewhere, and lines extracted from it are still widely used in source populations in selection programs.

The usefulness of the barley CC II and the maize Iowa Stiff Stalk Synthetic populations as sources of germplasm has stimulated the initiation of many different kinds of germplasm enhancement programs based on carefully selected materials in a number of different species. As enhanced germplasm has become more widely known and available, breeders have increasingly turned to such sources and away from traditional collections, in which variability is stored in a static state.

RECOMMENDATIONS

Germplasm collections are only valuable if they are used. The acquisition and compilation of a minimal set of data are required to locate the accessions most likely to contain the traits of interest. Efforts must also be made to enhance the ability to access important traits contained within exotic accessions.

Programs of genetic enhancement should be developed to make a diversity of germplasm resources useful to crop breeders.

Germplasm collections are of little value unless useful alleles for disease resistance and tolerance to stresses such as cold, heat, and drought, including alleles from primitive sources, are introgressed and assimilated into appropriate genetic backgrounds. Significant enhancement (prebreeding) programs have been undertaken in the past by publicly supported universities, especially in North America, and by the Agricultural Research Service of the U.S. Department of Agriculture. Some of the commodity-oriented international agricultural research centers, such as the International Rice Research Institute and the Centro Internacional de Mejoramiento de Maíz y Trigo, have also made large numbers of crosses involving exotic germplasm. Useful germplasm often reaches germplasm collections via regional and international testing networks of the international centers. Enhancement activities are also pursued in many large private-sector breeding programs. The products of these efforts are increasingly becoming generally available through national registration programs. Private groups should be encouraged to participate in these programs, making available obsolete lines of historical importance for possible use by other countries.

Increased efforts are needed to evaluate and document germplasm accessions.

Breeders and other researchers do not generally have the resources or facilities to screen large collections. There is, however, considerable information for many collections, but this information is often not readily available. Managers of collections must make increased efforts to gather this information and make it available with seeds or plant parts from accessions to aid in selection and use of germplasm. At a minimum, the basic information about the origins and habitat of accessions is needed. Such information is essential to germplasm use and can aid management by reducing redundancy and enabling the establishment of core subsets as outlined in the previous chapter. Germplasm workers should strive to become active partners in germplasm evaluation, enhancement, and use by contributing their knowledge on and experience with unimproved germplasm.

7

Biotechnology and Germplasm Conservation

B iotechnology requires germplasm, as both raw material and a source of natural variation. As a way of shaping and using genetic information, biotechnology has implications for germplasm conservation and use. This chapter discusses these opportunities and the allocation of resources.

Although biotechnology is commonly thought of as recombinant DNA (deoxyribonucleic acid) technology, it is used here in a broader sense to include tissue culture, cryopreservation, plant micropropagation, and animal regeneration from early embryos. Biotechnology influences germplasm conservation in several ways. First, it provides alternatives in some cases to conserving whole organisms. Second, it can assist with the exchange of germplasm. Third, the techniques of molecular biology can be applied to the problems of managing and using germplasm. The fourth influence results from the increased demand for germplasm and conservation services by the biotechnologists themselves.

Molecular biology provides a scientific framework that describes the elements of the genetic system as sequences of four nucleotide bases that make up DNA. Knowledge of how these DNA sequences are expressed and how expression is regulated and coordinated during development is growing rapidly. It is now commonplace to introduce foreign gene constructs into an organism, and the ability to add regulatory sequences that determine when and how strongly the introduced genes will be expressed to alter the phenotype of the recipient is often possible. The numbers of genes that have been isolated, cloned, and sequenced increase daily, and the information thus obtained already represents a genetic resource of considerable and growing scientific and commercial value.

ALTERNATIVES TO STORING SEEDS AND WHOLE ORGANISMS

Any living cell appears to have the genetic information needed to regenerate the complete organism, but among more complex and highly organized multicellular forms, relatively few have developed this capacity as a means of reproduction. In plants, clonal or asexual propagation occurs naturally in the form of propagules such as bulbs, tubers, runners, and stolons. Horticultural propagation techniques as well as in vitro propagation methods have been developed for many plants and have extensive commercial use. In vitro cell and tissues cultures are providing new approaches to multiply germplasm resources and open opportunities for long-term cryopreservation.

In Vitro Conservation of Plants

The germplasm of vegetatively propagated crops is normally stored and shipped as tubers, corms, rhizomes, roots, or in the case of woody perennials, as cuttings. Although some of these crops have seeds that could be stored, they are frequently highly heterozygous and thus do not breed true to type from seed. Perennials are usually maintained ex situ either in plantations (for example, temperate and tropical tree fruits, sugarcane, and strawberries) or are stored during the winter and planted the following spring (for example, potatoes, sweet potatoes). For these and other crops whose seeds are short-lived, there are a number of potential advantages in storing them as in vitro cultures. These include economies in space and labor and, provided that appropriate conservation methods are chosen, greater genetic stability. Disadvantages, however, include the need for special facilities and trained technicians and the small amount of experience in the use of in vitro methods for germplasm conservation on a large scale. The following summary is drawn from recent reviews of the subject (Withers, 1985, 1986, 1991a,b).

In vitro storage of germplasm was first suggested in the mid-1970s (Henshaw, 1975; Morel, 1975). Although whole plants can be regenerated from the cells of many plants (totipotent), the preservation of unorganized cultures, such as cell suspensions or callus, carries some risk of generating spontaneous somaclonal mutants. In contrast, cultures of organized meristems frequently are not only more stable but also propagate more rapidly, since meristematic areas do not have to differentiate after recovery from storage. Effective storage systems are often inexpensive and easy to maintain, and should reduce the overall work load in germplasm banks. Frequent monitoring of culture viability and for microbial contamination should not be necessary.

The full exploitation of in vitro genetic conservation is impeded if a species cannot be propagated from cultured tissue or cells. For example, a satisfactory mass propagation technique for coconut palm does not yet exist (unlike oil palm, which can be mass-propagated via callus culture). Coconut callus can be grown in vitro, but it cannot be reproducibly induced to form new plantlets. Plantlets can only be produced in vitro from germinating zygotic embryos. Each coconut embryo produces only one plant so there is no multiplication of the original material. Similarly, the impact of in vitro culture on conserving woody species has been less than for other species because of difficulties in the culture and regeneration of woody species. In these cases, insufficient research has been conducted to develop appropriate in vitro culture methods. A thorough investigation of

Advances in biotechnology provide tools for conserving and managing plant genetic resources. Here a small clump of pine tree plantlets, regenerated by individual plant cells in tissue culture, grow in a test tube. Credit: U.S. Agency for International Development.

the technical problems encountered in developing in vitro methods for woody species is urgently needed (National Research Council, 1991b).

In Vitro Storage

Two approaches to in vitro storage have proved successful, namely slow growth and cryopreservation. The slow growth approach involves applying retardant chemicals or reducing the culture temperature. Subculture intervals can be extended up to 1 or 2 years, thereby greatly reducing the time, labor, and materials required to maintain the cultures. Slower growth reduces the frequency of cell division and consequently the number of times a random mutation is multiplied in the culture. Such genetic changes that occur in tissue cultures are called somaclonal variations. Stress is an intrinsic factor in slow growth, and little is known about its effect on somaclonal variation. What began as a clonal culture may change into a population of cells consisting of the original genotype plus variant genotypes. Also, stress factors may act differently on such a population of genotypes, favoring some somaclonal variants. This could result in a changed population of cells and the failure to conserve the genetic integrity of the original clonal material. Undifferentiated callus cultures are more susceptible to somaclonal variation than organized tissue systems, such as shoot cultures. Only organized cultures are recommended for slow-growth storage. This technique of long-term root, tuber, or shoot tissue culture storage is well developed for some crops such as banana (*Musa*).

Cryopreservation involves suspending growth by keeping cultures at an ultralow temperature, typically that of liquid nitrogen (–196°C). It offers the prospect of storage for indefinite periods with minimal risk. However, certain cultures can suffer damage during freezing and thawing. Until recently, routine cryopreservation methods were available or under development only for cell suspension cultures. Larger, organized structures frequently suffered serious structural injury and loss of viability.

However, two new approaches to cryopreservation may lead to more widespread applications for genetic conservation. They focus on reducing cell damage from ice crystal formation. One approach is through vitrification of cellular water by a cryoprotectant mixture, and the other involves encapsulation of specimens within an alginate gel that is then dehydrated. For vitrification the specimen is infused with a cryoprotectant mixture that promotes the conversion of much of the cellular water into a noncrystalline, vitreous solid when rapidly cooled (Sakai et al., 1990). For encapsulation the specimen, such

Seed samples stored in liquid nitrogen are inspected at the National Seed Storage Laboratory at Fort Collins, Colorado. Cryopreservation cuts storage costs because seed life is extended. Fewer grow-outs are needed to replenish seed when it begins to lose its ability to sprout. Credit: U.S. Department of Agriculture, Agricultural Research Service.

as a shoot tip or somatic embryo, is encased in an alginate gel to form an artificial seed. This artificial seed is then dehydrated in the air before cooling (Dereuddre et al., 1990). The enveloping gel appears to minimize deleterious effects from dehydration and also protects the specimen from physical damage, being larger and more robust than an isolated shoot tip or embryo.

Despite these optimistic developments using plant tissues, much research is needed to bring the level of development of cryopreservation techniques for plant materials to that available for microbial systems and animal semen and embryos. Many technical barriers remain that prevent the routine use of cryopreservation for plant meristems, pollen, and plant cell cultures. No conservation collections or germplasm banks are yet using cryopreservation for non-seed germplasm storage, although several are involved in cryopreservation research.

Collecting Germplasm In Vitro

The laboratory facilities required for in vitro culture normally include a sterilizer, a laminar flow hood (to provide a clean, sterile work surface), incubators, growth chamber and greenhouse in addition to a pure water source, chemicals, glass and plastic labware, and other standard items of equipment. Some success has been claimed for collecting material in the field directly in vitro. The level of sophistication ranges from using a fully equipped local laboratory, for transfer of collected material as soon as possible, to working in the field in a portable glove box or on a clean table with a simple box-like cover to exclude contamination. Following surface sterilization, explants are removed and either inoculated to sterile culture media in the field or held in sterile buffer for later inoculation in the laboratory (Sossou et al., 1987; Withers, 1987). A cruder alternative is to sterilize tissue explants with nontoxic agents and inoculate them to media containing antibiotics and fungicides.

Although there is some potential for using in vitro collection for vegetatively propagated crops and those with short-lived seeds, there are serious implications for plant quarantine since the collected explants may carry pests or pathogens that might well be detected or excluded in material cultured by more rigorous methods.

In Vitro Exchange of Germplasm

During the past 20 years, advances in tissue culture technology have led to the development of commercial micropropagation, a relatively new industry that supplies young plants for a variety of horticultural, agronomic, and plantation crops (Constantine, 1986). One consequence has been the rapid development of in vitro exchange as a means of transferring germplasm between different laboratories. An International Board for Plant Genetic Resources data base assembled by a survey (Withers and Wheelans, 1986) indicated that some 135 plant genera were exchanged from 1980 to 1985. Some 94 percent of more than 480 attempts were successful. The same data show that 49 countries attempted 110 international exchanges.

Several of the international agricultural research centers now distribute in vitro cultures. For example, the distribution of potato germplasm in this form is now a routine procedure. Shoot cultures, which are inoculated into small tubes of semisolid medium, are cultured for 2 to 3 weeks to induce rooting and reveal any microbial contamination. Transfer to fresh culture medium for a passage before potting enhanced survival. In 1984 the Centro Internacional de la Papa (International Potato Center) exported 2,500 culture tubes to more than 30 countries. In that year more than 88 percent of certified-pathogen-free material was distributed as cultures rather than as field-grown tubers (Huaman, 1986). Cultures are now being replaced by the distribution of small tubers produced in vitro. These are more robust than plantlets. They can be stored in the dark for more than 4 months, and the recipient can plant them directly into pots or nursery beds without a further culture step. In modern potato breeding programs, in vitro cultures provide a disease-free reference collection during the years of field testing needed to select the most desirable clones.

In 1984 the Centro Internacional de Agricultura Tropical (CIAT, International Center for Tropical Agriculture) distributed more than 2,000 accessions of 238 cassava clones as cultures to 21 countries and received 240 clones as in vitro cultures from Costa Rica, Guatemala, Panama, and the Philippines. By 1985 the International Institute for Tropical Agriculture (IITA) had distributed in vitro cultures of sweet potato germplasm to 47 countries. IITA also imports sweet potato, cassava, yam, and cocoyam as in vitro cultures and is testing in vitro tubers of yam as a more convenient material for germplasm exchange (Ng, 1986).

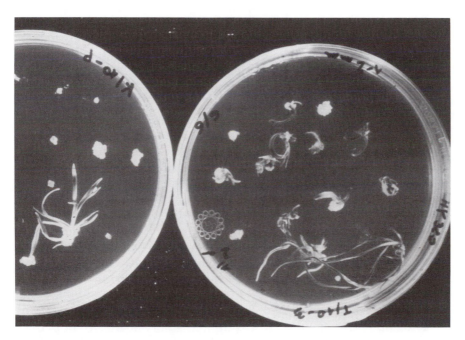

Pictured are haploid barley plants recovered from anther culture. Used to create diploids, they greatly accelerate the production of true breeding lines. Credit: Food and Agriculture Organization of the United Nations.

In Vitro Conservation of Animals

The in vitro conservation of livestock is discussed in greater detail in a separate report of the committee (National Research Council, 1993). The two principal methods for conserving animal germplasm in vitro are as frozen semen and embryos. To recover the germplasm, both methods require the maintenance of female animals for either insemination or implantation. However, embryos have an advantage over semen because they provide the complete genotype.

In some domestic species embryos for transfer or cryopreservation are generally collected from animals that have received hormonal treatment to induce an excess of eggs for insemination (superovulation). The resulting embryos are evaluated for their potential to produce a pregnancy or to withstand freezing. Embryos from cattle, sheep, goats, and horses have all been frozen in liquid nitrogen, thawed, and successfully implanted. Similar success has not been achieved for swine embryos.

The collection of embryos can be carried out surgically or nonsurgically, by flushing. Implantation may also be a surgical or nonsurgical procedure, depending on the species. Generally one embryo is transferred per recipient in cattle and horses, two in sheep and 20 in pigs. Related technologies allow in vitro fertilization, embryo sex determination, splitting of embryos into parts to produce identical clones, and the creation of chimeras by associating cells derived from embryos of different genotypes. Mapletoft (1987) recently reviewed embryo transfer technology.

Singh (1988) summarizes the potentials for disease transmission by embryo transfer. None of the disease agents tested replicated in the embryonic cells of embryos in which the zona pellucida was intact. However, because some agents can replicate in the cells of zona pellucida-free embryos, the integrity of this structure is important. Proper washing in the presence of antibiotics was shown to be effective in rendering embryos free of many disease agents (Singh, 1988). However, two or three bovine viruses and five of the porcine viruses tested adhered to the zona pellucida after in vitro viral exposure and washing. These agents could thus be transmitted by embryos if they are excreted in the reproductive tract. Only one disease agent (pseudorabies virus) was found to be transmitted when pig embryos were transferred from infected donors.

Two approaches are used to ensure that embryos are free of pathogenic organisms. The traditional method is based on testing donor animals over extended periods of time to establish that they are healthy. The other is based on the research on the transmissibility of agents via embryo transfer and requires that embryos be processed in accordance with international standards. Although the protocols have been shown to be reliable for certain diseases, the health of the embryo depends entirely on the care taken during embryo collection and processing. The second method is therefore likely to require additional safeguards, such as a period of quarantine for the recipients, particularly when testing for the more serious diseases.

MOLECULAR CONSERVATION TECHNOLOGIES

Within the past decade rapid progress has been made in developing and applying methods to identify, isolate, and characterize individual genes at the molecular level. At the larger scale of the genome, the DNA sequence can be cut up into fragments and reassembled in the form of a linkage map. While the task of characterizing germplasm at the molecular level has just begun, these molecular technologies can offer new approaches for both preserving and evaluating germplasm resources.

DNA as a Genetic Resource

The DNA sequences in the genomes of germplasm accessions are the sources of the genes required in breeding programs. When identified and isolated by cloning, these genes may be used to prepare transgenic organisms that express them. Transformation may avoid much of the genetic disruption that accompanies sexual hybridization and, since it is not limited by sexual compatibility, can make use of genes from other life forms. The genes used to breed crop plants that are resistant to various herbicides or insect larvae or to increase the growth rate of transgenic fish, are most conveniently conserved as cloned DNA. The extent of this kind of conservation is limited by the technical problems of identifying and cloning the commercially important genes that breeders would like to have and obtaining high rates of stable transformation in a broad range of recipient genera. Genes of this kind that have been cloned so far are also protected by patents and so are not freely available to anyone who wishes to exploit them commercially (see Chapter 12).

For economic and technical reasons it is unlikely, in the foreseeable future, that gene synthesis will make the physical storage of germplasm in the form of seeds, whole plants, or tissue cultures obsolete. The conservation of DNA and the assembly of sequence data bases are not alternatives to conventional germplasm conservation because they are not coordinated in a genome (Peacock, 1984). Recent progress in the synthesis of artificial chromosomes in yeast raises the possibility of conserving coordinated assemblies of genes that would allow more complex phenotypic changes to be engineered. To be of practical use, however, DNA and artificial chromosomes require the conservation of recipient organisms. Their genetic information is incomplete. Unlike a seed or an embryo in a surrogate mother, a DNA sequence cannot yet be used to regenerate a whole organism. Nevertheless, cloned genes, genomic libraries, and sequence data bases have significant potential uses in germplasm conservation and management. They may be more compatible with genetic stock collections, rather than national and international germplasm collections because of the specialty nature of the material and data.

In the broader context of conserving global genetic diversity, some (Adams et al., 1992; Adams, 1993) have proposed the collection of small samples of plant material (leaf tissue, seeds, and so on) for conservation as sources of DNA. Tissue samples from all taxa threatened with extinction would be collected and preserved in liquid nitrogen. These samples would, in theory, be available indefinitely. DNA extraction would be deferred until needed. At that time, the DNA could be isolated, immobilized on membranes, used as a source of specific genes or DNA sequences (see polymerase chain reaction technique below), and then returned to liquid nitrogen storage. Several conservation groups are now establishing a network of such DNA banks with duplicate samples as a safeguard against loss through extinction (Adams et al., 1992; Adams, 1993). This technique may be useful for the many undescribed or unstudied plant species with seeds that cannot be stored, but it is unlikely to be of direct value for those of recognized importance that are already conserved in germplasm banks.

DNA Sequence Data Banks

There is now so much DNA sequencing carried out in various laboratories worldwide that the ability to compare new sequences with those that have already been described and well characterized is of considerable importance. Comparisons may suggest unsuspected functions or may reveal useful homologies between unrelated organisms. Several organizations support the central storage and coordination of DNA sequence information. The two major data banks are the European Molecular Biology Laboratory and, in the United States, GenBank, which is operated by IntelliGenetics and the U.S. Department of Energy's Los Alamos National Laboratory.

The rate at which data on DNA sequences is accumulating is likely to continue to increase exponentially in the near term as a result of such major initiatives as sequencing the entire human genome and the genomes of rice and *Arabidopsis*. This rapidly growing body of data raises important problems of storage and access to facilitate rapid comparisons with newly obtained information.

Restriction Fragment Length Polymorphisms

When DNA is extracted from an organism and digested with one or more restriction enzymes and the products (called restriction fragments) are separated by electrophoresis on a gel, the result is a smear of many DNA fragments with different lengths. The longest fragments move slowly in the gel, staying near the origin; the smallest move faster and are farthest from the origin. If the smear is transferred to a membrane by blotting, it may be hybridized with a radiolabelled DNA probe. After autoradiography, the regions of hybridization are revealed as one or more distinct bands at constant and characteristic distances from the origin of the gel. Each band identifies a restriction fragment that contains a stretch of DNA complementary to the sequence of nucleotide bases in the DNA probe. The same digest tested with different DNA probes shows different band patterns that are also constant and characteristic for each probe.

When the DNA digests from separate organisms are compared, the differences in banding patterns are called restriction fragment length polymorphisms (RFLPs) and correspond to points, or very small regions, of physical difference on the chromosomes. The number of polymorphisms (RFLPs) provides some indication of the number of differences that exist between the genomes of the organisms being compared. The RFLPs segregate as allelic differences and may be tested for linkage with each other and with other characters of agronomic importance. RFLPs, in theory, provide an almost inexhaustible number of markers and should make possible the rapid construction of linkage maps even in species in which this has been hampered because of the lack of morphological markers or long generation times.

There are several practical limitations, however. The DNA extracts, digests, gel separations, blots, hybridizations, auto radiography, and pattern interpretation are expensive and time-consuming and require trained personnel. It may be necessary to test several hundred, or even thousand, different DNA probes to find RFLPs suitably spaced throughout the genome. The ideal is to locate one at about every 10 map units. Some of these difficulties may be overcome by automation of the procedures. The analysis of the resulting data can be complex. In some cases the results obtained for one segregating F_2 population cannot be directly applied to an F_2 generation from another, different cross because of major duplication at the DNA level that may even obscure the chromosomal assignments obtained with the first set of probes. This difficulty is proving to be an obstacle to the use of RFLPs as an aid to selection in maize breeding, because maize contains an excessive number of polymorphisms in the noncoding regions of its DNA that complicate the banding patterns and often make them difficult to decipher. In contrast, among some inbreeders, such as the tomato, there are not enough RFLPs among commercial cultivars, evidently reflecting their genetic similarity.

Recent research indicates that RFLPs can be used to map quantitative trait loci (Paterson et al., 1988). Linked RFLPs that identify relevant portions of chromosome arms may potentially be used to select characters that are inherently difficult to assay, such as resistance to insects that cannot be reared, drought tolerance, and processing quality of harvested product. In germplasm enhancement programs, linked RFLPs could be used to mark genes for resistance to pests or pathogens that do not occur when or where the work is being done and that cannot be introduced because of quarantine regulations. An example might be breeding U.S. maize for resistance to African streak virus.

A barrier to using RFLPs is the investment of effort needed to find useful probes. Probes could be used to screen germplasm accessions for alternative alleles for important disease resistance loci.

Polymerase Chain Reaction

The polymerase chain reaction (PCR) is an extremely sensitive and accurate method for recovering microgram amounts of single specific DNA sequences present in biological samples at very low concentrations. The reaction requires pairs of single-stranded primers (DNA template molecules) of 20 to 30 bases in length found on opposite strands at either end of the double-stranded DNA sequence of interest. These primers are added to a solution of DNA extracted from the sample, together with the four nucleotide bases that make up DNA and a DNA-polymerizing enzyme from the bacterium *Thermophilus aquaticus (Taq)*, which is stable at high temperatures. The reaction mixture is heated to 92°C for 30 seconds to denature the DNA into single strands, cooled to 50°C for 1 minute to allow annealing to occur, and then heated to 72°C for 2 minutes to allow DNA synthesis to fill in the missing bases from one primer toward the other, using the single strands of sample DNA as templates for the complementary strand copies. The preparation is then heated again to denature the newly formed double-stranded DNA molecules, and another cycle of cooling-incubation-heating is begun. Since the primers are present in excess at each cycle, the amount of double-stranded DNA selected by the primers steadily builds up until after 40 cycles the preparation is sufficiently pure for that segment of DNA that it can be readily sequenced.

The ability to use PCR to select and amplify specific DNA sequences from desiccated dead seeds and inviable frozen semen and embryos raises the question of whether these "dead" materials are useful genetic resources. Depending on their scientific interest, commercial value, or rarity, there may be a case for keeping them as a source of particular DNA sequences. Already collected, stored, inventoried, and evaluated, there would be little more to do with them until they were needed for DNA isolation. However, the fact that they had become inviable could mean that the original storage conditions were unsatisfactory and that, if left in place, further changes to the DNA might eventually reduce their usefulness. The DNA banks discussed above avoid these difficulties.

The use of PCR to recover sequences from herbarium specimens (Rogers and Bendich, 1985) may mean that specimens should be regarded as a form of germplasm. PCR analyses will be useful for comparative genetic studies, which, until now, have depended on morphological or chemical comparisons rather than functional genetic analyses using the tools of molecular biology.

As with isolated DNA, cloned DNA, and DNA sequence information stored in data banks, the DNA of inviable specimens and cryogenically stored tissues cannot be used directly to reconstitute an organism. These forms of DNA are presently only valuable as a germplasm resource to the extent that they can be incorporated into other living organisms or for research. For the long-term preservation of genetic variability, they are only useful to the extent that viable hosts in which they are readily expressed can also be maintained.

PCR is a useful technique, but the requirement for primers means that it can only be used to find sequences that are already known in some detail. PCR is useful, however, for recovering allelic forms of well-characterized genes from any source, whether or not it is viable.

Randomly Amplified Polymorphic DNA Markers

PCR can also be used with single, arbitrary, 9- or 10-base primers to generate probes for detecting RFLPs. These polymorphisms are called randomly amplified polymorphic DNA or RAPD markers (Welsh and McClelland, 1990; Williams et al., 1990). It is expected that any arbitrary 9- or 10-base sequence will occur at a sufficient frequency at points on opposite strands of DNA so that synthesis of the intervening segments will occur in PCR from points where the primer has hybridized. The optimum size range of these fragments can be controlled by prior digestion of the extracted DNA with one or more restriction enzymes. The several kinds of DNA fragments amplified in the PCR step can then be used as probes to detect RFLPs. The advantage of the method is that the original primers are relatively inexpensive and easy to make and that many probes are available that should cover all parts of the genome (Anderson and Fairbanks, 1990).

Other Uses for Probes

DNA probes of particular genes can be used to detect their chromosomal location by in situ hybridization and to detect the presence of those genes in segregating populations or new alleles in germplasm accessions. Although these methods are now used in plant breeding in material with known genetic backgrounds, they will likely also be of use as a research tool in germplasm conservation when simpler, cheaper methods are available. However, the same technology can detect the presence of pest and pathogen nucleic acids and could simplify quarantine procedures by greatly reducing the time needed to test imported plant materials for diseases or organisms, including viruses, of quarantine significance.

RECOMMENDATIONS

Recent advances in biotechnology provide powerful tools for conserving, evaluating, and using genetic resources. Cloned DNA fragments and synthetic DNA are unlikely to replace conventionally stored seeds and other germplasm in the foreseeable future. In any case whole organisms must be conserved as recipients for the expression of introduced DNA.

Transformation works for a number of plants and animals, and is now being applied to commercial cultivars or breeds. Many field tests of various transformed crop plants have been carried out since 1987. However, for the present, the impact of transformation on germplasm enhancement will likely be small because of the effort and resources now required to characterize, isolate, and clone useful genes.

PCR for recovering and amplifying selected DNA sequences makes possible comparative studies of DNA from herbarium and museum specimens and dead seeds and sperm. As a consequence, these materials may become much more useful for ecogeographic and evolutionary genetic studies of germplasm. They may eventually become a source of alleles at well-characterized loci.

Recombinant DNA biotechnology is rapidly evolving. Its techniques are expensive, time-consuming, and complicated and call for highly trained staff. Until it becomes more user friendly, it is best regarded as a potentially useful adjunct to germplasm conservation rather than a means of revolutionizing this important activity. This limitation applies in particular to the larger collections of plant germplasm, where current resources seriously limit activities.

Research is needed to apply in vitro culture and cryogenic storage methods to a broad range of plant and animal germplasm.

In vitro conservation of plant germplasm as growing tissue cultures has potential as a means of conserving forms that cannot easily be kept as seeds, and for some species, the maintenance of cultured plantlets in slow growth is proving more efficient and less expensive than maintaining whole plants. In vitro techniques are a useful means of exchanging and distributing some clonally propagated germplasm. Cryogenic preservation has potential for long term storage but the method is not as well developed for plants as it is for semen and embryos of some livestock species.

Biotechnology research efforts should focus on developing enhanced methods for characterizing, managing, and using germplasm resources. These efforts should include the urgent need for more effective data handling systems for storage, retrieval, and sequence comparisons, some of which may be by-products of the considerable investment in sequencing the human genome.

Biotechnological innovations, including computerized information handling techniques, heighten the utility of germplasm collections rather than obviating their need. The rapid development of DNA sequence data banks, plasmid libraries, and cloned DNA fragments has created a genetic resource of growing size and importance. However, with this expansion comes a vast amount of data and information. Without sufficient attention to the management of these data, the resources they describe will be of little value.

Biotechnology could improve access to collections by providing tools for characterizing their accessions. Information on the genetic similarity between accessions could aid in developing priorities for acquisition. By linking molecular markers to specifically desired genetic traits, biotechnology may provide a mechanism for rapid searching of large collections without the need for lengthy field trials. Molecular techniques for characterizing genetic material, such as restriction fragment length polymorphism analysis, appear likely to provide the breeder with greater efficiency in selecting and developing new breeding lines and varieties.

DNA sequences used as probes can detect viruses in germplasm bank materials and also reveal RFLPs that can be used to construct linkage maps and, as linked markers, to select desirable traits. Linked RFLP markers might be used to detect genes for resistance to pests and pathogens that themselves cannot be used because of quarantine restrictions. Information from RFLPs could potentially assist in the selection of core subsets of larger germplasm collections by providing another measure of genetic diversity, but only if the technology becomes much less expensive.

8

Documentation of Genetic Resources

In the first decades of genetic resources activities, emphasis was given to the study and conservation of genetic variation. Now, general attention is focused on the management and utilization of the germplasm materials held in collections (Frankel and Brown, 1984). As a result, information on the germplasm in collections has become nearly as important as the germplasm itself.

In the management of the collections, comprehensive information is required to decide which material should be included or excluded from a collection. Descriptive information about each accession is also required. Easily accessible information is needed to optimize the efficiency of the management and use of the collections. Readily accessible information is required to facilitate the exchange of materials and information among germplasm banks and to help the users in experimenting with conserved germplasm.

Increasingly, genetic resources activities are being coordinated on a global scale, leading toward the realization of a cohesive worldwide network. Eventually, this would make possible the more efficient allocation of available genetic resources. This can only be done successfully if information on existing germplasm collections is both available and compatible.

The distinction between the information and the systems that allow for the management of that information is important. Information is the raw material that can be used for decision making. Information systems are the mechanisms for storing and using large amounts of information.

This chapter provides an overview of the current status of the documentation of germplasm collections, the information and management of the information in the germplasm system, and some possibilities for future developments.

INFORMATION ON GERMPLASM COLLECTIONS

A distinction can be made between the information needed to manage and maintain the collection and the descriptive information needed for its utilization.

Management Information

The information needed for germplasm-bank management is the information produced during the acquisition, monitoring, regeneration, and distribution of accessions (Ellis, 1985). The amount and type of information that is processed for germplasm-bank management depend entirely on the procedures used for these basic operations. For that reason they are very specific to a particular germplasm bank. The types of management information are quality, quantity and type of the material, storage location, distribution to the users, and treatment of the material (Konopka and Hanson, 1985).

Descriptive Information

Since descriptive information should support all kinds of decisions concerning the choice of material for utilization, it should contain the information most appropriate for that task, including information that is available from other germplasm-bank collections as well. The main reason cited by breeders for not using material in germplasm banks is that the information provided with accessions is either irrelevant to needs or too incomplete to be of use (Peeters and Williams, 1984).

Users must be able to make rational choices of the germplasm that they want to use in their research or breeding programs. The information at the germplasm bank should support this choice. Users must make decisions about their activities on the basis of information about their own and other's collections, including passport, characterization, and evaluation data, described in Chapter 5. Together these basic data describe the smallest unit contained in the germplasm bank, the accession.

Once the basic data have been entered into the information system, the quality of the information can be enhanced. The integrity and plausibility should be confirmed. Then the quality of data can

These samples of different varieties of primitive barley are from the Plant Genetics Resources Center in Ethiopia. The germplasm bank keeps samples of different populations under refrigeration and tests them periodically. Credit: Food and Agriculture Organization of the United Nations.

be increased. For example, if the collection site is documented but the longitude and latitude are not available, these can be added. If the longitude and latitude are available, climatologic data on the collection site can often be added.

Sources of Additional Information

Besides the data base on the particular collection, many additional sources of information can be used in managing a germplasm bank. These can be data bases on collections at other germplasm banks, central crop data bases, and data bases on related subjects.

Information on Other Collections

The aim of the user is to select the material most suitable for a particular purpose from a wide range of genetic variability (Holden,

1984). However, a curator or user may want to know where other collections are located, the type of material they included, and what is available from those collections.

There are several ways that information on other collections can be gathered. Information on an accession level is available in printed material, such as catalogues and scientific papers. Information can also be obtained on magnetic media or on-line from the institute holding the collection. Another, probably more efficient, way of compiling information on other collections is the use of central crop data bases. Since many germplasm banks have computerized all or part of their collection data, it is possible to combine the data sets of individual germplasm banks for each crop into central crop data bases. This approach has great potential (Frese and van Hintum, 1989; Vanderborcht, 1988). Because duplicate samples unnecessarily aggravate the management of collections, from the curator's viewpoint, the collection data in central crop data bases would make it possible to trace duplicate samples more efficiently.

Central data bases could be used to set priorities for collection missions and regeneration programs. Depending on the information included, it can function as a source of data for many types of studies. In addition to the detailed information on the level of collection accessions, it can be useful to have summarized information, presenting a general overview on a species level or other convenient category. An important source of this kind of information is the crop directories of the International Board for Plant Genetic Resources (IBPGR). These directories, although sometimes not very discriminating, describe most major collections by crop (Bettencourt and Konopka, 1988; Lawrence et al., 1986). Other sources of information on collection locations are available from, for example, the United Nations Development Program and International Board for Plant Genetic Resources (1986) and Sgaravatti (1986).

Information on Related Topics

A multitude of scientific disciplines, such as taxonomy, genetics, and seed technology, must be applied to work with genetic resources. Therefore, it would be logical to use the information available from these disciplines to cross-check or supplement specific information on germplasm. For instance, in the field of taxonomy, there are a variety of information systems that could be adapted for use in genetic resources (Aleva et al., 1986; Allkin and Bisby, 1984). Meteorological data bases can be linked with collection information to enable a more in-depth study of the distribution of characteristics under environmental constraints. The results of these types of studies could be translated into more rational approaches toward collection, use, and management.

Standardization of Information

Information exchange is a vital part of genetic resources work. The compatibility between information systems determines the efficiency of this exchange. To allow an easy exchange of information, a certain degree of standardization is necessary, but to be functional it must be widely accepted.

Most problems concerning the exchange of information are caused by either technical or logical incompatibility. The problems concerning technical compatibility, those related to hardware and software, can be solved relatively easily and are discussed below. Logical incompatibility causes more serious problems. A first step in preventing logical incompatibility is to guarantee the interpretability of data files by always accompanying them with notes explaining the structure of the data files and the codes that were used in them. Before entering information from a foreign system, the format and coding must usually be converted, which apart from being very laborious sometimes causes loss of information. To minimize this effort and the loss of information, data-bank standards have been proposed. Early efforts on morphological and varietal characters of rice were made at the International Rice Research Institute (IRRI) that led to a computer-printed catalog in 1970 (Chang, 1985a). Seidewitz (1974–1976) proposed standards for terminology in his thesauri for names of descriptors and defined descriptor states. In the exchange of information workshop in Radzikow, Poland, in 1984 (International Board for Plant Genetic Resources, 1984b), a standardized list of passport descriptors and a format for data exchange were proposed. Since 1978 IBPGR and collaborators have published descriptor lists (International Board for Plant Genetic Resources, 1982; International Board for Plant Genetic Resources et al., 1985; International Rice Research Institute and International Board for Plant Genetic Resources, 1980) for various crops, giving passport, characterization, and evaluation descriptors, which generally follow those proposed during the workshop in Radzikow.

Efforts have been made to compile an international standard list with addresses and address codes. Currently, the only standard that is commonly used in germplasm-bank information systems is the three-letter country abbreviations supported by the Statistical Office of the United Nations (Anonymous, 1982). However, even this list of codes is sometimes supplemented with extra codes that serve a different purpose.

The problems in the standardization of the logical aspects of germplasm-bank documentation are mainly caused by the fact that germplasm banks differ. This makes it unlikely that two will use exactly the same information system.

Documentation systems can be fully compatible even if different equipment and software are used, but this requires consensus over standards for exchange of information. The Radzikow workshop tried to develop such standards (International Board for Plant Genetic Resources, 1984b), but only very few germplasm banks thought that it was possible and necessary to implement the proposed standards.

COMPUTERIZATION OF GERMPLASM COLLECTION DATA

Management of information about accessions in germplasm collections is an integral part of genetic resources efforts (Blixt, 1984). As a result of the growth of collections in recent years, the amount of data available has increased considerably. More powerful systems are needed to manage effectively and efficiently the mounting quantity of data. Computers have become essential to this task.

Hardware

In the early 1970s, many organizations recognized the need to computerize germplasm collection data to facilitate the maintenance and utilization of collections. Originally, plans were envisioned in which all major germplasm collection centers, as part of a global network, would be equipped with compatible computers and documentation systems to promote the free exchange of information (Williams, 1984). This approach proved to be impractical, because of the diversity of hardware and software available. As a result, a variety of computer systems is still used for documentation of genetic resources.

Since the early 1980s, the introduction of personal computers has greatly expanded the capacity to store and retrieve information. This equipment has been introduced in many germplasm banks and has the potential to greatly enhance data management capabilities.

Software

A variety of software systems are used for documentation systems. Generally, they can be grouped into three types: software developed in house, commercial data-base management systems, and commercial application software (International Board for Plant Genetic Resources, 1988a).

Software developed in house that uses third-generation programming languages (Pundir et al., 1988) like PASCAL or FORTRAN is well-suited to the specific needs of the germplasm bank. Its development requires sophisticated programming knowledge to develop it initially and to update it to add new functions or to address other problems as they arise. Thus, it can be both expensive and time-consuming.

Commercial data-base management systems are becoming increasingly popular (Perry et al., 1988). They are specifically developed to enable the efficient and safe management of large quantities of data. Such software packages do not generally require sophisticated programming skills and are flexible enough to be used to create suitable applications. Updating of the data-base management software can be done by the manufacturer and thus is less expensive and time-consuming. In addition, manufacturers offer assistance to users in developing applications for their software.

Commercial application software is available for purposes other than data-base management. This is software that was not intended for data-base management, such as word processors, spreadsheets, text editing, or statistics. Although these packages can be used for data storage, the facilities for data-base management functions, such as selective retrieval, can be limited or absent. These types of software packages are suitable only for data sets with a limited size and are not recommended for use as primary data management tools. They can, however, be quite helpful for preparing data for entry into a data base.

Standardization of Systems

A large variety of hardware systems and software packages are used in germplasm collections worldwide. Difficulties in the technical compatibility between systems are bound to arise. Problems with hardware compatibility can usually be addressed, however. It may be necessary to transfer the data from one medium or format to another to enable its transfer to another computer system. File transfer centers for this purpose were recommended by the workshop on the exchange of information in 1984 (International Board for Plant Genetic Resources, 1984b).

While standardization of all documentation software would be the most straightforward solution, the variety of computer systems, software, and their availability make this impractical. As an alternative, a standard for data exchange, such as the American Standard Code for Information Interchange (ASCII) fixed-format files, could be adopted for data exchange. ASCII fixed-format files can be created, read, and processed by virtually all commercial software and require few adjustments by individual data systems.

Standardization also assumes that data need to be exchanged extensively between germplasm banks. This may be true more for large collections, such as those of the international genetic resources centers, some national programs, and regional collections. In these collections, however, there are usually individuals with expertise to convert data to a format appropriate to the recipient. This is the current practice of the Germplasm Resources Information Network (GRIN) of the U.S. National Plant Germplasm System. Data in the GRIN system are stored in a large, sophisticated data management system that is incompatible with the systems of other manufacturers. However, an individual requesting information can receive data from GRIN that have been reformatted to be compatible with most major commercially available data-base management software or that have been reformatted in ASCII code. The problems of technical incompatibility can thus be minimized.

Information Supply to Users

The classic manner in which germplasm banks present their collection directly to the users has been through the listings of passport and characterization data. With the current information technology, this seems an inefficient way of making the available information accessible to users. These lists are usually very incomplete selections of the available material and information. More extensive catalogues are often very bulky, expensive to produce, and difficult to update. Other media, such as microfiche (Porter and Smith, 1982) and compact disks (Centro Internacional de Mejoramiento de Maíz y Trigo, 1988), have been used to bring the information to the attention of the users.

The media most commonly used for the exchange of computerized information are floppy diskettes and magnetic tapes. They provide a low-cost means of storing massive amounts of information. The magnetic and optical storage media also permit an interactive retrieval of, sometimes preselected, information. Suitably equipped users are able to process these data into the most suitable form and to find precisely what they are looking for with a minimum of effort. Printed media are not totally obsolete, however. They are still an effective way of presenting collection information in scientific articles or booklets containing a collection overview with summarized information (Pundir et al., 1988). Infrequent users of large collections or users without computer facilities may find printed listings to be sufficient and more appropriate for their limited needs.

An alternative way to provide information to users is through the use of information service units at germplasm banks. The incorporation of stand-alone applications into national and international computer networks provides users with increasing possibilities to consult remote data bases using on-line data communication. In the United States, GRIN is an example of such an information network used for genetic resources (Perry et al., 1988).

CURRENT STATUS OF GENETIC RESOURCES DOCUMENTATION

The current status of genetic resources documentation can best be typified as one that is rapidly evolving. The situation has changed since the comprehensive paper of Knuepffer (1983) and status report of the International Board for Plant Genetic Resources (1984c). Many institutions have altered their data management systems with respect to hardware, software, or both; the results of massive screening activities have been documented; and for many crops, central data bases have been organized, as have useful data bases on related topics. The situation is different for every germplasm bank and for every crop. In this diverse and rapidly changing situation, it is impossible to describe the current situation accurately or comprehensively, but some trends can be observed.

Collection Data Bases

Small collections, usually working collections, are just being computerized, usually using personal computers and a commercial database management system. Many breeding institutes and small germplasm banks use this relatively new option, for example, the University of Birmingham in the United Kingdom and the French Institut National de la Recherche Agronomique in Montpellier. Even some larger germplasm banks, such as Zentralinstitut für Genetik und Kulturplantzenforschung in Germany (formerly, East Germany), the Nordic Gene Bank in Sweden, and the Center for Genetic Resources in The Netherlands (van Hintum, 1988), use a commercial data-base management system on stand-alone personal computers or personal computers linked to a minicomputer.

Most large germplasm banks computerized their collections some time ago. The very large ones, such as the IRRI, use internally developed programs on mainframe computers or minicomputers (International Board for Plant Genetic Resources et al., 1985). The smaller ones often use less customized packages, also on large computers. Consiglio Nazionale delle Ricerche in Bari, Italy, uses a statistical package on a mainframe computer. The technology of data-base management has evolved very quickly; the technological solutions applied in the first germplasm information systems have become outdated. The older systems often lack the flexibility and performance of modern data-base management systems. Therefore, many large germplasm banks have already replaced, or are in the process of replacing, their older systems with a commercial system, for example, the IRRI and Institut für Pflanzenbau und Pflanzenzüchtung der Bundesforschungsanstalt für Landwirtschaft in Braunschweig, Germany (formerly, West Germany).

In 1988 the International Board for Plant Genetic Resources (unpublished data) performed a survey for preparing crop inventories. Responses were received from 52 germplasm banks maintaining mostly minor collections of cereals or vegetables; these centers were located in Africa (2 centers), Asia (5 centers), South America (8 centers), Australia (2 centers), and Europe (35 centers). The following conclusions concerning documentation can be drawn.

• Personal computers are increasingly being used at germplasm banks. They either stand alone or function in combination with mainframe or minicomputers. This development considerably decreases the problems associated with the exchange of information.

• Germplasm banks are rapidly computerizing their data. Only 2 of 52 respondents to the survey mentioned above did not use computers for documentation purposes, but they were planning to start using them in the near future.

• The software used for documentation is mainly applications of commercial data-base management system software (66 percent) or in-house developed software programs (30 percent). Few applications are based on other software products, such as spreadsheets (4 percent).

The extent to which the collection data are computerized is difficult to estimate since detailed information on this point was scarce. Overall, it appears that active and working collections in particular are in the process of computerizing characterization and evaluation data. Many of the respondents did not mention the status of passport data in their collections. Although many respondents expressed their willingness to exchange data, it appears that the major concern of most germplasm banks at present is computerization of their own collection data.

Central Crop Data Bases

The importance of international central crop data bases was stressed above. For a number of crops, these data bases have been established, usually by germplasm banks that hold one of the collections. The extent to which these data bases are complete varies enormously, as do the type and quality of the information they contain. Often only passport data are included (for example, International Board for Plant Genetic Resources, 1985a). Sometimes, this is extended with characterization, evaluation (Chang, 1985a; International Board for Plant Genetic Resources, 1986; Vanderborcht, 1988), or seed management data (Frese and van Hintum, 1989; International Rice Research Institute, 1991). Problems in compiling central data bases are generally caused by lack of interest, lack of funds, lack of computer facility, or incompatibility of data (International Board for Plant Genetic Resources, 1985b, 1986, 1989b; International Rice Research Institute, 1991).

Other Sources of Information

The number of other sources of information is large. As a result of data computerization, the accessibility of those sources has increased considerably in recent years. There is a multitude of on-line and off-line data bases (Anonymous, 1989) and other possible sources of information (Food and Agriculture Organization, 1987a).

FUTURE DEVELOPMENTS

Developments in information technology have occurred rapidly. This has made it difficult to predict their impact on genetic resources data management activities. Nevertheless, it is useful to look at some of the major trends and how these could affect the documentation of genetic resources.

Four major developments in information technology have occurred:

- Decreasing prices and physical sizes and the increasing quality and performance of hardware;
- Increasing quality, performance, and user friendliness of commercial software;
- Growing popularity and possibilities of computer networks; and
- Growing familiarity of the public with computers.

These developments have several unavoidable consequences for germplasm-bank documentation:

• The limitations in the documentation of germplasm collections will shift from problems with hardware and software to problems in using the information in germplasm information systems and related data bases.

• All data on genetic resources will be computerized by using user-friendly information systems. This will stimulate the use of the information, for example, for the compilation of central crop data bases.

• Improvements in computer networks and standardization of commercial data-base management software packages will allow for much easier communications between germplasm banks and between germplasm banks and users, leading to the flow of more information.

RECOMMENDATIONS

Information about the accessions in a germplasm collection is essential. The seeds in a collection are of little value if separated from the information that will enable researchers to select those appropriate to specific needs. Once obtained, this information must be recorded in data bases in a form that is accessible to the broadest possible range of potential users.

Germplasm collections must minimally make available passport information for the materials they hold.

Too often, loading data on accessions has been accorded less importance than design and development of the computer software. An earlier report of the committee (National Research Council, 1991a) recommended, for example, that the U.S. National Plant Germplasm System direct resources more to the addition of existing information to the data base than to software and hardware development. For materials in collections, basic passport, characterization, and evaluation data should be obtained from collectors, breeders, researchers, and germplasm curators. For materials being added to collections, strict policies against adding undocumented accessions should be enforced.

Compilation of data for germplasm applications should follow easily exchanged, readily available protocols.

The availability of computerized information in a collection data base, in central crop data bases, and on-line in other data bases will change the management of genetic resources collections considerably and should be anticipated. Central crop data bases, combined with an infrastructure for easy exchange of information, will allow for a genetic resources strategy with networks of dispersed, rationalized collections of freely available material (Perret, 1989). This strategy would create many possibilities in the field of genetic resources management. The choices of data bases to be used will be less critical than the quantity and quality of the data and the implementation of the procedures to acquire, enter, and distribute the data. However, the protocols implemented for data management should follow those used in most widely available commercial data base applications.

International crop germplasm data bases should be developed.
National and international centers should collaborate in developing a compatible and interlinkable data base for every major crop. National research programs will be greatly benefitted by having access to an internationalized documentation and exchange system. The international agricultural research centers and other well-established national centers, such as the U.S. Department of Agriculture, may serve as international data-base centers.

9

The Conservation of
Genetic Stock Collections

Genetic stocks are lines that contain various mutant alleles, chromosomal rearrangements, or other cytological abnormalities that are useful only for genetic and basic biological experimentation. They rarely, if ever, have any direct commercial value, but they are essential to much basic and applied research. The term *genetic stocks* is occasionally used (for example, in some of the Food and Agriculture Organization [FAO] debates) in a much broader context to include the advanced lines of breeders that are largely unrelated to genetic research endeavors. In this chapter, however, genetic stocks refers solely to stocks used strictly for research purposes.

This chapter first considers the species of agriculturally important plants, animals, and microorganisms for which significant genetic stock collections have been established. It reviews the locations of major collections and the types of institutions that maintain them and then the compositions of genetic stock collections and how they differ from the compositions of other germplasm collections. Finally, the special problems that genetic stock collections pose in terms of long-term conservation and the steps that need to be taken to ensure their long-term security are examined. Genetic stock collections of other classes of organisms that have no direct economic benefit but that are used solely for such purposes as research and teaching (for example, the mouse *Mus domestica*, the fruit fly *Drosophila melanogaster*, the coliform bacterium *Escherichia coli*, and the pink bread mold *Neurospora crassa*) are important and may face many similar problems, but they are not considered here.

IMPORTANCE AND USE OF GENETIC STOCK COLLECTIONS

Genetic stock collections are an important component of the total conserved genetic resources both of economically important species, such as corn (*Zea mays*) and yeasts (*Saccharomyces* sp.), as well as of organisms used purely for research, such as fruit flies (*Drosophila* sp.) and bread molds (*Neurospora* sp.). Such collections have been of critical importance to the rapid progress made in the science of genetics and its applications to plant and animal improvement over the past few decades. From this viewpoint, genetic stock collections and their contributions to scientific progress are underappreciated outside the genetic research community. Furthermore, the importance and value of genetic stock collections will find increasing use in the identification, location, and isolation of specific genes and their manipulation and transfer into economically important plants and animals for breeding programs using molecular genetics tools. Consequently, the perception that genetic stock collections are primarily a research tool of no direct economic benefit is likely to change quickly as the impact of the new biotechnologies permeate commercial agriculture.

In addition, advances in molecular genetics will provide new challenges and opportunities for genetic stock collections. Of particular importance in this regard is restriction length fragment polymorphisms (RFLPs) (see Chapter 7). To exploit this technology, however, breeders and geneticists require not only known genetic strains of the organism of interest but also the DNA (deoxyribonucleic acid) probes necessary to detect RFLPs. The quality, maintenance, and continued availability of publicly developed genetic stocks and DNA probes for RFLP detection will become an issue of growing concern to geneticists and breeders alike in both developed and developing countries.

GENETIC STOCK COLLECTIONS

International agencies and most national programs have placed little emphasis on conserving genetic stock collections, even those of economically important species. There are several reasons for this. First, genetic stock collections have usually been regarded as serving entirely different purposes—genetic research and teaching—that require different inputs and management strategies compared with those required by cultivar collections of crops or breed collections of domestic animals. Second, genetic stock collections are usually seen as providing no direct economic benefits to developing countries, and hence to be of little interest to international agencies, especially those whose programs and funding are directed toward the poorer developing countries. Third, the development of genetic stock collections is an expensive and time-consuming process and depends on close interactions with sophisticated research programs. As a result, extensive genetic stock collections of most organisms are confined to developed countries.

Furthermore, genetic stock collections are generally regarded as the responsibility of individual research groups that have developed and maintained them on a purely ad hoc basis. Only when the demand for access to genetic stocks and the size of the collections exceed the capacity of the individual researchers have there been moves to develop integrated national or international collections. Efforts to develop such recognized genetic stock collections have usually been initiated through scientific societies such as the Genetics Society of America and the Organizing Committees of the International Wheat and Barley Genetics Symposia.

Significant genetic stock collections exist for the major crop plants and microorganisms of economic importance in agriculture and medicine (Table 9-1). For the other groups of species, small to modest genetic stock collections have often been established, but they have not been developed to the same degree as those for crop plants, because of the lack of a long-term research effort in genetics and breeding or the lack of appropriate long-term maintenance technologies.

In the case of microorganism collections, it is debatable whether they should be strictly regarded as genetic stock collections. Such collections often contain genetically well defined strains of microorganisms for particular traits (for example, in the case of wheat rusts, virulence against specific crop cultivars). However, they are perhaps more akin in composition and use to cultivar collections for crop plants rather than genetic stock collections. On the other hand, in terms of problems of maintenance and long-term security, they have more in common with genetic stock collections.

Most genetic stock collections have been developed and maintained in the industrialized countries of the north, which have a history of strong scientific research programs in genetics, biology, and medicine. However, even in these developed countries, genetic stock collections have not received the recognition they deserve and frequently lack adequate financial support. Consequently, many such collections face problems of management, availability of information and materials, and long-term security of conserved stocks.

Development and Location

Genetic stocks have been developed by scientists as aids to basic research, therefore, they therefore differ dramatically in origin from

TABLE 9-1   Agriculturally Relevant Genetic and Microbial Stock
Collections

Agricultural Group	Summary of Genetic Stock Collections
Major agricultural crops	Maize, barley, tomatoes, wheat, peas (some small genetic stock collections also exist for many others).
Minor crops, pasture and ornamental plants	Includes some horticultural and ornamental species with an impressive array of mutant forms governing traits such as flower color, flower shape, and growth habits (for example, in the sweet pea, *Lathyrus odoratus, Antirrhinum majus,* and *Petunia hybrida*). Most of these, however, have been isolated and retained for their direct economic importance rather than for their value in genetic studies.
Domesticated livestock	Genetic stock collections have not developed for domesticated animals to the same degree as they have for crop plants. In general, the collections are relatively small, and they are usually closely integrated with germplasm maintenance programs.
Economically important aquatic animals	Significant genetic stock collections have not developed for fish or other aquatic animals.
Economically important insects	Modest genetic stock collections have developed for several species, such as mosquitoes, Mediterranean fruit fly, sirex wasp, and the sheep blowfly, and especially in the case of agricultural pests for which attempts have been made to reduce the population size of the pest below the economic threshold level by using lethal genes or chromosomes or by releasing irradiated males.
Beneficial and harmful microorganisms of animals	Significant laboratory collections of animal pathogens and microbial flora of the digestive tract have been developed worldwide for diagnostic and animal research purposes.
Beneficial and harmful microorganisms of plants	Significant collections of many of these organisms are maintained worldwide because of their great economic importance and the availability of appropriate technology for the long-term maintenance of microbial collections.

NOTE: No substantial genetic stock collections have been developed for forestry and
fuelwood species.

the usual components of germplasm collections, which principally come from collecting missions or applied breeding programs. The development of a major genetic stock collection usually takes many decades and involves research and publication of results by many individual scientists. Initially, in the evolution of such collections, individual scientists develop and maintain particular stocks or classes of stocks of direct relevance to their personal research.

Development of Genetic Stock Collections

As investigators require the same or similar genetic materials for research, certain researchers or institutions usually become recognized as the primary sources of those stocks. If demand continues to grow, such individuals or institutions may develop into nationally or internationally recognized genetic stock centers. However, once the demand for genetic stocks has grown past the point at which individual researchers are willing or able to maintain sufficiently well-characterized material to meet the demand, efforts are usually made to develop a formal genetic stock center. One approach is to develop a center at one institution that handles all stocks of a species. An alternative approach is to develop a completely decentralized system in which scientists from many institutions agree to maintain and supply a designated class of genetic stocks.

Since genetic stocks are principally developed and maintained by research scientists, most are held in universities or research institutes. This is in contrast to germplasm collections for crop plants, which are usually maintained by national governments because of their strategic importance to national economics.

Maintaining Genetic Stock Collections

The fact that genetic stock collections are largely maintained by individuals within universities and research institutes has important implications in terms of continuity of funding and long-term security. Initially stock collections are usually funded as part of a research project, but once they grow in size, funds for maintaining the collection must compete with other aspects of the research program. Even when collections achieve national or international status, financial support may not be assured, because maintenance of collections is usually regarded by research funding bodies to be mundane and unproductive, in terms of publishable research results, and of low priority.

The maintenance of genetic stock collections by part-time curators in research institutes, where they must compete with other activities for funds, is seen by some to have significant advantages. It is impossible to conserve all genetic variants of any species. The present system allows flexibility and forces the genetic stock collections to evolve in response to the needs of the user community. Under this view, the development of a more centralized system, in terms of funding and management, carries with it the danger of developing large collections with mainly historic entries of limited value to current researchers. The disadvantage of this view is that stocks of little interest today may become extremely important tomorrow. For example, the transposable element stocks from the work of Barbara McClintock were mere historical curiosities in the late 1950s, but they have become the center of attention for gene isolation and cloning today.

Components of Collections

Many different types of genetic stocks are used in biologic and agricultural research. Genetic stocks can be conveniently divided into three groups: single-gene or single-trait variants, cytogenetic stocks, and other genetic stocks.

Single-Gene or Single-Trait Variants

In the group of single-gene or single-trait variants, there are three subgroups. The first is morphologic and physiologic variants. Although the importance of some of these stocks may have been reduced with advances in modern molecular genetics, many are still of value as genetic markers and in studies of gene function—more valuable today for molecular research than they are for classical genetic studies.

The second subgroup is electrophoretically detectable protein variants. The use of electrophoretically detectable variants at genetic loci governing the production of specific proteins has increased dramatically over the past two decades in many spheres of genetic research. Consequently, genetic stocks carrying well characterized variants have become an increasingly important component of genetic stock collections.

The third subgroup is made up of variants with different RFLPs. The use of RFLPs as genetic tools in applied agricultural research will undoubtedly increase markedly in the near future. The identification and maintenance of genetic stocks carrying known genetic markers as well as the probes (DNA segments) used to detect them will become priority tasks.

Cytogenetic Stocks

There are three types of cytogenetic stocks. The first is variants with different chromosome structures. Variants whose chromosome structures differ from the norm because of deletion, duplication, inversion, or translocation of specific DNA segments from one chromosome to another are important in gene mapping, determination of the chromosomal location of specific genes, and studies of chromosome structure and function.

The second type is variants in chromosome number. These may include, particularly in plants, changes in whole genomes (haploids or polyploids) or changes in the number of a single chromosome, chromosome arm, or chromosome segments (nullisomics, monosomics, telosomics, trisomics, or tetrasomics). Such variants are also valuable in genetic mapping and in basic genetic studies, such as in the study of genetic architecture of chromosomes.

The third type of cytogenetic stock is alien addition and substitution lines. The wild and weedy relatives of the major crops are sometimes used as germplasm sources in breeding programs. A first step is often the transfer of individual chromosomes, or chromosome segments, carrying identified genes of potential economic importance into the crop species. The three types of alien chromosome segments used to transfer the desired gene into commercial crop varieties are addition lines, substitution lines, and translocation lines.

Other Genetic Stocks

There are also three types of other genetic stocks. One type is multiple gene stocks in which a particular chromosome carries multiple genetic markers that are particularly useful in linkage studies.

A second type is near isogenic lines that carry alternative alleles at specified loci. In near isogenic lines, alternative alleles at a locus are transferred by repeated backcrossing into the same genetic background. Such lines have been used principally in studies of the comparative effects of different alleles in similar backgrounds, but more recently, they have attracted increased attention from scientists interested in identifying and isolating genes and gene products.

Wild relatives of crop plants are a third type. Wild relatives of crop plants that are difficult to grow or manage within the framework of a normal cultivar collection or germplasm bank may be more readily maintained in genetic stock collections. These stocks need special management because seeds scatter at maturity. Most wild relatives of wheat, barley, and maize belong to this group. Other examples are the wild species of *Solanum* and *Lycopersicon* held in the tomato genetics stock collections.

Breeding Lines Versus Genetic Stock Collections

A central theme of the FAO International Undertaking on Plant Genetic Resources (see Chapter 14) is that all plant germplasm should be available for exchange without restriction. Many countries, especially those that have significant privately funded plant breeding programs dependent on some form of proprietary protection such as patenting, are unable or unwilling to guarantee the free availability of privately held germplasm.

This controversy has focused on genetic stocks. The reason is that FAO's undertaking defined plant genetic resources as the reproductive or vegetative propagating material of the following categories of plants: (1) cultivated varieties (cultivars) in current use and newly developed varieties; (2) obsolete cultivars; (3) primitive cultivars (landraces); (4) wild and weedy species, which are near relatives of cultivated varieties; and (5) special genetic stocks (including elite and breeders' advanced lines).

This system of classification was developed in the mid-1960s (Frankel and Bennett, 1970) and emphasized crop cultivars and wild and weedy relatives of crops because of their importance in plant introduction and breeding. The category of special genetic stocks (category 5 above) includes all plant germplasm not included in the other categories, hence it includes genetic stocks and breeders' advanced lines. Genetic stocks and breeders' lines have several features in common that underlie their inclusion in category 5. For example, both are usually developed by individuals, often at considerable expense and personal effort. During development, the individual may choose to maintain exclusive control of such germplasm. Moreover, neither genetic stocks nor advanced breeders' lines are commonly held in most germplasm collections. Their status differs substantially from those of released cultivars or wild and weedy relatives, in that for the germplasm of wild or weedy relatives there is either no single identifiable originator or the originator, in seeking public release, has voluntarily forfeited exclusive control, at least for further breeding and scientific research purposes.

There are substantial differences between advanced breeders' lines and genetic stocks, however. Advanced breeders' lines are the elite materials from which a breeder selects new cultivars for commercial release. Consequently, they are often of substantial economic value, which is why their control and availability have generated controversy. Genetic stocks, on the other hand, are seldom of direct economic importance, and their value lies in their role in scientific research. Except in their early days of development, genetic stocks are usually held in the public domain and are freely available (usually in very small quantities) on request. As defined here, genetic stock collections do not include advanced breeders' lines. Rather, they contain genetically defined stocks principally of interest to researchers rather than to plant breeders.

MAINTAINING GENETIC STOCK COLLECTIONS OF AGRICULTURAL CROPS

The conservation of genetic stock collections requires technical and scientific inputs and physical resources broadly similar to those required for maintaining germplasm collections or agricultural crops. In the case of seed crops, for example, these broad requirements include the following:

- Suitable low-temperature, low-humidity facilities;
- Access to back-up storage facilities;
- Adequate seed-handling facilities for drying, cleaning, packaging, and viability testing of seed samples; and
- Computerized information storage and retrieval systems.

Special Needs and Requirements

Genetic stocks require highly specialized knowledge, procedures, and care for adequate regeneration and maintenance. For example, many stocks are weak, semilethal, or completely or partially sterile in homozygous condition, and in practice, they must be maintained in a heterozygous condition. Albino mutants are one obvious example. Some stocks may not routinely survive in the field and need to be grown in a greenhouse, whereas others may need special treatments (long days or vernalizing temperatures) before they flower or may require genetic or cytological screening. Genetic stock collections are difficult, and sometimes impossible, to maintain as part of national or international germplasm collections.

Curators of germplasm collections are reluctant to accept responsibility for genetic stocks if they lack the skills or resources to provide the specialized maintenance procedures required. Offers of additional training, staff, or resources seldom accompany requests for germplasm collections to take responsibility for the maintenance of genetic stocks. Similarly, users of genetic stocks are reluctant to have such materials incorporated into germplasm collections, because experience has indicated that they quickly become contaminated or lost.

The maintenance of genetic stock collections by active researchers overcomes the problem of the need for specialized skills in the regeneration of genetic stocks but might create other difficulties with respect to long-term conservation. Institutes that take responsibility for or house genetic stock collections often lack suitable medium- to long-term, low-temperature seed storage and related drying, packaging, and computing facilities routinely available in first-class germplasm banks. The curators of genetic stock collections that lack such facilities are forced to regenerate their stocks repeatedly, despite the difficulties and dangers that this implies, to avoid the loss of viability. In such situations, the total number of stocks that can be effectively maintained by a single curator is relatively small, and the cost per accession is relatively high.

The major source of funds available for maintaining genetic stock collections in universities and research institutes is competitive research grants. This support is vulnerable to short-term changes in funding policies as well as changing perceptions of what constitutes competitive front-line research. Indeed, the maintenance of genetic stocks is already regarded by some funding bodies as a low-priority area that should be funded on a long-term basis by appropriate government departments.

Commitment to the maintenance of genetic stocks in universities or research institutes is usually made by the individual researcher, not by the university or institute as a whole. Consequently, the continuation of collections may be jeopardized if the research interests and commitment of the curator change or the institution changes the responsibilities of the position if the current curator resigns, retires, or dies.

Genetic stock collections have periodically come under serious threat of disruption, loss, or closure. Many important genetic stock collections have been maintained indirectly through support for other areas of research and by agencies not charged with the duty of their maintenance. In view of the importance of genetic stock collections in underpinning genetic and biotechnological research and the investment that has gone into the development of such collections, a number of actions have been taken to put at least some of them on a firmer long-term footing.

In the United States, the Genetics Society of America established the Committee for the Maintenance of Genetic Stocks (CMGS) in 1958. This committee served previously to alert geneticists to situations in which valuable collections of stocks were endangered by the lack of adequate support or supervision. In the recent past, however, the CMGS has sought to play a greater role in developing objective criteria to be considered in determining which genetic stock collections warrant priority support and in promoting coordinated national support for priority centers.

The U.S. National Plant Genetic Resources Board has also considered the issue of long-term funding for genetic stock collections and has supported the concept that the U.S. Department of Agriculture (USDA) should be the agency that assumes the lead role and responsibility for the maintenance of genetic stocks of commercially important plants as part of the U.S. National Plant Germplasm System. USDA has provided full or partial support for several important genetic stock centers, for example, those for soybeans, peanuts, and tobacco; and in recent years, it has provided partial support for several others, including barley, maize, and tomatoes, as part of the U.S. National Plant Germplasm System (NPGS). Largely as a result of the efforts of USDA, it can be argued that, at least with respect to the major economic plants, genetic stock collections in the United States are as well maintained as they are anywhere in the world. Discussion of the role of the NPGS in maintaining important U.S. collections of genetic stocks can be found in an earlier report of this committee (National Research Council, 1991a). Nevertheless, significant problems of continuity and adequacy of funding and the management of genetic stocks of these species remain.

Two examples follow to illustrate the types of problems confronted by genetic stock collections.

EXAMPLES OF GENETIC STOCK COLLECTIONS

Management of genetic stocks collections presents a variety of technical, managerial, and financial challenges. The following profiles of two large genetic stock collections illustrate the need for competent, technically sound management, and the importance of cooperative efforts from the broad scientific community.

The Charles M. Rick Tomato Genetics Resource Center

This center, referred to here as the Tomato Genetics Stock Center (TGSC), was originated and developed by Charles M. Rick of the Department of Vegetable Crops at the University of California, Davis, principally as a by-product of his research on the genetics, evolution, and breeding of this crop. Davis is near the center of the largest tomato-growing area in the world, and in 1987 nearly one-quarter of the world's processing tomatoes were grown in eight counties within a 129-km radius of Davis. As a result, tomato genetics research and improvement has been carried out at the University of California, Davis, and in nearby private farms over many years to meet the needs of this large and economically important industry.

The TGSC arose from Rick's early investigations of chromosomal and genetic variants found in spontaneous unfruitful plants in commercial crops. The early surveys of commercial crops yielded a wide range of stocks such as haploids, triploids, and tetraploids as well as sterile diploid mutants that included many different monogenic mutants. This research subsequently led to a program of linkage mapping that generated a large array of additional mutants, linkage testers, and other stocks in which various markers were combined. In addition to collecting spontaneous mutants, genetic variants were generated by chemical mutagens, X rays, and fast neutron bombardment. Over time, the collection was also enriched by acquisitions of valuable genetic stocks from other research centers in the United States and other countries (Rick, 1987).

The accessions of wild taxa, the second major group in the TGSC, can also be traced largely to Rick's research activities. These include samples of the nine known species of *Lycopersicon* and four closely allied species of the genus *Solanum*. Rick's first trip to South America, where the tomato and its relatives are native, was in 1948. Since that time he has made 12 major expeditions to various parts of the world to collect wild germplasm. Other investigators have contributed accessions from their plant exploration and collection activities.

The TGSC collection has about 2,900 accessions. The principal components include some 700 lines carrying monogenic traits such as morphological variants, male sterility, resistance to pests and diseases, 800 linkage testers, chromosomal stocks and other miscellaneous lines, and 1,000 accessions of wild species.

For most of its existence, TGSC was supported largely by its host institution, the University of California, as part of its overall research and teaching program. Additional support came from grants from the National Institutes of Health and the National Science Foundation for research on the acquisition and maintenance of relevant genetic stocks. As TGSC increased in size additional resources were required. Initially, funds were sought from the National Science Foundation, and a grant of $96,500 was awarded for the 3-year period from 1976 to 1979. A second grant of $109,369 was received from the National Science Foundation for the period from 1979 to 1982. Subsequently, support was sought from the Agricultural Research Service (ARS), U.S. Department of Agriculture, which has provided an average of $40,000 per year from 1982 to the present. Beginning in 1985, the Genetic Resources Conservation Program of the University of California has provided an average of $6,700 per year. An endowment, donated primarily by industry, supplies additional funds.

Operations of the TGSC

The TGSC is responsible for the maintenance, regeneration and increase, dissemination, cataloging, and data management of the accessions it holds.

Maintenance of Stocks   All accessions are stored as seed at 8°C under 35 percent relative humidity. Entries are maintained in a working collection and in a duplicate collection maintained under the same conditions as a conservation measure. While the longevity of the accessions varies with species and genotype under these conditions, most can be stored for 20 to 25 years. As a further safeguard, reserve samples are also stored at the National Seed Storage Laboratory (NSSL) of the ARS at Ft. Collins, Colorado. Replacements and additions to accessions already in the NSSL are made each year.

Seed Increase   About 300 stocks, on average, must be increased each year. There is no uniformly applicable procedure for seed increase because of the genetically diverse nature of the lines in the collection. For wild species it is TGSC policy to make a large seed increase of each new accession as soon as possible after collection or accessioning. In this way unnecessary regeneration with the attendant risk of altering the original genetic composition of the accession is avoided.

Dissemination of Seed   Over a recent 5-year period, about 3,000 seed samples were distributed annually in response to about 225 requests from 175 scientists. Almost half of the requests came from scientists outside the United States. TGSC supplies seed free of charge and attempts to meet all requests for stocks. With the growth of molecular genetics and its potential impact on plant improvement, demand for genetic stocks will probably increase.

Accessioning of Stocks   It is the practice of TGSC to maintain all stocks of bona fide aneuploids and mutants, that is, those for which convincing data have been published. Many compound-mutant stocks, developed for such purposes as gene mapping, are likewise accessioned. From 1983 to 1987, about 90 new stocks were added to the collection each year, of which about 55 percent came from outside the University of California, Davis, program. It is expected that the collection will continue to expand.

Data Management The growing number of accessions and the increasingly detailed information available on each accession means that data management is a substantial and growing problem for TGSC. TGSC currently lacks a computerized information storage and retrieval system. Passport data are lacking and inadequate for the TGSC accessions stored at NSSL, and the TGSC collection inventory has not been entered into the Genetic Resources Information Network (GRIN) of the U.S. National Plant Germplasm System. A computerized data base compatible with GRIN is planned to service the management of TGSC and to make the updating of GRIN more efficient. It will also allow wider and easier access to the TGSC inventory by users.

Future Operations

The TGSC collections, like many other genetic stock collections, were developed primarily from the research of one person, C. M. Rick. Over the past 40 years the collection has evolved from a small number of accessions primarily used in Rick's personal research program to a very substantial number of accessions that are collectively of national and international importance. During this evolutionary process the TGSC changed radically in its responsibilities, facilities, and size. The evolution of TGSC was relatively smooth because of Rick's standing in both the scientific and industrial communities and the support he has received from the University of California and other sources.

Nevertheless, the TGSC faces ongoing problems. These include lack of long-term financial support, an ill-defined curator succession policy, and inadequate physical facilities.

A task force involving members of the University of California Department of Vegetable Crops, the University of California Genetic Resources Program, and private companies concerned with tomato improvement was established to review the present and future funding, personnel, and space needs of TGSC. Based on this review, the task force sought to outline a suitable management structure, to identify sources of stable long-term funding, and define optimum facilities for TGSC. The report of the task force has been published (Genetic Resources Conservation Program, 1988).

Barley Genetic Stock Collections

Barley geneticists and breeders have established an internationally coordinated network of centers for the maintenance of barley genetic stocks. Under this scheme, individuals are appointed to co-ordinate maintenance of each of the different categories of genetic and chromosomal variants commonly found in barley. As of this writing, coordinators have been appointed for 17 different classes of genetic stocks.

The coordinators' responsibilities include the maintenance of as complete a collection of their particular type of stock as possible. They also supply seeds to other researchers on request, compile up-to-date information on the stocks in their collection, and prepare an annual report for the *Barley Genetics Newsletter*.

Individuals are also appointed to coordinate work on each of the seven barley chromosomes. These coordinators are responsible for updating the linkage map of their designated chromosome using pub-lished data, information provided directly by barley researchers, or data from their own studies. The revised linkage maps are also pub-lished regularly in the *Barley Genetics Newsletter*.

In addition to the coordinators for the different genetic stocks and barley chromosomes, there is an overall network coordinator.

The development of a dispersed, but coordinated, network of ge-netic stock collections for barley occurred because of the widespread economic importance of this crop and because of the long-standing genetic and breeding programs on barley that are in place in many countries. Barley is an excellent subject for genetic studies because it is an annual plant, predominantly self-pollinated, with a small num-ber of chromosomes ($N = 7$), and has a large number of easily classifi-able marker loci. Thus, substantial genetic stock collections were established in different research institutes throughout the world. As the collections grew in size, efforts were made to increase coopera-tion among them to minimize duplication of effort and to ensure that those with a special interest and expertise in any given group of genetic stocks were closely involved in their maintenance. The present system was formally put in place by the Committee for Nomencla-ture and Symbolization of Barley Genes at the Second International Barley Genetics Symposium held in Pullman, Washington, in August 1969 (Robertson and von Wettstein, 1971).

It is not possible to consider here all the elements of the interna-tional system for the maintenance of barley genetic stocks. Conse-quently, this discussion is limited to the collection held by the De-partment of Agronomy, Colorado State University, Ft. Collins. This collection, designated the World Barley Genetic Stock Center by the First International Barley Genetics Symposium, Wageningen, The Neth-erlands, in 1963, plays an important role in coordinating the mainte-nance of barley genetic stocks in the international network.

The Barley Genetic and Aneuploid Stock Collection

The Barley Genetic and Aneuploid Stock Collection (BGASC) in the Agronomy Department at the Colorado State University, Ft. Collins, was established by D. W. Robertson and his colleagues as a resource for genetic linkage studies in barley. These studies began in the 1920s and emphasized simple morphologic marker genes of value in early linkage studies. Over the years the collection was enriched by the addition of a wide range of naturally and artificially generated variants, linkage testers, and more recently, of cytogenetic stocks from the local program and others in the United States and overseas. Unlike TGSC, BGASC does not maintain stocks of wild relatives. These are maintained as part of the USDA barley germplasm collection.

The size of BGASC has varied over the years as new materials are sent to the collection, tested for allelism, and discarded or accepted into the collection. There are now over 3,000 accessions in the collection.

Throughout its existence, BGASC has been supported principally by Colorado State University, Ft. Collins. Additional support came from grants from the National Science Foundation from 1961 to 1979. USDA has provided financial support since 1979 to ensure the continued operation of the center.

The center also received partial outside support for the publication of the *Barley Genetics Newsletter* as an informal international communication medium in barley genetics. Initially, in 1971, three European countries, Denmark, the Federal Republic of Germany, and Sweden, supported 50 percent of the cost of publishing and distributing this newsletter. However, this arrangement was terminated in 1986. Consequently, in 1987 the American Malting Barley Association took responsibility for the publication and distribution of the *Barley Genetics Newsletter*.

Operations of the Barley Genetic and Aneuploid Stock Collection

The BGASC is responsible for the maintenance, regeneration, dissemination, accessioning, and data management of its materials. As the designated world collection, it seeks to maintain a sample of all accessions held in the network.

Maintenance of Stocks All accessions are maintained in a working collection at Ft. Collins, Colorado. As a safeguard, additional samples of some stocks are also stored in the National Seed Storage Laboratory, USDA, which is located on the Ft. Collins campus of Colorado State University. A program is also under way to duplicate the entire collection.

Some 300 genetic stocks that have been well studied genetically are considered to be active stocks. These are frequently requested by researchers throughout the world and are regenerated more often to maintain enough seed for distribution.

The list of these stocks is published in most of the issues of *Barley Genetics Newsletter* and *Genetic Maps* (National Institutes of Health, Bethesda, Maryland).

Seed Increase   Much of the collection was increased during the period from 1969 to 1971. Stocks have been increased since then according to demand. However, recently a systematic growing out of the entire collection was begun. The stocks are mainly grown in the USDA-ARS greenhouse in Ft. Collins. About 300 to 500 stocks are grown each year, and a sample of all seed is supplied to the National Seed Storage Laboratory for long-term conservation.

Trisomic and other aneuploid stocks are routinely increased as part of the research program. In the case of trisomic stocks, chromosome counts are made for every plant, because only approximately 30 percent of the seeds harvested from trisomic plants are trisomics; the rest are normal diploids.

Accessioning of Stocks   BGASC maintains stocks of at least two alleles of all defined loci as well as aneuploid stocks and stocks with multiple genetic markers useful in gene mapping. Although it would be desirable to maintain stocks of all reported variants, the numbers are so large that it is impossible. Because of limited resources, the policy of the BGASC is to maintain only one mutant or variant stock per locus. Allelism tests are carried out to identify those stocks to be retained in the collection. With some exceptions, no stocks are accepted at present because of uncertainty about the long-term future of BGASC.

Data Management   Updated lists of the stocks held by BGASC are published periodically in the *Barley Genetic Newsletter* and *Genetic Maps*. The center encourages geneticists who identify new mutants to publish detailed descriptions, according to a standard format, in the *Barley Genetic Newsletter*. This ensures that an adequate description is readily available and frees the BGASC from unnecessary record keeping. However, if new mutants are not acquired by the BGASC, they are unlikely to be maintained other than by the people describing them.

No description has been published for the majority of stocks maintained at the center. Some information was obtained when an entire collection was grown during the period from 1969 to 1971. Detailed records are again being acquired during the current program of stock regeneration and increase. These records may be added to the GRIN network. Similarly, if funds are available, these data could be published as a special issue of *Barley Genetics Newsletter*, together with a listing of the complete collection.

Future Operations

Like TGSC, the BGASC has grown in size and scope over the last half century. It now represents a resource of considerable national and international importance and plays a pivotal role in coordinating the international network of genetic collections for barley. BGASC has also faced a major funding crisis in recent years because of the loss of National Science Foundation funding and the loss of European support for the *Barley Genetics Newsletter*. It has been able to arrange alternative funding from the USDA and private industry, respectively, so prospects for the immediate future are promising.

TOWARD A MORE SECURE FUTURE

Genetic stock collections are an important complement to genetic resources collections. Genetic stocks are essential to an understanding of the genetic structure of a species, and this basic knowledge is of importance in enhancing the effectiveness of plant improvement programs. As a consequence, it is likely that demand for well-characterized genetic stocks from geneticists, biochemists, physiologists, and biotechnologists will increase in the future.

RECOMMENDATIONS

The requirements for conserving genetic stocks and germplasm collections are similar. Nevertheless, the maintenance of genetic stocks poses a number of unique problems because they are often difficult to regenerate and frequently require specialized expertise and extraordinary care in terms of their genetic identity and characterization. As a consequence, the maintenance of genetic stocks and regular germplasm collections have usually followed separate paths.

Important collections of genetic stocks must be provided a secure base of financial and technical support.

Germplasm collections have generally been maintained by the major public institutions involved in crop improvement. In contrast, the maintenance of genetic stocks, even those of the major crops, were undertaken on a voluntary basis by individual scientists or groups of cooperating scientists, usually with the support of research institutes or competitive research grants. This system has generally worked well. There are, however, major problems, particularly with respect to the continuity of financial support, data storage and management, and the relationship between genetic stock and germplasm collections.

It needs to be emphasized, however, that the changes made should not be at the expense of flexibility. The prime function of genetic stock collections is to provide materials for research. Therefore, the composition of the collection must evolve with research needs. The changes also should not be at the expense of stock quality which is critical for reliable research results. To accomplish this, management of genetic stock collections must continue to be the responsibility of specialist scientists.

For many genetic stock collections, the provision of dependable long-term funding remains the single most important issue threatening their future security. The situation in the United States has improved in recent years, principally because an increasing number of genetic stock centers have been brought under the umbrella of USDA, as illustrated above for TGSC and BGASC.

Directors of germplasm centers often do not recognize the length of time and long-term apprenticeship necessary to become familiar with genetic stocks. Even reasonably complete documentation cannot cover the intricacies of growing, crossing, and maintaining the temperamental stocks that are found in genetic stock collections. Thus, the costs of arranging transfers of genetic stocks to germplasm centers are almost always greatly underestimated in terms of both time and effort. Only together can genetic stock and germplasm collections claim to represent the total spectrum of genetic variation in a species, and while there are good and valid reasons for their separate existence, there are no good reasons for the lack of cooperation and coordination that would enhance support for both types of collections.

With better long-term funding, improved data management facilities, and greater coordination with germplasm banks, genetic resources centers would provide genetic stock collections with the stability and security they need to meet the challenge of the rapid advances in molecular genetics likely to occur into the twenty-first century.

Major collections of genetic stocks should be overseen by an advisory board of qualified experts in genetics, breeding, and seed storage.

There are problems in deciding which collections deserve priority support as national or international resources. The important factors in reaching such decisions are the extent of and potential for use of the collection, the quality of material supplied to the users, and the costs of its maintenance. Advisory boards can help with policy development in the areas of stock acquisition, discarding unnecessary or redundant materials, and coordination with other genetic stock or genetic resources collections. Such boards can also play an important role in developing a succession policy for the center that can be put in place, allowing it to continue to operate effectively when key professional staff leave or retire.

Storage and retrieval of data for genetic stocks must be computerized.

Data collection, management, and distribution also remain problem areas for genetic stock collections. The major genetic stock centers generally produce periodic newsletters, one of the functions of which is to provide users with up-to-date information on the status of the collections. However, many centers have been slow to computerize storage and retrieval systems for data management. Modern facilities could lead to substantial improvements in access to information and stocks. At genetic resources centers, computer software packages have been developed and are in use internationally.

Coordination of the activities of genetic resources centers and genetic stock collection have often been poor. The lack of duplicate samples deposited to long-term seed stores remains a major deficiency. Documentation is almost always inadequate.

10

The Genetic Resources of Microorganisms

Microorganisms generally receive scant or no attention in overall reviews of biological diversity and global genetic resources, perhaps because they are often studied by different methodologies and scientists based in laboratories rather than herbaria, museums, botanic gardens, or germplasm banks (Cronk et al., 1988; Office of Technology Assessment, 1987a; Plucknett et al., 1987; Wilson and Peter, 1988; Wolf, 1987). This is disproportionate to the key roles microorganisms play in the biosphere and is despite the extent to which they are already exploited commercially. Similarly, the genetic resources conservation movement has tended to underemphasize microorganisms, perhaps partly because of the common misconception that these are ubiquitous and so do not merit consideration in a conservation context. In ecosystems, microorganisms are important as symbionts (endophytes, mycorrhizae, and in insect guts), in nitrogen fixation (rhizobia, cyanobacteria, cyanobacteria-containing lichens), in the biodegradation of dead animal and plant material, and in controlling the size of populations of plants and insects through natural biocontrol.

Microorganisms are not readily circumscribed. They include algae, bacteria (including cyanobacteria), fungi (including yeasts), certain protistan groups, tissue cultures, and viruses. Tissue cultures and animal and plant cell lines are not discussed here.

This chapter assesses the extent of the maintained gene pool of microorganisms and focuses on the problems of preserving genetic stability in the long term and constraints to the development and enhancement of microorganisms. The actions needed to safeguard this massive resource so that future generations can investigate its utility are also considered.

ORGANIZATION OF MICROBIAL CULTURE COLLECTIONS

The World Federation of Culture Collections (WFCC) is the key international coordinating body on the culture collections of microorganisms. This organization, established in 1970, is recognized by both the International Union of Biological Sciences and the International Union of Microbiological Societies and promotes liaison between individuals and organizations responsible for the maintenance and development of culture collections (Kirsop and DaSilva, 1988). A series of handbooks describing the resources available to culture collections has been initiated by WFCC; volumes concerned with filamentous fungi (Hawksworth and Kirsop, 1988) and yeasts (Kirsop and Kurtzman, 1988) have already been issued and others covering bacteria and animal cell lines are in preparation.

The World Data Center for Microorganisms (WDC), now the responsibility of WFCC (Komagata, 1987), produced the first *World Directory of Collections of Cultures of Microorganisms* in 1972; the third edition of this directory (Staines et al., 1986) includes information on the names of organisms in 327 culture collections distributed throughout 56 countries. This information is complemented by the Microbial Strain Data Network (MSDN), sponsored by the Committee on Data for Science and Technology, WFCC, and the International Union of Microbiological Societies (Kirsop, 1988a). The MSDN was established in 1985 to provide on-line information on the data elements coded for strains by individual collections using a controlled vocabulary (Rogosa et al., 1986); it is accessed primarily by electronic mail.

The United Nations Educational, Scientific, and Cultural Organization started to identify a series of microbial resources centers (MIRCENs) in 1974; (DaSilva et al., 1977; Kirsop and DaSilva, 1988; Zedan and Olembo, 1988). Sixteen MIRCENs are currently recognized worldwide and are expected to provide the infrastructure for an international network geared to the management, distribution, and utilization of the world's microbial gene pool.

The European Culture Collections Organization, founded in 1982, has provided synopses of the resources available in Europe and arranges annual conferences. The Commission of the European Community commenced an ambitious program to establish a Microbial Information Network Europe (MINE) in 1985. Culture collections in Belgium, France, Greece, Germany, Italy, The Netherlands, Portugal,

Spain, and the United Kingdom enter data in a common format (Gams et al., 1988) through national centers to produce an integrated machine-readable catalog. In the United Kingdom, more extensive strain data, including biochemical and physiologic characteristics, have been compiled into the on-line Microbial Information Service (MiCIS); MiCIS and MINE are expected to become fully integrated. The objectives and key features of these systems have been compared by Allsopp et al. (1989). For the Scandinavian countries, the Nordic Register of Culture Collections was initiated in 1984, but this is not yet generally accessible.

Nationally, culture collection liaison is, in some cases, coordinated by Federations of Culture Collections, as in Brazil, Japan, the United Kingdom, and the United States. These activities have been reviewed by Krichevsky et al. (1988). Only in the case of Brazil, however, has a single integrated national catalog been produced (Canhos et al., 1986).

Individual culture collections can be divided into four main, but not mutually exclusive, categories.

• *Service collections*, which have as their primary objective the supply of authenticated cultures to all who request them. These are generally listed in publicly available catalogs; such collections are invariably directly or indirectly supported by government funds and supply cultures at less than full economic cost.

• *In-house collections*, which are established to meet the requirements of particular organizations, institutions, or individual companies. Collections of this type can be very substantial, but catalogs are not generally available and cultures are supplied on a discretionary basis.

• *Research collections*, which are built up by individual scientists or teams as a part of their research programs. Such collections are often unique resources, because they include novel and unusual strains in restricted groups, but long-term storage facilities are rarely adequate and resources permit cultures to be made available only to close colleagues. Research collections are often endangered or lost when individual scientists change positions, retire, or pursue different lines of research.

• *Laboratory suppliers*, which make available limited numbers of organisms, generally single strains of species, which are commonly used in teaching or research. Prices are generally below those that service collections are obliged to levy.

The funding arrangements for culture collections vary markedly, but apart from some service collections sponsored by national governments or international coalitions of collections, these are rarely adequate or sufficiently long-term to enable collections to optimize their value as a resource to the scientific community. Further information on particular collections and their services has been compiled by Hawksworth (1985), Hawksworth and Kirsop (1988), Kirsop and Kurtzman (1988), Malik and Claus (1987), and Staines et al. (1986).

MICROBIAL RESOURCES IN CULTURE COLLECTIONS

The only overview of the microbial resource maintained in culture collections worldwide is that found in the *World Directory* (Staines et al., 1986). Although this compilation cannot be considered comprehensive, particularly with regard to in-house and research collections, it includes all major service culture collections and does provide a reasonable approximation to the numbers of species of microorganisms available publicly. Because this catalog accepts whatever name is used for a particular organism by each collection (but cross-references synonyms), an allowance must also be made for synonymy in interpreting this compilation. Table 10-1, which was constructed to allow for this factor, indicates that about 18,500 species of microorganisms are currently available from culture collections.

Species numbers alone do not, however, provide an adequate representation of the gene pool, in view of the considerable range of genomic variation known to occur even within single species. The number of microorganisms in which the genetic diversity has been analyzed in depth and experimentally is small. In those cases in which this has been accomplished, the numbers of genotypes distinguished can be substantial. For example, in the alga *Chlamydomonas reinhardtii*, 159 mutant lines are known (Harris, 1984; Harris et al., 1987); in the filamentous fungus *Neurospora crassa*, this figure is about 3,000 (Fungal Genetics Stock Center, 1988); in the yeast *Saccharomyces cerevisiae*, about 850 genetic strains are maintained at the Yeast Genetic Stock Center (Kirsop, 1988b); numerous special forms and races of plant pathogenic fungal species are kept at the collection of fungal pathotypes of the Institut voor Plantenziektenkundig Ondersoek in Wageningen, The Netherlands; 6,000 strains of *Escherichia coli* with various genetic markers are maintained at the *E. coli* Genetic Stock Center (Bachmann, 1988), and some 1,500 serological types of *Salmonella* are maintained at the World Health Organization's International Salmonella Center (Staines et al., 1986). Nitrogen-fixing bacteria of the genus *Rhizobium* have been the focus of many collections; 73 collections in 38 countries have significant strains of the six species of this genus from about 220 different host legumes (McGowan and Skerman, 1986). Furthermore, many filamentous fungi (for example,

TABLE 10-1 Numbers of Species of Microorganisms Maintained in Culture Collections Compared with the Numbers Described and Probable World Species Totals

Group	Species in Culture Collections[a]	Described Species	Estimates of Total World Species	Proportion of Species Maintained, in Percent	
				Described	World Estimate
Algae	1,600	40,000[c]	60,000	4	2.5
Bacteria (including cyanobacteria)	2,200[b]	3,000	30,000	73	7
Fungi (including lichenforming fungi and yeasts	11,500	64,200[d]	800,000[g]	18	1.5
Viruses (including plasmids and phages)	2,900	5,000[e]	130,000	58	2
Protoctists (including protozoa, but excluding algae and "fungal" protoctists)	300	30,800[f]	100,000	1	0.3
Total	18,500	143,000	1,120,000	13	2

[a]Staines et al. (1986) rounded figures allowing for 25 percent synonymy in fungi and 10 percent in bacteria and algae.

[b]Totals for species with valid names are given here.

[c]P.C. Silva (in Hawksworth and Greuter, 1989).

[d]Hawksworth et al. (1983).

[e]700 Plant viruses (Martyn, 1968, 1971), 1,300 from insects (Martignoni and Iwai, 1981); those from other hosts are estimated.

[f]Wilson and Peter (1988).

[g]In a more recent study, Hawksworth (1991) conservatively estimates the total number of fungi species in the world as 1.5 million.

Coprinus and *Schizophyllum* species) have different, or in some cases multiple, mating type alleles, meaning that additional strains require preservation.

Most collections do not specialize to the extent of individual genera or species, but to varying degrees, they concentrate on particular systematic or biological groups. Such specialization is usually in response to the requirements of the funding bodies of the collections or the particular research interests of the staff or institutions where they are located. Most also have strains predominantly isolated from the geographical area in which they are located.

In view of these factors and bearing in mind the extent of infraspecific genetic variability that may well occur in many species, it is prudent to consider the existing culture collections as being essentially complementary rather than duplicatory. Even when strains derived from the same original isolation are retained in more than a single collection, they cannot be assumed to be genetically identical. This phenomenon results from the often inherent genotypic variability, so different genotypes may be deposited in different collections; inadvertent selection during maintenance for survival under the different preservation methods used; or contamination with other strains (Bridge and May, 1988; Bridge et al., 1986; Lawrence, 1982).

The present efforts have developed on an ad hoc basis over the past 80 years, and at present, they do not appear to be capable of adequately conserving this vital world resource. Informal and often unwritten agreements to specialize in different areas exist between collections, as their curators are generally alert to the need to maximize their combined efforts to adequately preserve the microbial gene pool. Although the WFCC, through the WDC and MSDN, compiles data on the contents and coverage of collections, it has no executive authority over collections and lacks the resources needed to encourage and enable particular collections to expand to establish new collections where gaps in the world's coverage are identified.

CONSERVING MICROBIAL DIVERSITY

To make a realistic assessment of the efficacy of the current system of culture collections in conserving microbial diversity, it is useful to obtain estimates of the numbers of species of known or potential economic importance. Microorganisms, with the exception of some larger fungi, lichen-forming fungi, and larger algae, in general lack a history of inventory production on a regional basis equivalent to those of the floras and faunas covering, for example, vascular plants, bryophytes, birds, mammals, arthropods. In addition, when surveys are carried out, only rarely is there any attempt to study systematically all the different habitats available even to a single group of microorganisms. This is an inevitable consequence of the labor intensiveness of sampling, culturing, and identifying massive numbers of microorganisms.

Numbers and Richness of Microbial Species

The numbers of species described and currently accepted in most groups of microorganisms worldwide can be estimated with some confidence from available catalogs of names (Table 10-1). However, about 120 new species of bacteria and 1,500 new species of fungi are described as being new to science each year, clearly demonstrating that knowledge of these groups is grossly inadequate. It is hazardous to extrapolate from such figures to the numbers that can reasonably be expected to occur in nature, particularly because the numbers of newly described species can be taken as an indication of the limited numbers of microbiologists actively engaged in systematic work rather than of the numbers of species to be found.

An analysis of the information presented in Table 10-1 indicates that 1,120,000 is a reasonably conservative estimate of the world's microorganism species. With respect to culture collections, this indicates that, overall, only 2 percent of the species expected to be found are currently preserved in them. For all groups of microorganisms other than bacteria and viruses, only 1 to 18 percent of the described and currently accepted species are represented in culture collections, representing a mere 0.3 to 2.5 percent of the estimated number of species actually in the world. The higher proportions for bacteria, 73 and 7 percent, respectively, may be due at least in part to an underestimate of the known but now not validly published species and a consequently low value for the estimated figure worldwide.

MAINTENANCE OF GENETIC STABILITY IN CULTURE

Preservation methods of major service culture collections that aim to conserve genetic resources must be capable of maintaining genetic stability and viability in the long term. This is of major concern not only from the conservation standpoint but also to industry, agriculture, and medicine, for which particular biochemical, pathogenic, or other attributes need to be safeguarded. The problems are particularly acute with respect to plasmids and viruses, which must be preserved within the cells of the host organism. In the case of plasmids in bacteria and yeasts and of viruses in filamentous fungi, there is a danger that the genomic material of the plasmid will be incorporated into that of the host (Chater, 1980). Strains cited in patents are deposited in specially recognized collections (Bousefield, 1988); such collections have a responsibility to maintain patent strains in an unaltered state for a minimum of 30 years (Crespi, 1985).

Traditional Preservation Methods

Traditional preservation methods involving regular subculturing of growth at low temperatures or maintenance under mineral oil are subject to inadvertent selection and contamination, especially when collections lack sufficient numbers of specialists to operate adequate quality control protocols. While growth is continuing, albeit at a much reduced rate, normal sexual processes involving changes in the genome can occur; these can include sexual and parasexual recombination, aneuploidy, and polyploidy.

Freeze-Drying and Liquid Nitrogen Preservation

In bacteria, yeasts, and filamentous fungi with discrete propagules, freeze-drying (lyophilization) in a cryoprotectant is currently the preferred long-term storage method; regular viability checks are still needed, however, as the long-term value of the method is not considered to be fully established. Freeze-drying gives very low survival rates for algae and protozoa and is not yet applicable in these groups. Liquid drying (drying without freezing) is used in some bacterial and yeast collections for strains that are difficult to maintain by freeze-drying (Banno and Sakane, 1979) and is adequate for at least 10 years. Bacteriologists sometimes also maintain strains as suspensions frozen onto glass beads and stored at $-60°C$ to $-70°C$, but this cannot be recommended for resource centers when long-term viability must be assured. The ideal long-term preservation method for microorganisms is in the vapor of or immersed in liquid nitrogen at $-150°C$ to $-196°C$ when viability is expected to be indefinite. Rates of freezing and the cryoprotectant used can affect survival of the freezing process, but it is possible to tailor procedures to ensure optimal survival for particular groups (Morris et al., 1988). The development of protocols for cooling and thawing are also critical for many groups of algae and protozoa. At present even genetically important stocks of algae must be maintained by subculturing on agar plates, in liquid media, or in water. Light is essential for photosynthetic algae maintained by subculture techniques.

The main disadvantage of the use of liquid nitrogen storage systems is the cost of both refrigerated containers and the supply of liquid nitrogen; at least one major collection manufactures its own liquid nitrogen on site. In the case of microorganisms that cannot be grown on artificial media, such as many plant pathogenic fungi (for example, *Puccinia* rust) and plant viruses, they can sometimes be successfully maintained by placing the infected host tissue in liquid nitrogen. At present, the major service culture collections lack the substantial resources necessary to take advantage of the enormous potential for such on-host preservation.

In both freeze-drying and liquid nitrogen storage techniques, as all metabolic processes are suspended, the possibility of genetic change taking place during extended periods of storage is precluded. However, instability can arise from differential survival of propagules during the freezing and thawing processes, and there is consequently some risk of inadvertent selection during the process. This is unlikely to be significant when a high percentage of the propagules survive the freezing process. Current research aims at improving protocols to maximize survival rates of individual propagules (Morris et al., 1988). A new generation of ultra-low-temperature mechanical refrigerators able to operate below –130°C are becoming available and may provide an alternative to liquid nitrogen storage in the future.

Various aspects of stability have been discussed by Kirsop (1980); for further information on the preservation methods for microorganisms see Gherna (1981), Kirsop (1988c), Kirsop and Snell (1984), Malik and Claus (1987), Smith (1988b), and Smith and Onions (1983).

MAINTENANCE IN NATURAL HABITATS

Although the conservation of habitats for other groups of organisms inevitably safeguards the environment for those microorganisms that are already present, the range of potential habitats to be safeguarded is immense and these do not always coincide with those environments that are important for other groups of organisms. Furthermore, because of the current state of knowledge of the world's microbial groups, habitats previously unexplored and considered unimportant repeatedly yield novel organisms (see above).

The concepts of the classification of plant communities and centers of plant diversity that are proving of value in the conservation of vegetation types (International Union for the Conservation of Nature and Natural Resources, 1987) are difficult to translate into microbiology. Some of the problems encountered with respect to the fungi have been reviewed by Apinis (1972).

In addition, while a scientist wishing to collect a particular flowering plant, tree, bird, or mammal known to be in a conserved habitat should have a high expectancy of success, the isolation of microorganisms is both a time-consuming and an uncertain task. Interesting and novel strains are not uncommonly found at low frequencies. Furthermore, many of the larger fungi, rusts, smuts, and other host-specific species are strongly seasonal in occurrence, may not be visible in every year, or may have fruiting bodies that mature and disappear within a few hours. The probability of a research worker being able to obtain another isolate of a strain from a particular habitat in situ is, consequently, often extremely low. It could also be both exorbitantly expensive, time-consuming, and cost-ineffective. In addition, because the ecological roles and extent of functional redundancy among microorganism populations are unknown, management schemes to preserve then in situ cannot, in most cases, be made with confidence. Once isolated in culture and found to be new or to have new properties, the only realistic option available to ensure that it continues to be available is in most cases ex situ conservation in a culture collection. Only in the case of perennial lichen-forming fungi is habitat conservation a realistic measure for the maintenance of microbial diversity (Seaward, 1982).

Although the conservation of unmodified natural habitats should be supported by microbiologists, as they are sources of numerous novel organisms, in general, in situ conservation is not a viable option for the supply of already isolated and characterized microbial genetic material to researchers.

POTENTIAL OF MICROBES IN THE AGRICULTURAL, BIOTECHNOLOGICAL, AND INDUSTRIAL SECTORS

The current uses to which microorganisms are put in the agricultural (including food), biotechnological, and industrial sectors are manifold and difficult to summarize succinctly (Table 10-2). Worldwide, the positive economic value of microorganisms must be calculated in at least many tens of billions of U.S. dollars, bearing in mind their role in the pharmaceutical and fermentation-based industries. However, the economic potential of only a small percentage of the microorganisms already present in culture collections has been investigated.

Exploitation of Metabolites

An indication of the richness of microorganisms is by the number of secondary metabolites from these organisms that are exploited. By

1974, 3,222 antibiotics were known from microorganisms (Berdy, 1974), about 60 percent of these being from actinomycete bacteria. The filamentous fungi are particularly promising from this standpoint: while β-lactam antibiotics continue to be of major importance, significant new drugs are being obtained from fungi, including ones that act as immunosuppressants and those that lower cholesterol levels. Cyclosporin, first approved for use in 1983 and obtained from a fungus regarded for almost a century as unimportant, helps prevent the human body from rejecting organ transplants (Winter, 1985). Secondary metabolites have been reported from only about 2,000 species of fungi, or 3 percent of the known species, and most species have been represented by single strains studied when grown under a single set of conditions. Numerous biologically active metabolites remain to be isolated and characterized. When a desired property is located, production can be increased dramatically by strain improvement; *Penicillium* strains that produce over 25,000 units of penicillin per ml are now used, early *Penicillium* strains produced only 300 units per ml (Kristiansen and Bu'Lock, 1980).

Genetic Engineering

Genetic engineering methods increase the potential applications of the beneficial products that are discovered. Gene sequences producing desired enzymes or other compounds from one organism can be introduced via plasmids into the cells or, in some cases, the genomes of hosts, which can be cultured in bulk. Examples are the transfer of the insecticidal crystal protein gene from *Bacillus thuringiensis* into *Escherichia coli* (Qi and Yunliu, 1988); manufacture of human insulin, alpha interferon, and other products from similarly engineered *E. coli* strains (Primrose, 1986); transfer of enzymes from filamentous fungi to yeasts (van Arsdell et al., 1987); transfer of human tissue plasminogen activator genes into filamentous fungi (Upshall et al., 1987); and the expression of human immunodeficiency virus enzyme into yeasts (Barr et al., 1987). The range of possibilities is immense, and the technology is rapidly advancing (Bennett and Lasure, 1985). Hybridization between protoplasts of different species of fungi and the subsequent production of recombinants has been effected and has opened an area with tremendous promise (Peberdy, 1987).

Application of Microorganisms in Agriculture

The range of applications of microorganisms in agriculture is already substantial and is rapidly developing into new areas. The Ti

TABLE 10-2 Applications of Microorganisms in the Agricultural, Biotechnological, and Industrial Sectors

Uses	Examples of Microorganisms Used	Selected References
Cheese production	Bacteria (Lactobacillus, Streptococcus) and fungi (Penicillium)	Marth (1982)
Beers, wines, brandy, and distilled alcohols	Fungi (Saccharomyces)	Benda (1982), Brandt (1982), Helbert (1982)
Biocontrol of grasshoppers	Protozoa (Microsporidium moscema lacustris)	R. J. L. Muller, International Institute of Parasitology, St. Albans, United Kingdom, personal communication, September 1989
Biomass conversions (waste biodegradation)	Bacteria (Clostridium) and fungi (Sporotrichum, Trichoderma)	Birch et al. (1976), Rolz (1984)
Breads	Fungi (Saccharomyces)	Ponte and Reed (1982)
Fermented foods	Bacteria (Pediococcus), fungi (Monascus, Neurospora, Penicillium, Rhizopus, Saccharomyces)	Steinkrause (1983), Wang and Hesseltine (1982)
Food (direct usage)	Fungi (Agaricus, Auricularia, Lentinus, Pleurotus, Tremella, Volvariella)	Chang and Hayes (1978), Wu (1987)
Organic acids (e.g., citric, oxalic)	Bacteria (Bacillus), fungi (Aspergillus, Candida, Penicillium, Pichia)	Meers and Milsom (1987)
Vitamins (e.g., riboflavin)	Fungi (Aspergillus, Blakeslea, Nematospora)	D. L. Hawksworth, International Mycological Institute, personal communication, January 1989
Enzymes (e.g., cellulases, lipases)	Bacteria (Bacillus, Escherichia), fungi (Aspergillus, Auerobasidium, Candida, Rhizoctonia, Trichoderma)	Böing (1982), Moo-Young et al. (1986)

Application	Organisms	References
Antibiotics	Bacteria (*Streptomyces*), fungi (*Acremonium, Penicillium*)	Blanch et al. (1985), Berdy (1974), Goodfellow et al. (1988)
Amino acids (e.g., glutamic acid)	Bacteria (*Corynebacterium, Brevibacterium, Escherichia*)	Nakayama (1982)
Fuel production (acetone-butanol, ethanol, biogas)	Bacteria (*Clostridium*), fungi (*Saccharomyces*)	Blanch et al. (1985), Moo-Young et al. (1986)
Herbicides	Fungi (*Collectotrichum*)	Templeton et al. (1979)
Nitrogen fixation	Bacteria (*Rhizobium*)	
Pesticides	Bacteria (*Bacillus*), fungi (*Beauveria, Trichoderma, Verticillium*), viruses (*Baculoviridae*)	Aronson et al. (1986), Brady (1981), Entwistle (1983), Wood and Way (1988)
Blood products (e.g., factor 8, interferon)	Bacteria (engineered *Escherichia coli*)	Primrose (1986)
Pickling (lactic acid fermentation)	Bacteria (*Lactobacillus, Leuconostoc, Pediococcus*)	Vaughn (1982)
Protein (single-cell)	Algae (*Chlorella, Scenedesmus, Spirulina*), bacteria (*Methylophilus, Pseudomonas*), fungi (*Candida, Chaetomium, Fusarium, Paecilomyces, Trichoderma*)	Birch et al. (1976), Kristiansen and Bu'Lock (1980), Olsen and Allermann (1987), Reed (1982)
Sausages (fermented; e.g., salami)	Bacteria (*Lactobacillus, Micrococcus, Pediococcus*)	Haymon (1982)
Mycorrhizae	Fungi (*Boletus, Rhizoctonia, Russula*)	Harley and Smith (1983)
Vinegars	Bacteria (*Acetobacter*)	Ebner (1982)
Waste detoxification	Algae (*Prototheca, Scenedgumm*)	Robinson et al. (1988)
Yogurts (and other fermented dairy products)	Bacteria (*Lactobacillus, Leuconostoc, Streptococcus*)	Chandan (1982)

plasmid of *Agrobacterium tumefaciens* is important to biotechnology and plant breeding for use in introducing cloned genes into a variety of crop plants (Fillatti et al., 1987). The potential for introducing gene sequences from bacteria or fungi capable of producing natural insecticides into the genomes of crop plants is especially exciting. Following the recognition of the role of bacteria on plant surfaces in ice nucleation and, therefore, frost damage, new horizons for the control of this damage have opened (Lindlow, 1983); application of *Pseudomonas syringae* strains that lack a crucial protein favoring ice nucleation reduces frost damage.

The applications of microorganisms in the biocontrol of pests and weeds are becoming increasingly recognized. The bacterium *Bacillus thuringiensis* can control a wide range of insect pests, especially lepidopteran larvae (Aronson et al., 1986), as can strains of some fungi, particularly *Metarrhizium* species (Brady, 1981), and insect viruses (Entwistle, 1983). Work on the control of weeds by using fungal pathogens is arguably one of the fastest growing areas of biocontrol, with narrowly host specific *Cercospora* and *Collectotrichum* strains and rusts proving to be particularly efficacious (Wood and Way, 1988).

There is also major potential for new applications in the conversion of agricultural, forestry, and other wastes to usable products such as cattle feed; in the detoxification of harmful compounds in situ by microorganisms (Sahasrabudhe and Modi, 1987); and in the control of environmental pollution (Hardman, 1987). Immobilized microalgal cells show potential for a wide range of applications, including the accumulation of phosphate ions from effluents and chlorinated hydrocarbons (Robinson et al., 1988). A new strain of *Pseudomonas* tolerant of toluene also promises to be of value in pollution control (Inoue and Horikoshi, 1989).

In developing countries, major short- and medium-term benefits can be expected from improved inocula for mycorrhizae and nitrogen-fixing *Rhizobium* strains; these improve tolerance to environmental stress and reduce the need to apply artificial fertilizers, respectively (Mantell, 1989).

In addition to new and developing fields described above, it must not be forgotten that living collections of microorganisms are also relevant to breeding programs, because authentic sources of pathogens are prerequisites for resistance testing in both plant and animal breeding programs. They are also essential for raising monoclonal and polyclonal antibodies and developing other diagnostic methods for the rapid identification of pathogens; the strains from a wide geographic or host spectrum essential to developing such systems are only likely to be supplied quickly through service culture collections.

RECOMMENDATIONS

The microbial genetic resources base currently maintained in culture collections worldwide is scarcely representative of the global genetic resource. This situation is fully recognized by the WFCC; indeed, the Sixth International Congress of Culture Collections held at the University of Maryland in November 1988 passed a resolution calling for "appropriate financial and material support for research on the isolation, characterization, systematics, ecology and conservation of natural and genetically engineered organisms to enable the Collections to competently and professionally fulfill their research and service potentialities" (Canhos, 1989:501).

The need to conserve genetic resources was stressed in 1972 at the United Nations Conference on the Environment (United Nations, 1973). Unfortunately, the participants did not appear to have appreciated the enormous diversity of microorganisms not preserved in culture collections. The recommendations mentioning microorganisms focused on the production of inventories of collections, although it was suggested that governments "cooperatively establish and properly fund a few large regional collections" (United Nations, 1973:15). This last proposal has not been taken up internationally to any significant extent, with notable exceptions being the MIRCEN-sponsored collections in developing countries (Kirsop and DaSilva, 1988).

Frankel (1975a) considered the following actions to be necessary to conserve the world's genetic resources of crop plants: (1) linking existing institutions by agreements for the exchange of material and information; (2) the designation of base collections for particular crops or regions operating under agreements on technical standards; (3) the establishment of a cooperative network of data banks; (4) institution of a program of emergency collecting of threatened genetic resources; and (5) training in genetic resources work. All these actions are equally applicable to microorganisms, with only minor modifications and additions. International and national action is needed in several areas to adequately conserve the world's microbial gene pool.

A much greater proportion of the microbial gene pool must be captured in world collections.

The number of species in service culture collections worldwide needs to be increased substantially. Increased acquisitions of strains from previously unexplored habitats and regions will inevitably lead to the discovery of species not previously known in culture and species that have not been described previously. The priority must clearly be species and strains already proved to be of value to humans. The second priority should be those species that have already been described, that is, biotic material about which something is known. This would provide a cost-effective opportunity to build on what is already known without the need to attempt to repeat original isolations or characterizations. Further, retaining these is preferable to the costs and uncertainties of reisolation.

To achieve these two objectives, a substantial increase in existing capacity would be required (Table 10-1). The system of base collections specializing in particular regions developed by the International Board on Plant Genetic Resources appears to be a model for developing a strategy for the world conservation of microorganisms.

Secure, long-term international funding is needed for conserving, managing, and using the world's microbial diversity.

Because service culture collections must necessarily take a long-term view of their role in conservation, it is imperative that they preserve as a wide a range of the world's microbial genetic resources as possible. Funding must be sufficient both to maintain a sufficient volume of strains and to enable long-term preservation methods and adequate quality control procedures to be implemented. If the model of world base resource collections is eventually adopted for microorganisms, international rather than national funding will be required.

The figures given in this chapter demonstrate that the global genetic resource of microorganisms is not adequately conserved in existing culture collections. The extent to which increased resources should be devoted to isolating additional species and maintaining them must reflect their potential value. In the first instance, the priority in the case of service collections should be to make available known species and strains whose activities have been documented.

Isolation, culture, preservation, and documentation methods are needed for conservation and to ensure the long-term viability and usefulness of microbial resources.

Particular attention needs to be given to the preservation of obligate pathogenic and symbiotic microorganisms, together with infected hosts or their natural symbionts, especially since many of these microorganisms can be expected to have potential in biocontrol and in enhanced productivity.

Researchers often need to locate strains with particular combinations of physiological, biochemical, and other attributes. To do this, computerized data bases readily accessible over telephone lines or electronic mail networks are required; significant progress in this direction is being made through the MSDN and MINE/MiCIS initiatives (see above), but progress has been slow because of limited resources.

PART II

POLICY
ISSUES

11

Exchange of Genetic Resources: Quarantine

Quarantine is a strategy of control to prevent the spread of pests and diseases. It covers all regulatory actions taken to exclude animal or plant pests or pathogens from a site, area, country, or group of countries. For example, when animal or plant genetic resources are imported from another country or region, there is a risk that they may contain or carry pests or pathogens that could be damaging to agriculture. For this reason, countries use quarantine practices to protect their agriculture and living natural resources from potential damage or destruction.

Quarantine is usually a government responsibility, and the manner in which quarantine is executed differs among nations. National agencies responsible for plant quarantine may have other responsibilities, such as domestic pest control; research; pesticide registration, safety, and residue monitoring; or seed quality and labeling.

Significant among the pest or pathogen introductions into North America that have been damaging to agriculture are chestnut blight (*Cryphonectria parasitica*), white pine blister rust (*Cronartium ribicola*), Dutch elm disease fungi (*Ceratocystis ulmi*), Mediterranean fruit fly, European corn borer (*Ostrinia nubilalis*), potato golden nematode, gypsy moth, cotton boll weevil, San Jose scale, field bindweed, Johnson grass (*Sorghum halepense*), kudzu (*Pueraria lobata*), and witch weed. Cassava bacterial blight (*Xanthomonas manihotis*) is believed to have been introduced to Africa and Asia from tropical America by way of infected planting stakes (Plucknett and Smith, 1988). Rinderpest, a significant animal health problem for some countries of sub-Saharan

Africa, was introduced from Central Asia during the late nineteenth century (Acree, 1989).

REDUCING THE RISKS FROM PESTS AND PATHOGENS

Quarantine practices in most countries have at least three common functions. The first is exclusion or regulatory actions to prevent or reduce the risk of entry of exotic pathogens, pests, or parasites along artificial pathways. Second is the containment, suppression, or eradication of pests or pathogens that have been recently introduced. Third is the assisting of exporters to meet the quarantine requirements of importing countries.

The general concepts and objectives of plant and animal quarantine are similar; but differences in biology, agricultural production, marketing, exporting, and importing necessitate a variety of quarantine procedures. Animal and plant quarantine programs are intended to protect agriculture from the threat of entry of exotic hazardous organisms. In some countries this objective may be extended to the protection of natural domestic flora and fauna. Both types of programs regulate the importation of living individuals (for example, animals or plants or plant parts capable of propagation) and various commodities or unprocessed agricultural raw materials. Plant materials subject to restrictions may include seeds, straw, cereal hulls,

---

THE PRINCIPLES OF SUCCESSFUL QUARANTINE

Few reports have examined the challenges and opportunities of developing an effective and efficient quarantine program that addresses the needs and constraints posed by the increasing international movement of germplasm. One recent study (Plucknett and Smith, 1988) describes six principles of successful quarantine. They are summarized as follows:

1. Sound scientific and technical principles should form the foundation of a quarantine program. This should included expertise from a diverse array of scientific disciplines, including, for example, virology, mycology, microbiology, nematology, malacology, entomology, taxonomy, pathology, weed science, and genetic resources.

2. Pests and pathogens should be ranked by quarantine services according to the potential danger they pose to crops and the potential for success in excluding them. For example, germplasm from centers of diversity should receive a high priority because of the potential for such accessions to harbor coevolved pests or pathogens.

lumber, logs, bark, fruit, vegetables, cut flowers, fibers, gums, and spices. Animal materials of quarantine interest may include live animals, semen, eggs, or embryos; fresh, frozen, processed, or canned meat; milk and milk products; raw hides; and biological reagents or other compounds (protein, hormones, sera) extracted from animals. In many countries, garbage from other nations also is of quarantine interest.

Animal and plant quarantine regulations are similar in that they may:

• Require import permits issued by the quarantine service of the importing country (these may require the exporting country to certify that specified conditions have been met prior to shipment);
• Specify things that are prohibited from entry;
• Grant exceptions to the prohibitions for scientific purposes;
• Require inspection of imported materials upon arrival;
• Require appropriate treatment, if warranted, as a condition of entry; and
• Require, after arrival, quarantine or isolation in an approved facility.

Plant quarantine is concerned with a far greater number of species than is animal quarantine. Currently more than 240 crops or plant species are prohibited from entry to one or more countries (Kahn,

---

3. When germplasm must be planted and grown for the purposes of quarantine testing, it should be done in an area geographically and ecologically separated from the major growing areas for that crop, to prevent the establishment of crop-specific pests or pathogens.

4. When germplasm is endangered or the need for particular accessions is particularly urgent, some discretion should be possible on the part of quarantine officials in allowing exceptions for controlled entry, despite existing regulations to the contrary.

5. Decentralized quarantine services are generally more efficient because they enfold a wider range of expertise in germplasm assessment. However, decentralization places a greater burden on the national system to ensure high standards and to promote accurate and efficient information exchange.

6. Because delays in transit can be detrimental for many germplasm accessions, access to good communication and transportation services is essential for quarantine.

1982; Plucknett and Smith, 1988). More than 1,600 pests or pathogens of plants are the objects of quarantine worldwide (Kahn, 1988). In the United States, more than 1,300 pests or pathogens of pests are a significant threat to crops (Mathys, 1977). In addition, a worldwide survey of 124 countries lists more than 8,000 plants as weeds. Some, such as Bermuda grass (*Cynodon dactylon*) and wild oats, were already significant problems for many nations (Council for Agricultural Science and Technology, 1987; Holm et al., 1977, 1979).

Importations of animal germplasm consist primarily of living individuals, embryos, or if admissible, eggs (birds, fowl), all of which are developed through sexual reproduction. In addition, semen may be imported, subject to regulation.

Importations of plant propagative material include seeds, plants, various asexually produced propagative units, or the plants developed from them. These include asexually produced seed (nucellar seeds of *Citrus* species); underground or aerial storage organs such as bulbs, corms, stems, crowns, or roots; rhizomes; rooted and nonrooted cuttings; suckers; various types of grafts; and tissue cultures. Tissue cultures may include cells, protoplasts, embryoids, embryos, undifferentiated tissues, or plantlets produced from tissue culture. Pollen may also be imported, but this is probably done less frequently and with less overall known risk than that from the importation of animal semen.

QUARANTINE AND GLOBAL TRANSFER
OF PLANT GENETIC RESOURCES

The transfer of genetic resources has, at times, resulted in the unintended introduction of serious pests or pathogens (International Board for Plant Genetic Resources, 1988b; Plucknett and Smith, 1988). In the 1940s more than 20 million citrus trees were lost in Argentina and Brazil because of tristeza disease from virus-infected imported nursery stock (Knorr, 1977). As the international transfer of germplasm has increased, so has the potential hazard to crop production around the world (Karpati, 1981, 1983; Plucknett and Smith, 1988). Movement of wheat germplasm into the United States, for example, has been increasingly hampered over the past decade by the spread of Karnal bunt disease (Plucknett and Smith, 1988).

The increasing emphasis on germplasm from wild species is a particular problem. Little may be known about the pathogens that may be present. Wild species are frequently from the geographical region where a crop originated or was domesticated and may harbor coevolved pests or pathogens of potential significance (Plucknett and Smith, 1988).

Poor-quality maize cobs harvested in Yemen. Credit: Food and Agriculture Organization of the United Nations.

Although hundreds of pests and pathogens have already been introduced along natural and artificial pathways, there are hundreds of other crop pests and pathogens that have not yet been introduced (Holdeman, 1986; Kahn, 1979, 1989a,b; MacGregor, 1973; Schoulties et al., 1983; Yarwood, 1983; Zimdahl, 1983). Analyses of the life cycles of these exotic pests show that many organisms do not have a natural means of spreading medium to long distances. These organisms, however, could readily and quickly be moved on imported articles. Quarantine services and the enforcement of biologically based regulations are the only means available to reduce the chances that such organisms might enter a country by artificial pathways. The importation of animals and plants for any purpose, including the transfer of germplasm resources, can be identified as a low- to high-risk pathway, depending on the host, various biological factors, and the occurrence of transferrable exotic pests in other countries.

When the potential risk is perceived to be low, germplasm resources usually move with minimal delays under the principle of the least drastic action. When the risk is perceived to be high, there may be an adverse impact on genetic resources transfer when more drastic actions are taken to exclude pests. Some countries are willing and able to provide safeguards such as testing, isolation, or treatment that can lower the risk to an acceptable level. In other cases, the safeguard mechanism chosen is to exclude the material from entry. Rubber tree (*Hevea brasiliensis*) germplasm, for example, must be passed through an intermediate quarantine site outside of the American tropics before transfer to Southeast Asia because of the threat of South American leaf blight (*Dothidella ulei*) (Turner, 1977). Indonesia excludes any vegetative *H. brasiliensis* germplasm from entry (Plucknett and Smith, 1988).

Quarantine as a Pest Control Strategy

Quarantine is often combined with plant protection to include all regulatory activities carried out by local, regional, national, and international government agencies or organizations. Two components of quarantine that affect the exchange of plant genetic resources on a global scale are (1) exclusion of plant parts or taking of regulatory actions that will reduce the chances that pests and pathogens might enter a country along artificial pathways; and (2) phytosanitary certification or providing of assistance to a country's exporters to meet the quarantine requirements of importing countries.

Both importers and exporters of plant germplasm are affected by these two functions. Importers are subject to the regulations of their own country, which might not only require that exporting countries meet certain phytosanitary standards but might also place some restrictions on the imported germplasm after entry. Although the quarantine service of the exporting country may assist in meeting the requirements, it is the importing country that sets quarantine standards.

Quarantine and Genetic Resources

Plant quarantine practices are frequently viewed as an impediment by those working in research, breeding, or genetic resources management. This has been particularly true for germplasm that is transferred in a vegetative form (Chiarappa and Karpati, 1984; International Board for Plant Genetic Resources, 1988b). The timely transfer of some, but not all, species may be adversely affected by quarantine-imposed delays.

For example, *Prunus* germplasm is prohibited from entry into the United States because (1) the plum pox virus (among other exotic viruses) does not occur in the United States; (2) the virus, which is transmitted by several species of aphids (but only within 30 minutes of its acquisition by the aphid), does not move on natural pathways beyond the range of its aphid vector; (3) several species of the vector already occur in the United States; (4) the virus can cause severe economic losses; (5) the virus could become established in U.S. nurseries and, subsequently, in commercial orchards if imported infected bud wood or root stocks were used in nursery stock production; and (6) several U.S. *Prunus* species and varieties are known to be susceptible. A period of 5 to 10 years or more may be required to test introduced *Prunus* germplasm for the presence of viruses (National Research Council, 1991a), and a longer period of time may be required to propagate material for release from tested plants.

Wheat breeders in many countries may be confronted with prohibitions or postentry safeguards that may limit research or breeding programs if the Karnal bunt fungus (*Tilletia indica*) is found to be present in germplasm samples. The perception of risk or potential damage from this fungus is highly controversial in some regulatory and plant breeding circles. It is known to occur in Mexico and India as well as a few nearby countries. However, damage in Mexico is considered by many to be extremely minor. In India, in the past, damage has been considered to be minor, but in recent years heavy losses of some susceptible varieties have been reported in limited areas (Lambat et al., 1983; Plucknett and Smith, 1988).

Legume seeds (beans, soybeans, cowpeas) from some areas may carry latent bacteria and fungi. Restrictions may require that the seeds be subjected to seed health testing and virus indexing. Some countries may have restrictive policies to prevent the entry of specific pathogens, and testing or indexing may not usually be considered adequate safeguards. Breeders may receive only a portion of the seeds they sought to introduce. Bean breeders at the Centro Internacional de Agricultura Tropical (International Center of Tropical Agriculture) are severely limited in the number of seeds they can draw from the center's germplasm bank because of Colombian national quarantine restrictions (Plucknett and Smith, 1988).

Legal Basis of Quarantine

The legal foundation that supports national quarantine regulations and actions is usually either legislation passed by national governments as acts, statutes, orders, decrees, or directives or enabling legislation that authorizes a minister or secretary of agriculture to issue regulations. In addition, regulatory actions may be taken based on policy, guidelines, or instructions for quarantine officers (Kahn, 1977).

State or provincial governments may enact legislation or promulgate rules and regulations to exclude or reduce the chances for entry of pests. State or provincial quarantine services often assist the national government in quarantine activities such as survey or export certification. The international exchange of plant genetic resources is not usually affected by state quarantines, except insofar as state officials must often be used to obtain phytosanitary certificates for export.

Some countries are also bound by quarantine regulations that are promulgated by regional parliaments, such as the European Community or Andean Pact (Kahn, 1989a). In addition, a number of regional plant protection organizations exist to provide member countries with nonbinding regulatory recommendations (Plucknett and Smith, 1988).

The International Plant Protection Convention of 1951 provides an international mechanism for harmonizing most international plant quarantine activities. By 1987, 89 countries were signatories to the convention, and many others followed (Plucknett and Smith, 1988). The convention has been important to efforts to standardize quarantine practices among nations. It has encouraged the establishment of regional organizations (Table 11-1).

The quarantine services of most importing countries require that plant genetic resources be accompanied by a phytosanitary certificate issued according to the standards set by the Food and Agriculture Organization of the United Nations. The certificate must be addressed to the quarantine service of the importing country and signed by an authorized officer of the quarantine service of the exporting country. The certificate must include the plant's place of origin and botanical name. The certificate must also contain a statement certifying that the plants or plant products have been inspected and found to be free from quarantined pests.

Unfortunately, a country's printed regulations may have little to do with actual practice. A phytosanitary certificate does not ensure that the plant material will be able to enter a country. Countries with developed quarantine services usually do not rely on these documents as a sole safeguard, even though they may be required for entry.

Biologic Basis of Quarantine

The biologic foundation of plant quarantine rests on knowledge about the identity of pests, pathogens, and hosts; their geographic

TABLE 11-1   Regional Plant Protection Organizations

Organization	Number of Member Countries	Headquarters
Asia and Pacific Plant Protection Commission (APPPC)	23	Bangkok, Thailand
Caribbean Plant Protection Commission (CPPC)	18	Port of Spain, Trinidad
Comite Tecnico Ad-Hoc en Sanidad Vegetal para el Area Sur (COSAVE)	6	Montevideo, Uruguay
Junta del Acuerdo de Cartagena (JUNAC)	5	Lima, Peru
Organismo Internacional Regional de Sanidad Agropecuaria (OIRSA)	7	San Salvador, El Salvador
Inter-African Phytosanitary Council (IAPSC)	48	Yaounde, Cameroon
North American Plant Protection Organization (NAPPO)	2	Washington, D.C.
European and Mediterranean Plant Protection Organization (EPPO)	36	Paris, France

SOURCE: Plucknett, D. L., and N. J. H. Smith. 1988. Plant Quarantine and the International Transfer of Germplasm. Study Paper No. 25. Washington, D.C.: Consultative Group on International Agricultural Research, World Bank. Reprinted with permission, ©1988 by The World Bank.

distribution; and the life cycles of these organisms as they are influenced by the environment, climate, and farm practices. An understanding of how these factors influence colonization or establishment of exotic pests is important for determining the entry status of imported germplasm. Quarantine is intended to prevent the release of an exotic pest under conditions favorable to its establishment. The nature of the organisms of quarantine interest affect greatly the complexity of quarantine activities. For potatoes, methods of detection for the major bacteria, fungi, and nematodes of quarantine significance existed in 1975, when the Centro Internacional de la Papa (International Potato Center) began distribution of germplasm. Methods were needed, however, for detecting several viruses and a viroid in potato germplasm (Kaniewski and Thomas, 1988). For temperate fruit trees, lack of knowledge about some pathogens means that sophisticated serologic, nucleic acid, or electron microscopic detection methods are unavailable. For these pathogens, grafting to sensitive indicator plants may be the only detection method available and can greatly lengthen quarantine-imposed delays.

Geographic Basis of Quarantine

The distribution of exotic pests and pathogens in various countries, regions, or continents is the geographic basis for promulgating quarantine rules and regulations. Information about the occurrence of such organisms is found in the scientific, agricultural, and regulatory literature. Reports (Plant Health Division, Agriculture Canada, 1978–1986; Plant Protection and Quarantine, 1980–1986) of pests intercepted at ports of entry provide information about the detection of exotic pests, but the information is useful only if the country of origin is identified.

The center of diversity of a crop is also likely to be the center of diversity for many of its pests and pathogens (Kahn, 1977). Plant collectors often import germplasm from areas where pest, pathogen, and host species have evolved. These can make germplasm from these regions a particularly rich source of genes for resistance or tolerance traits, but it can also constrain its importation (Neergaard, 1977, 1984; Plucknett and Smith, 1988).

Some crops have been successfully imported without their co-evolved pests and pathogens. Their continued productivity, however, depends, in part, on the effectiveness of quarantine in maintaining their isolation. Examples include rubber in Africa and southeastern Asia, which still escapes from South American leaf blight; coffee in Latin America, which escaped coffee rust (*Hemileia vastatrix*) from Asia and Africa until recently; and banana in Latin America, which avoided the black sigatoka fungus until the 1970s (Plucknett and Smith, 1988).

Organisms from outside a center of diversity are not necessarily innocuous. Although potatoes originated in the Andes of South America, the late blight fungal pathogen, *Phytophthora infestans,* has origins elsewhere (Kahn, 1977).

Pests and Pathogens of Quarantine Significance

An organism is considered to be of quarantine significance if it does not occur in the country of concern and is known to cause economic damage elsewhere or has a life cycle that suggests that it is capable of causing damage under favorable conditions. It also may be of significance if it does occur but is not widely distributed; if it is under a national suppression, containment, or eradication program; if there are more virulent strains that do not occur; or if it has the potential to cause economic damage to crops. Finally, an organism may already be a common pest but regulations require that growers plant only pathogen-tested nursery stock. In such cases, the government may take steps to ensure that imported stocks meet domestic standards.

Pathways for the Entry of Pests and Pathogens

Plant pests and pathogens can move or be moved along natural or artificial pathways (Table 11-2). Some organisms have stages or life forms that enable them to survive biologic, chemical, or physical stresses and physiologic, morphologic, or anatomic characteristics that facilitate active or passive dispersal.

Some organisms (most nematodes, bacteria, scale insects, mites, and snails) move, or are naturally moved, from plant to plant or from plants in one field to plants in another, but they have no means of natural spread over long distances. Such organisms may be dispersed longer distances by the accumulation of a series of short-distance natural spreads over time until a natural barrier, such as the absence of a host, is reached. Some pathogens (certain fungi, viruses, bacteria) are moved much longer distances by their insect or vertebrate animal vectors. Still others travel long distances as airborne spores. A few insects, such as locusts and monarch butterflies, are migratory.

Many pests and pathogens of quarantine significance have very inefficient means of natural movement or spread or life cycles that are not conducive to natural spread. They may be moved along artificial pathways, however (Table 11-2). Quarantine is most effective when humans, or the activities of humans rather than those of nature, are the prime long-distance movers of pests and pathogens.

TABLE 11-2   Natural and Artificial Pathways for the Movement of Plant Pests and Pathogens

Natural Pathways	Artificial Pathways
Winds, storms, jet streams	Cargo (agricultural and
Air and convection currents	nonagricultural)
Ocean currents	Mail
Surface drainage	Baggage
Natural seed dispersal	Common carriers (ships,
Fliers (insects and mites)	vehicles, airplanes)
Migratory species (locusts)	Dunnage, crates, packing
Self-locomotion (zoospores)	materials
Vectors (insects, nematodes)	Smuggling
Other carriers (birds and other	Farm practices (irrigation,
higher animals)	used farm equipment)

Pests and pathogens have the best chance of surviving along artificial pathways when they are on or in their hosts. They are then protected from adverse environmental stresses such as extremes of temperature and relative humidity, prolonged exposures to sunlight or ultraviolet light, or their own parasites or predators. Some, such as viruses and other obligate parasites, may not be able to survive in the absence of a host.

The importation of infected, infested, or contaminated plants or animals presents the greatest opportunity for pest and pathogen survival and entry. The chances of survival of the pest or pathogen usually depend on the survival of the host.

Pest Risk Analysis

Pest risk analysis is a determination of the entry status of any imported article, including propagative material, based on the known or perceived risk (chances) of inadvertently introducing hazardous pests or pathogens by artificial pathways (Kahn, 1979, 1985). It is based on data about known pests or pathogens of the germplasm and the ease with which they could gain entry, colonize, and become established. Other factors include the effectiveness of inspection methods; the availability and effectiveness of treatments if a pest is found; the existence of the diseases or disorders in the country of origin for which the causal agent is unknown; the availability in the importing country of technical support in the form of pathologists, entomologists, and other specialists; and the ability to monitor pathogens, provide diagnostic services, and apply timely control measures.

A risk analysis is an essential precursor to decisions about the importation of germplasm. The risk associated with a particular species may vary with the nature of the material being imported (Figure 11-1). Seeds, for example, generally present a lower risk than living plants do. Risk analysis necessitates access to a wide array of accurate and timely information about crops and their pests and pathogens.

If the potential benefits exceed the risks, importation of germplasm is justified. An example is the benefit to agriculture from the importation of germplasm when balanced against the risk (that a pest will escape and become established) coupled with the use of a quarantine station as a safeguard. However, the benefits from the importation of large amounts of the same material for commercial purposes cannot justify either taking the risk of importation nor the expense involved in passing material through quarantine. For example, the importation of potato tubers, which are considered by many countries to

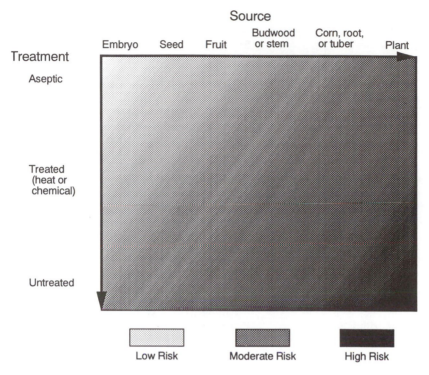

FIGURE 11-1 Factors considered in pest risk analysis and the relative degree of risk they pose. Source: International Board for Plant Genetic Resources (IBPGR). 1988. Conservation and Movement of Vegetatively Propagated Germplasm: *In Vitro* Culture and Disease Aspects. Report of the IBPGR Advisory Committee on In Vitro Storage. Rome: International Board for Plant Genetic Resources. Reprinted with permission, ©1988 by International Board for Plant Genetic Resources.

present a high risk, can only be justified by passing small amounts (a few tubers) through a quarantine station. However, allowing large amounts to enter for commercial planting or consumption may not be as easily justified.

Germplasm collected from a site with a higher incidence of pests and pathogens has a higher chance of being infested, infected, or otherwise contaminated. Plants or animals from such locations may present a higher risk. Risk and hazard factors related to various germplasm collection sites are given in Table 11-3.

In general, vegetative propagations of genetic resources constitute a higher level of risk than do seeds. Almost all vegetative propagations taken from mother plants that are systemically infected with

TABLE 11-3 Some Plant Germplasm Sources

Germplasm Source	Risk- or Hazard-Related Factors
Collected in the wild	Pest and pathogen occurrence and incidence is often unknown; no survey or control
Tubers, roots, or seeds collected in marketplaces	Health of mother plant is unknown; there is no foliage; material is difficult to inspect; even so, pathogens may be latent; country of origin is uncertain
Farms	On average, pest control activities are relatively low or absent
Orchards, plantations	Often, a high level of inspection, pest survey and management, or control
Experimental fields	May have a high level of pest survey and management or control; personnel are often aware of pests, pathogens, symptoms, and signs
Experimental plots isolated from commercial plantings or located where pests are not known to occur	Risk is lowered by the absence of specified pests or lower inoculum levels; survey and control are practiced; personnel are aware of pests, pathogens, and symptoms
Greenhouse with floor-level planting beds	Risk is lowered by isolation and improved phytosanitation, but soilborne pests could create sanitary and management problems
Commercial greenhouses with raised benches	Risk is lowered by plant isolation and high levels of phytosanitation, survey, and control
Research greenhouses with raised benches	Similar to commercial greenhouses with raised benches; highly trained personnel
Plant tissue cultures, aseptic plantlet culture	Propagations usually have the same health status as their mother plant, insofar as obligate or fastidious pathogens are concerned, but other pests and pathogens are usually eliminated during processing; isolated from contamination
Approved certification	Pathogen-tested plants; approved procedures; survey; phytosanitation; precautions against recontamination
From certain quarantine stations, third-country quarantine, other high-containment locations	Plants grown under the highest levels of pest detection, eradication, phytosanitation, treatment, and isolation; no recontamination

NOTE: Germplasm sources are listed in order, with the highest perceived pest risk given first.

pathogens (for example, viruses and many bacteria and fungi) are also infected. Few of these pathogens are transmitted in the seed. Thus, seeds are considered to be safer for transfer, although there are important exceptions (Kahn, 1977; Plucknett and Smith, 1988). Even when pathogens are known to be seedborne, frequently not all seeds are infected or contaminated. For seedborne viruses, the percentage of transmission may be as low as 1 infected seed in 50,000 (for example, lettuce mosaic virus in lettuce) to more than 50 infected seeds. Not all germinating seeds known to be infested or contaminated with certain fungi or bacteria produce infected plants.

In general, size, age, and volume of propagules of genetic resources can affect the level of risk. High-risk plant genera are imported in the smallest quantity necessary to establish the base material (for example, a few tubers, a small packet of seeds, or a sample of semen). For vegetative propagations and plants, smaller and younger propagules are less likely to harbor pests. For animals, carefully collected and appropriately treated embryos serve a similar purpose (see Chapter 7) (National Research Council, 1993).

The absence of signs or symptoms does not necessarily mean the pathogen is absent. Some nematode-transmitted viruses may not produce observable symptoms in infected germplasm (Bos, 1977). Many cassava seeds infected with cassava bacterial blight (*Xanthomonas campestris* pathovar *manilotis*) do not exhibit symptoms (Plucknett and Smith, 1988).

IMPORTATION OF GENETIC RESOURCES

The quarantine regulations of most countries (Plant Protection and Quarantine, 1960–1988) require that plant and animal materials, including genetic resources, enter through named ports or points of entry. If an airplane or ship has several ports of call in the importing country, inspection usually takes place at the first one. The authorized ports must have inspection stations or facilities. Material that arrives by mail may be held at a post office until it has been inspected by the plant quarantine service. The entry of genetic resources that are prohibited except for scientific purposes may take place through either an inspection station or some other facility named in the special permit authorizing entry. There are, however, many ways in which the inspection of germplasm may be unintentionally or intentionally circumvented (Plucknett and Smith, 1988).

A range of safeguard options can be used by a quarantine service in issuing special permits for the importation of small amounts of germplasm. The most drastic is to prohibit importation of the host with no exceptions even for scientific purposes. For example, some countries prohibit coconut plants and seeds from anywhere because of diseases of unknown etiology. Others allow entry but require varying degrees of quarantine isolation.

Countries without quarantine stations may allow entry of items that are exempted from prohibitions after passage through intermediate or third-country quarantine (Berg, 1977; Kahn, 1983). Under third-country quarantine, a country imports high-risk material from another country but the material first goes through quarantine in a third country. The first two countries are usually in the tropics or subtropics, but the third country is usually in a region with a temperate climate. The third country may accept the risk because the crop is not grown there; the organisms of concern have narrow host ranges and are not likely to attack crops in the third country, or the organisms are not likely to become established because the environment is unfavorable.

Several industrial countries have provided intermediate quarantine services in the past for specified tropical cash crops such as coffee, banana, and rubber (*Hevea* species). Unfortunately, support has declined or been withdrawn for many such efforts, despite urgent needs (Plucknett and Smith, 1988). It has been suggested that renewed support must come from a consortium of donors that includes industrial and developing nations, commodity and consumer groups, and bilateral and multilateral aid organizations (Plucknett and Smith, 1988).

RECOMMENDATIONS

Quarantine is essential for protecting a nation's agricultural system. However, regulations and practices must continually balance the potential for release of harmful pests or pathogens with the needs of germplasm scientists, research efforts, and breeding programs. Protection of agriculture is not, however, solely the responsibility of quarantine services. Efforts are needed on the part of national and international germplasm programs to reduce the potential for harboring pathogens in the collections they manage. For quarantine, opportunities to improve present efforts exist in the areas of cooperation, information, and technology.

Research should be directed toward the development of improved and efficient methods of pathogen detection, germplasm treatment, and safe transfer.
Many of the molecular technologies described in Chapter 7 have the potential to provide powerful tools for the detection of pathogens

in plant materials. Fundamental research into the biology of pests and pathogens is an essential part of assessing both the risk of inadvertent transfer and potential methods of detection or elimination. There is a particular need for increased studies of viruses and viruslike diseases and to address the pests and pathogens of forage species and the wild species related to cultivated crops. Tissue culture has been successfully used to eliminate pathogens from a number of crop species and enable their safe transfer (International Board for Plant Genetic Resources, 1988b).

New and emerging molecular technologies promise to provide the capacity for rapid detection of pathogens that may currently be severely restricted in their transfer because of potential quarantine difficulties.

Movement of germplasm through quarantine could be made more efficient through increased cooperation between nations and international institutes, and between quarantine officials and the users of germplasm.

Many regional quarantine programs already exist, and these should continue to be supported. In many cases, it may be both practical and efficient to establish regional centers for the quarantine testing of particular materials and, thus, share responsibilities, costs, and benefits.

A greater effort is needed to establish sites for intermediate (third-party) quarantine sites. It has been suggested that support for such efforts must come from a consortium of donors that would include governments, commodity and consumer groups, and bilateral and multilateral aid organizations. Such facilities provide a relatively safe mechanism for the transfer of germplasm that might otherwise be restricted because of pests or pathogens with which it might be contaminated.

As stated above, quarantine services should be flexible in decisions regarding the fate of germplasm that is endangered or of particular value. Germplasm scientists with specific expertise in a crop can, with appropriate safeguards, receive and test germplasm that might otherwise be detained in quarantine or denied entry into a country. For such cooperation to succeed, germplasm scientists and quarantine officials must acknowledge a mutual goal of efficient and safe transfer of germplasm.

Quarantine services in developing nations are in particular need of facilities, supplies, and trained personnel (Plucknett and Smith, 1988). International centers and established national programs can cooperate through the provision of appropriate training of germplasm and quarantine scientists and technicians. Investments in equipment, facilities, and training by international centers, existing national programs, and aid agencies will greatly benefit both the receiving nations and the programs or nations with which they exchange germplasm.

The potential to introduce exotic pests or pathogens through the distribution of contaminated germplasm from national and international collections must be reduced.

Maintenance of germplasm in a national or international collection does not mean that the material is a priori free of pests or disease. This is especially true for material maintained and exchanged in the vegetative state (for example, plants in the field that are exchanged as bud wood or cuttings). Methods exist to free a number of vegetatively propagated crops, such as potato, sweet potato, and citrus of viruses or viroids that might infect them (International Board for Plant Genetic Resources, 1988b). Once tested and shown to be free of disease, such material can often be maintained and exchanged as in vitro cultures, a practice now routine with potato germplasm (International Board for Plant Genetic Resources, 1988b).

Germplasm collections should, to the extent possible, be tested for at least the most potentially significant pests or pathogens they may contain. It has been suggested that, for vegetatively propagated germplasm collections, information about whether the plants have been tested for pathogens and the results of those tests should be part of the data available to quarantine officials (International Board for Plant Genetic Resources, 1988b). Such information should be available for seed collections that are likely sources of seed-borne pathogens as well.

It may not be technically feasible to eliminate easily some pathogens from some accessions. Although this might limit the distribution of those accessions, it should not be cause to remove them from collections. Wild species, for example, might possess potentially serious pathogens, but their potential contribution of new genes may justify their continued maintenance in a collection. The technique of reducing or eliminating pathogens by selection of disease-free plants during growouts should also be applied cautiously. Although this may yield a disease-free sample, in many cases the selection of a few plants as seed sources for a heterogeneous accession could result in significant loss of genetic diversity (Plucknett and Smith, 1988).

Quarantine should not be used as a mechanism to further economic or political objectives.

Sound quarantine policies and practices should be biologically based and in accordance with known or potential pest risk (an estimation of the chances that a hazardous pest or pathogen will gain entry along artificial pathways). They should be executed through the least drastic actions that will provide the safeguards and reduce the risk to an acceptable level in a timely fashion. Finally, considerations of the costs of accidental release of a harmful pest or disease versus the benefits to be gained from the imported material should be considered. Usually, the benefits to be derived from the importation of genetic resources justify taking the risk, provided that safeguards are in place, whereas the benefits derived from commercial importations may not necessarily support taking the same risk.

12

Exchange of Genetic Resources: Proprietary Rights

Plant variety protection has been an intense source of controversy in the debates surrounding the free international flow of germplasm. Other forms of proprietary rights have become equally controversial as plants and animals, as well as individual genes, are given protection under the regular patent system.

Studies of proprietary rights in agriculture, primarily of plant variety protection, have found that such rights bring somewhat increased private breeding activity but have not found significant effects on the genetic diversity of the varieties that are marketed (Butler and Marion, 1985; Council for Agricultural Science and Technology, 1985; Lesser, 1987; Mooney, 1983; Perrin et al., 1983). Nevertheless, there has been international political concern over the growth of proprietary rights in the biological area and particularly over the possibility that such rights benefit the developed world at the expense of the developing world.

One argument that is commonly made is that the developing world should or will respond to the growth of biological proprietary rights by restricting access to its genetic resources. Although the committee is unable to make the political judgments necessary to evaluate the likelihood that the governments of developing nations will adopt such a policy, it notes that, for the most part, this has happened only with industrial crops. Moreover, the actual flow of new germplasm from developing nations to the developed world is probably now quite limited; the major current flows are probably of advanced material in the opposite direction.

The committee attempted to evaluate the underlying concern: what will be the actual effect of proprietary rights on the international flow of genetic resources, the practical terms of exchange, and particularly on the access of developing nations to genetic material? These affect all nations' access to germplasm resources and may also influence the willingness of developing nations to permit scientific access to their own genetic resources. The committee did not consider the issue of whether developing countries should strengthen their own intellectual property systems, either to encourage domestic innovation or to facilitate or limit the licensing of technology and genes from abroad.

In evaluating the effect of property rights on the exchange of genetic resources, the committee reviewed the growth of proprietary rights legislation and use, explored the actual barriers to exchange that have arisen or may arise for several species of importance to developing countries, examined several alternatives currently under discussion, and developed relevant recommendations.

PROPRIETARY RIGHTS ON LIVING BEINGS

Proprietary protection—whether patent, copyright, trade secrets, or other legal instrument—is a legal right conferred on the inventor or creator of a particular concept or thing. The legal right is that of excluding or restricting others from using the concept or thing, except in certain conditions or with the permission of the inventor. Such permission is usually given in return for a royalty or the equivalent of a royalty included in the price of a product. This special monopoly return, typically for 17 to 20 years, is designed to provide a reward for the innovation and to induce investment in such innovation (Bent, 1987; Office of Technology Assessment, 1989).

In most nations and for many forms of intellectual property, there is an obligation to disclose the innovation in return for obtaining this monopoly right. This concept of disclosure—with its implication that the patent be a building block to further knowledge—is generally taken quite seriously by the courts. Moreover, it is recognized that patent protection sometimes provides a socially beneficial alternative to other methods of proprietary protection that provide less public disclosure. Thus, patents sometimes provide an alternative to trade secret protection in which data are held confidentially within a firm, as exemplified by the use of hybrids in which parental lines are strictly controlled.

For the forms of intellectual property protection considered in this chapter, the monopoly conferred by national statutes is usually effective only within the nation involved. If an invention is patented in the United States but not Canada, it is not a violation of U.S. or Canadian law to practice the invention in Canada, although U.S. law typically restricts import from Canada of the product or of products made through a patented process. There is a network of international conventions covering intellectual property; for those forms of intellectual property considered here, the conventions protect the right of nationals of one signatory to obtain protection in other participating nations, and some also establish minimum standards for the level and form of protection to be accorded by each participating nation.

Developed nations are strongly pressing for a new international intellectual property code beginning with the Uruguay Round of international trade negotiations (General Agreement on Tariffs and Trade). Under the drafts likely to be accepted if the Uruguay Round succeeds, nations will commit themselves to "provide for the protection of plant varieties either by patents or by an effective sui generis system." Most developing nations will have 10 years (and the least-developed countries even longer) before they must comply with this obligation. In the meantime, the obligation is to be reviewed ([Draft] Agreement on Trade-Related Aspects of Intellectual Property Rights, including Trade in Counterfeit Goods, MTN.TNC/W/FA, Articles 27, 65 and 66). Moreover, technology transfer is to be facilitated under the new Rio Convention on Biological Diversity on terms "which recognize and are consistent with the adequate and effective protection of intellectual property rights" (see Chapter 14).

Plants

A form of protection for vegetatively propagated plants has long been granted in the United States as a plant patent under the Plant Patent Act of 1930 (codified at 35 U.S. Code [U.S.C.] Sections 161–164). Such a patent conveys rights similar to those of any other patent and particularly that of prohibiting others from reproducing the plant asexually, but it could still be used for breeding. In 1970, similar legislation was adopted for sexually propagated plants, the Plant Variety Protection Act (codified at 7 U.S.C. Sections 2321–2583). A certificate under this act, which is the United States version of plant variety protection, conveys the right to keep others from selling the seed of the plant for seed purposes or for use in producing a hybrid (7 U.S.C. Section 2514). It does not, however, convey the right to keep others from using or making limited sale of the seed or plant for breeding purposes, nor does it prohibit a farmer from using, or

This improved variety of Mexican wheat is proving to be very successful for its yield and quality. Credit: Food and Agriculture Organization of the United Nations.

making limited sales of, seed from his or her own crops for seed purposes in following years.

Although few other nations distinguish vegetatively propagated and sexually propagated varieties in the way that the United States does, most developed nations and a very few developing nations have adopted bodies of law providing such protection for plants (Barton, 1982; Berland and Lewontin, 1986; Commission on Plant Genetic Resources, 1986; Mooney, 1983). Several developing nations are considering adopting such systems (Commission on Plant Genetic Resources, 1986). These laws generally meet the standards of the International Convention for the Protection of New Varieties of Plants (Union for the Protection of New Varieties, generally referred to as UPOV, December 2, 1961, as revised on November 10, 1972, October 23, 1978, and March 19, 1991). These standards are generally similar to those of the United States, but they vary in detail. For example, some cover both vegetatively and sexually propagated plants, and some provide for national authorities to test the variety to determine that it is actually new and has value for cultivation and use. (Under separate seed catalogue legislation, many nations also permit only tested and approved varieties to be distributed to farmers.)

In an extremely important decision in 1985, the U.S. Patent Office Board of Appeals indicated in *Ex parte* [on the application of] *Hibberd* (227 USPQ 443 (1985)) that plants might also be protected under the regular patent law, in contrast to the special plant variety protection legislation (Byrne, 1986). A regular patent conveys somewhat broader rights than plant variety protection does; it might not recognize, for example, the farmers' exemption, under which a farmer can use one year's harvest as the next year's seeds without infringing a plant breeding rights certificate. Recent changes to the UPOV convention will allow nations to eliminate the farmers' exemption. It is possible that the member nations of the European Community will permit regular patents on various forms of plants (European Commission, 1989). Certain exemptions on protecting plant varieties, as opposed to plants incorporating particular genes or traits, may be included. They are modeled on provisions of the UPOV convention, which allows nations to use both regular patents and PVP rights for the same genus or species. Canada, however, has previously rejected the application of regular patents to plants in *Pioneer Hi-Bred Ltd. v. Commissioner of Patents* (11 C.I.P.R. 165 [Fed. Ct., 1987]) for a soybean variety produced through cross-breeding. The decision was sustained on appeal to the Supreme Court of Canada ([1989] 1 Sup. Ct. Rev. 1623).

In those few nations that use regular patents for plants or plant components, the exact scope of the patent holder's rights is still unclear and will probably depend on the particular claims included in the patent. Thus, a patent on a gene will probably be held to cover all plants that contain the gene. Whether a patent on a plant or a seed (as opposed to a particular gene) will reach crosses containing only a portion of the genetic material from the patented variety depends on the precise claims of the invention. It is also unclear whether a patent on the seed or the plant will be held to convey the right to keep a farmer from reusing seed, a right which the plant variety protection certificate holder does not currently have, although nations will be able to grant that right under the new UPOV convention. The European proposal would grant this right (although the right would be extremely difficult to enforce, and a patent holder would probably often choose to pass on the right to farmers and to price the seed accordingly).

As noted above, a patent conveys no rights beyond national boundaries. A nation's patent system can reach foreign products only if these products are imported. Many nations maintain protection against the import of products that would have infringed a patent had they been made in the importing nation. In many cases, these rights even permit the exclusion of products manufactured legitimately under foreign license.

Finally, it should be noted that trademarks or certified seed systems can often provide an effective equivalent of proprietary protection. A particular trade name may obtain such recognition by farmers that its protection is equivalent to protection of a variety. If the right to sell certified seed under a particular name is restricted to specific firms, that right becomes a basis on which competitors can effectively be excluded from selling the same variety.

Animals

By logic parallel to that of *Ex parte Hibberd* and its well-publicized predecessor, *Diamond* [The Commissioner of Patents and Trademarks] *v. Chakrabarty* (447 U.S. 303 (1980)), the U.S. Board of Patent Appeals has also decided that animals may be patented (*Ex parte Allen*, 2 U.S.P.Q.2d 1425 (1987)). This particular case involved a claim for polyploid oysters produced through a hydrostatic process. Although the possibility of patenting such animals was upheld, the claim for these particular oysters was rejected because of the existence of previous chemical processes of inducing polyploidy in oysters. The hydrostatic process was patentable under traditional principles. This decision has become extremely controversial and the subject of congressional hearings; its overruling by legislative action is not at all inconceivable. A bill to exempt farmers from patent infringement suits for breeding of patented animals passed the House in 1988 (HR 4970) but did not reach the Senate floor.

Hungary, the Commonwealth of Independent States (under an inventor's certificate), and Germany, are the only other nations known to allow patents on animals (Byrne, 1985; World Intellectual Property Organization, Committee of Experts on Biotechnological Inventions on Industrial Property, 1987; Rote Taube, 1991). At the regional level, under the European Patent Convention, as determined by a 1990 appellate decision with respect to the Harvard mouse, animals are also patentable.

Microbes

Many nations have routinely permitted process patents, that is, patents on the process of using a particular microbe to produce a specific chemical. Even nations that prohibit the patenting of pharmaceuticals often permit the issuance of process patents on specific microbial processes of producing pharmaceuticals (World Intellectual Property Organization, Committee of Experts on Biotechnological Inventions on Industrial Process, 1987). Canada, however, is probably not one of these countries, at least this is the implication of the *Pioneer Hi-Bred Ltd.* case discussed above, which criticized an earlier case, *Re Abitibi Co.* (62 C.P.R.(2d.) 81 (1982)), that supported the patenting of microorganisms.

The United States, along with some other nations, has gone further and now permits patenting of the microbe itself. This is the basic holding of *Diamond v. Chakrabarty* (see above). The microbe must, however, be novel. This novelty requirement is likely to be satisfied if the microbe is genetically modified. It is a matter of debate whether the requirement is satisfied by the isolation of a strain found in nature. In such a case, virtually any microbe isolated from nature could be patented. The older leading case, *In re Mancy* (499 F.2d 1289 (CCPA 1974)), presumed that isolation was not enough. However, the same court ruled to the contrary later in *In re Berger* (563 F.2d 1031 (CCPA 1977)). This lower-court proceeding ultimately became *Diamond v. Chakrabarty* on appeal; the issue of the adequacy of isolation was abandoned before the case reached the U.S. Supreme Court.

For the patent disclosure to enable others to practice the invention and build on it for research purposes, the United States and a number of other nations require that a sample of the claimed microorganism typically be placed in a depository so that it can be stored and be made available for use by other researchers (*In re Lundak,* 773 F.2d 1216 (CAFC 1985); *Feldman v. Aunstrup* 517 F.2d 1351 (CCPA 1975); *In re Argoudelis* 434 F.2d 1390 (CCPA 1970); *Rabies Virus* (FRG Fed. Sup. Ct, 1987), 18 *I.I.C.* 396 (1987); *Deposit of Biological Materials,* 37 CFR Ch. 1, Subpart G, Section 1.801–1.809). Great Britain imposes a similar requirement (Patent Rules 1990, Schedule 2 (S.I. 1990 No. 2384), 29 November 1990) (American Cyanamid Co. (Dann's Patent) 1971 RPC 425 (H of L, 1970)); see also Cadman [1985]). There is also an international convention governing such depositories (Budapest Treaty on the International Recognition of the Deposit of Microorganisms for the Purposes of Patent Protection, April 28, 1977; 32 U.S.T. 1241, T.I.A.S. No. 9768).

The United States pattern requires essentially free access; thus, there is nothing to keep a competitor from obtaining material deposited by U.S. patent holders and then using it in nations that do not provide patent protection. Japan and Europe, however, generally permit patent holders to require that anyone obtaining the material agree to use it for experimental purposes only and not to provide it to third parties. Industry and many patent experts are urging the global adoption of such stronger restrictions upon access to materials in depositories (Beier et al., 1985; World Intellectual Property Organization, Committee of Experts on Biotechnological Inventions on Industrial Property, 1987).

Related Rights

Of perhaps equally great importance is the growing body of patent rights available on specific biological molecules, materials, or processes. Although it is probably too early to attempt a complete list, there are patents on at least the following types of biological entities and processes:

• Specific genes. The patent office is regularly issuing patents for specific genes. What are actually claimed in a gene patent are typically the isolated gene sequences, novel plasmids incorporating the sequence, and plants transformed through such a plasmid; hence such a patent reaches the use of the gene for genetic engineering but leaves breeders free to work with the gene in its natural context (that is, as it occurs naturally in a plant).

• Traits. In at least one case (*Ex parte Hibberd*, 227 U.S.P.Q. 443), the U.S. Patent Office has issued a patent for any plant (of a range of species) that is claimed to contain a novel trait like that produced through the technology described in the patent.

• Specific processes. A number of processes of genetic engineering and tissue culture have been patented. There is the possibility of controversy as to how far protection extends to remote products of these processes.

• Diagnostic probes. A variety of diagnostic tools are patented. It should be noted that the basic monoclonal antibody concept has not been patented, although a sandwich assay process for making antibody reactions into practical diagnostic tools has been patented (*Hybritech, Inc. v. Monoclonal Antibodies, Inc.*, 802 F.2d 1367 (Fed Cir. 1986)).

There are also questions over the rights of people who contribute material (for example, one of the person's own human cell lines) that is used in a commercial application (see Office of Technology Assessment, 1987b). The property right concept was rejected by the California Supreme Court (*Moore v. Regents of the University of the University of California*, 57 Cal. 3d 120 (1990)). Proposals for the copyrighting (as opposed to patenting) of specific gene sequences (Roberts, 1987) continue to be made, although they have not gained much support.

PLAUSIBLE IMPACTS OF PROPRIETARY RIGHTS ON THE TERMS OF INTERNATIONAL GERMPLASM EXCHANGE

Several possible effects of the growth of proprietary rights (and, more broadly, of the increased commercialization of biology) can be identified. These include price effects, decreased willingness of private and public entities to transfer genetic materials, and increased privatization of research activity.

Price Issues

Privately developed patented material is generally more expensive than publicly developed unpatented material. This may sometimes be a problem; in particular, the difficulty of paying the price may leave farmers in developing nations less well able to compete on world markets. It should be remembered, however, that the patent holder's ability to raise the price is limited by the existence of competition. Private or public competitors offering equivalent products can keep the price down. Moreover, even if the new material costs more than the previously used material did, the incremental cost must be less than the incremental benefit to the farmer; otherwise, the new material will not be bought. At least in the short term, then, new developments in the developed world are unlikely to hurt the access of farmers in the developing world to genetic material because old material is also available. As new material becomes essential to meet new pathogens, however, the terms of access to the material becomes critical, and farmers in the developing world might be hurt by the evolution of the world plant breeding structure.

Issues of Increased Control of Germplasm

In both the public and the private sectors, the increased value of germplasm (a value that derives from the underlying commercial opportunities in plant breeding and biotechnology) is leading to hesitation in sharing this material. The patent system plays a mixed role in this hesitation. Under most nations' laws, it is impossible to obtain a patent if a description of the invention was published before the patent application; circulation of the material to be patented is thus risky and is generally avoided until after patent applications are filed. (The United States permits publication up to 1 year before filing, but business practice normally reflects the need to obtain patent coverage abroad as well as in the United States.) Thus, the initial impact is to slow the circulation of material, as well as publication. The long-term impact, however, is to permit both publication and circulation. Once the material is patented, it can be exchanged without affecting commercial rights or it could be used for a fee. In the absence of a patent system, it might well be protected through secrecy instead. Examples of trade secrets are found in the hybrid corn (maize) and hybrid poultry industry where companies control the breeding stock and sell only the hybrids.

This effect operates in the public sector as well as in the private sector. For many plant species, advanced traditional breeding is conducted primarily in the public sector, for example, in land-grant universities in the United States, in research universities and publicly funded research centers in other developed nations, in a variety of international research institutions, such as those operated by the Consultative Group on International Agricultural Research (CGIAR), and in the more advanced universities and public research institutions in developed nations. Although these institutions sometimes obtain legal protection for their working materials and research products, they have, at least until recently, generally made such materials and products freely available throughout the world.

This flexibility is changing, however, and public sector institutions are becoming less free with their innovations and materials (Johnson, 1987; National Institutes of Health, 1984). More and more, the public sector has been using confidentiality or patents as a way of supporting its own research or as a defensive step to respond to fears that ideas or material (including specific cell lines) will be patented or misappropriated by others. Restrictive distribution agreements are becoming very common, as universities require recipients of genetic materials to certify that the materials will be used for noncommercial purposes only. More and more, public entities are entering informal or formal affiliations under which information goes preferentially to small firms that are being spun off from public institutions or to large firms that provide support. Moreover, as genes become patentable under the regular patent laws, institutions have been known to restrict their distribution of improved materials out of fear of patent infringement.

In at least one case, that of the People's Republic of China, international access to germplasm has been limited since the nation gave rights to particular materials to specific international firms. There have also, it should be noted, long been specific governmental rules restricting the export of certain specialized and industrial breeding materials, for example, pepper from India, oil palm from Malaysia, coffee from Ethiopia, and tea from Sri Lanka. These restrictions are in addition to practical restrictions deriving from inadequate funds or staffing for the supply of materials, and the practical restrictions are frequently more important than the formal legal restrictions. (It should also be noted that the practical scientific network frequently ignores all restrictions, except perhaps for those that serve quarantine purposes.)

The impact of patenting on the research use of material also deserves attention. Part of the idea of the patent system is to place information about patented technologies in the public sector so as to provide a baseline for further technological advance. This is the reason for publication and for depositories. It also implies a right to use patented materials without permission for certain research purposes; this right is explicit in many nations' patent laws but is only judicially mentioned in U.S. law. The exact scope of the right is unclear, however; under some U.S. statements of the law, academic experimentation is permissible, but experimentation by a profit-making institution might not be (for example, *Pfizer Inc. v. Int'l Rectifier Corp.*, 217 U.S.P.Q. 157 (CD Cal, 1982)).

While plant variety protection permitted the use of a protected variety as breeding material for further varieties, the regular patent system is less likely to permit such use. Thus, a patent on a gene is likely to cover a variety of uses of the gene and, particularly, to cover varieties that are derived from patented materials and still contain the patented gene. This increases the patent holder's rights. Legal control, however, is limited to those nations in which the genes are patented; it is also limited by the fact that naturally occurring genes are unlikely to be patentable unless they are moved outside their normal host range.

Issues of Allocation of Research Between Public and Private Sectors and Developed and Developing Nations

The growth of new biotechnology firms, fostered in part by proprietary rights, is one of a number of factors that has led to a general transfer of human and financial resources to the private sector. Although some private sector firms are likely to participate quite fully in the international scientific material exchange network, some are not. Few firms are willing to supply their near-commercial materials, at least until they have obtained proprietary rights protection for those materials. Thus, privatization itself slows exchange to some extent (but proprietary rights permit some exchange that would not occur if the only protection available were secrecy).

There are also questions of the allocation of research capability between the developing and the developed world, which, so far, is unclear. This may be an industry in which the combination of high labor costs and (relatively) low research facility costs give the nations of the developing world a major opportunity. Conceivably, however, the opposite will prove to be true, and there will be a concentration of research in multinational corporations in the developed world with less opportunity for developing nations to enter the economically successful biotechnology sector. If the latter proves to be true, proprietary rights may make it harder for firms in developing nations to obtain access to markets in the developed world—and thus give these firms less opportunity to expand and to conduct further research.

SELECTED SPECIES

To examine the relative strengths of various possible effects of the patent system, it is useful to consider species for which the actual economic situation can be examined. The examples given in this section were chosen to provide a range in propagation form (hybrid, seed, vegetative, and animal) and economic focus (of interest to developed and developing nations or to developing nations only).

Maize

Maize is the classic example of a crop that was originally domesticated largely in the New World (North, Central, and South America and the Caribbean). Public sector breeding efforts have been substantial in both the developed world (for example, the U.S. land-grant system) and the developing world. There are base collections within the International Board for Plant Genetic Resources designated network in the Commonwealth of Independent States, Japan, Portugal, Thailand, and the United States (International Board for Plant Genetic Resources, 1987).

The Centro Internacional de Mejoramiento de Maíz y Trigo (International Maize and Wheat Improvement Center) has a substantial ongoing breeding program that involves the incorporation of germplasm from many nations into elite germplasm. It distributes its improved material through national programs and by sharing its material with private seed firms. About 35 million hectares are said to be planted in improved maize from the international system (Hawkes, 1985).

The private sector, however, is by far the most important supplier of seeds. It includes both large companies with their own breeding programs and smaller firms that depend on the development of inbred lines by the public sector (Hawkes, 1985). In a number of nations, there are public analogues, such as the French Institut National de la Recherche Agronomique. In general, these entities retain their own inbred lines and typically sell only hybrids. Some of the seed is sold internationally; some is produced abroad by multinational seed companies, local firms, or the local public sector. These seed sources are responsible for the majority of the supply of seeds in the developed world and for some portion of the supply in the developing world. The exact portion of this trade in the developing world is unclear; in some developing nations, however, a substantial portion of the seed used is hybrid, most, but not all, of which may be bred and produced locally but with financial support from international firms. The international firms often possess the best local distribution systems.

As suggested earlier, the use of hybrid seed provides the seed company with a biological equivalent of patent law. Unlike open-pollinated corn, farmers cannot use the hybrid crop as seed the following year without substantial loss of productivity. They are constrained to return to the seed producer. The seed producer thus has a strong incentive to use and invest in research in hybrids. Thus, by retaining control over certain of their parental lines, the firms can ensure effective protection of the crop. Such physical control can be as effective as (if not more effective than) intellectual property protection. It also implies that developing nations are not likely to be able to obtain the parental material, except through the process of developing new lines and inbreds from the segregating hybrid materials. (Note that even when suppliers are willing to provide the material, there may still be the economic cost of importing or paying royalties on breeding material. Foreign materials are often of little direct agricultural use.)

Because of the use of hybrids, plant variety protections are essentially irrelevant for maize. It must be noted, however, that some 80 plant variety protection certificates had been issued for maize in the United States as of February 1, 1988. This compares with 436 for soybeans and 1,924 for all crops combined. In contrast, regular patent law has played—and may continue to play—an important role with respect to maize. This is a crop for which there will be significant biotechnological research in the United States. The *Ex parte Hibberd* patent is itself for corn that contains more than a specified percentage of tryptophan, although it is hard to visualize how such a claim can be justified as being novel or as being properly enforceable against methods of achieving maize with a high-tryptophan content other than that disclosed. Other potential impacts include the development of varieties resistant to specific herbicides (maize is one of the crops for which there appears to be an especially large number of alliances between chemical firms and seed firms [Mooney, 1983]), insects, or specific pathogens. There are also a number of recent patents on tissue culture techniques for corn. Moreover, there is strong interest in the development of new growth regulators, which could be patented under traditional principles (Office of Technology Assessment, 1986).

If traditional maize varieties or comparable hybrid varieties continue to be available (both in a market sense and in the sense that they have not become susceptible to an evolving pathogen), and if breeding materials are shared reasonably within the public sector, the existence of other new maize varieties cannot hurt the access of farmers in the developing world to genetic material. At worst, it can hurt farmers' economic competitive position compared with that of farmers in the developed world. Fundamentally, it is only if the traditional materials become unavailable that farmers in the developing world can actually lose access to genetic material (and even then there is a possibility of reconstruction from commercial material). To maintain this access, developing countries should continue to produce or introduce a reasonable number of hybrid materials in either the public or the private sector, which may then be used for further selection. Note that this analysis does not address the questions of the economic competitiveness of agriculture in developing nations, which declines as developed nations adopt more advanced varieties and improves as the developing nations themselves adopt more advanced varieties. Nor does it address the effects on the environment or pest management of introducing genes for herbicide resistance or a single gene for disease resistance, both of which are among the early likely targets of maize research.

Rice

Rice, which, along with wheat, is probably the most important crop to much of the developing world, presents a completely different breeding pattern. In general, breeding is done by the public sector. Even in the United States, for example, breeding in the west is done through a network involving the land-grant universities, the Rice Research Board operating under the authority of the California Department of Food and Agriculture, and the California Cooperative Rice Research Foundation, Inc. (Rice Research Board, 1987). This network, which has been working on crosses with African and Asian varieties (Rice Research Board, 1986), produces a number of varieties;

the leading four make up over 90 percent of the area planted to registered seed used in California (Rice Research Board, 1987).

The leading breeding work of importance to the developing world has been done at the International Rice Research Institute (IRRI). In some cases, IRRI has provided new lines that are directly used; in others, it has provided materials that are used in national breeding systems (Hawkes, 1985). These materials are provided by IRRI without charge. The resulting cultivars are planted in over 50 percent of the developing world's rice-growing area (Dalrymple, 1985, 1986b). Within the rice network, there are base collections in Japan, Nigeria (International Institute for Tropical Agriculture), the Philippines (IRRI), and the United States. The International Rice Germplasm Center collection at IRRI in the Philippines is undoubtedly the most important of these (Chang, 1984b; International Board for Plant Genetic Resources, 1987). In addition, a number of nations (Colombia, India, Indonesia, Japan, Korea, Taiwan, Thailand, and the People's Republic of China) have substantial rice breeding programs.

Rice is a crop for which the activity of the private sector and proprietary rights have been largely irrelevant to the developing world. However, the development of an effective system for the production and use of F_1 hybrids by scientists in the People's Republic of China means that this situation may change quickly and radically in the near future. Hybrid varieties occupy more than 40 percent of the total area planted to rice in China. If these varieties become widely used in the rice industry outside China, they will undoubtedly encourage private investment in rice breeding, especially in such wealthy areas as Japan, the Republic of Korea, Taiwan, and the United States. One of the leading research goals for rice, however, is the use of apomixis (asexual seed production) to bring down the cost of producing hybrid rice seed (Rutger, 1986). Should this prove to be possible, the economics of the industry will quickly revert to its pattern before the introduction of F_1 hybrids.

Potatoes

The potato is a vegetatively propagated crop that is important in both the developed and the developing worlds. (True seeds are used in certain breeding processes, while the vegetative propagating material is, in practice, called a seed potato.)

In the United States, new varieties are developed by the public sector, and propagating material is multiplied and marketed by the private sector. Intellectual property is irrelevant because of restrictions in the Plant Patent Act of 1930 that exclude potatoes from intellectual property protection. Although the history is not entirely clear, the law was oriented toward asexual reproduction of fruits and flowers, and there was evidently believed to be difficulty in patenting breeding material that was itself the marketed edible product.

In most Western European nations, however, potatoes are covered by plant variety protection. Among these, The Netherlands is a major exporter of seed potatoes. For potato exports to nations without plant variety protection, there is no legal method to prevent further multiplication of the varieties. At present, that is not a serious problem for the exporter, since disease-free multiplication is difficult. One or two generations of multiplication can be taken into account in the pricing; after that, new seed material will be required because of the accumulation of virus disease in stocks. As more nations realize the ability to produce disease-free seed potatoes, it is possible that these exports will be restricted or subject to contractual protection.

The global base collection of potatoes is in Peru (International Board for Plant Genetic Resources, 1987, 1990), and there is a substantial public sector breeding program of great importance for the developing nations. Thus, the Centro Internacional de la Papa (International Potato Center) has provided germplasm to more than 70 nations in forms ranging from clones to seeds (Bofu et al., 1987; Hawkes, 1985).

Additional major collections of wild potato relatives are held in the German-Dutch germplasm bank at Braunschweig, Germany, and in the United States at Sturgeon Bay, Wisconsin. Both adhere to the policy of making their collections freely available to bona fide users.

Given the relative magnitudes of the research programs, it is unlikely that proprietary rights will significantly affect the access of developing nations to germplasm. There are sales of potato-propagating materials from developed to developing nations. Mooney (1983) emphasizes the role of Dutch firms and states that they export about 300 million guilders (US$105 million) a year in seed potatoes, of which 100 million guilders (US$35 million) goes to developing nations. The yield in the developing world is about half that in the developed world, but the rate of yield improvement in the developing world has exceeded that in the developed world (Horton and Fano, 1985).

Oil Palm

The oil palm is included as a representative of a tropical industrial crop. Its share in the total annual production of the leading edible oils has been increasing over the past decade. Among the oil palm producers, Malaysia is by far the leader, producing 4.1 million metric tons in comparison with about 1.8 million metric tons each in

Indonesia and in the rest of the world. Extensive new plantings in many countries of Latin America, Africa, and the Pacific pose a long-term threat to this position, however.

On the research side, Malaysia has taken a decisive lead. The plantation companies have traditionally played a major role; but since the 1970s, their activity has been supplemented by the Palm Oil Research Institute of Malaysia (PORIM), a government entity. The fruits of such research are considered as a national asset that is to be guarded carefully from others.

In particular, since the early 1970s, Malaysia has banned the export of oil palm seeds, a ban whose effect is supported by the fact that commercial planting material is based on a hybrid. Thus, the owners of the parental breeding lines are able to exercise control with little need for proprietary protection.

Hybrid protection, however, is being threatened by the development—after long research—of methods of clonal reproduction of oil palm through tissue culture. The research, started in Malaysia by the Oil Palm Genetics Laboratory, a group financed by a number of plantation companies, was brought to a conclusion by Unilever (United Kingdom), followed by the Institut sur Recherche de Huile et Oléagineaux (France) and a number of other research organizations, including PORIM of Malaysia. There have been problems of mutations in the cloned offspring, but once these problems have been worked out, any owner of a population of hybrids will be able to multiply selected palms in such populations.

Plant breeder's rights have not, so far, been used with oil palms. This is partly due to the lack of uniformity. The hybrids do not satisfy the requirements of uniformity, and their parents are typically chosen as individual palms from a heterozygous population. Moreover, a buyer tends to select among suppliers on the basis of trust and reputation rather than on variety name.

For this crop, the actual restrictions on the flow of material, which is likely to be decreasingly effective over time, derive from developing nations themselves rather than from the developed world. Moreover, the role of proprietary rights in the breeding material has so far been minimal; it is likely to continue that way. What may prove to be important, instead, are new technologies of working with—or finding substitutes for—palm oil. Some of these technologies may be patentable.

Cattle and Poultry

For cattle, breeding material is typically supplied through the sale of live animals, embryos, or semen. In the case of dairy cattle, there has been relatively extensive use of the new technologies, and there has been a trend to spread the Holstein breed from Europe and the United States to other areas. This narrows genetic diversity; the use of artificial insemination reduces the number of effective sires and embryo transfer also reduces the number of females. For beef cattle, the transfer of genetic material is in the opposite direction, with imports of new breeds from Europe through Canada and directly from Brazil (Council for Agricultural Science and Technology, 1984b).

So far, formal intellectual property protection plays little effective role here, and those working in the field see their objective as one of upgrading the global animal stocks. Naturally, the price of the breeding material reflects the economic benefit of the improved material. And, in appropriate cases, for example, the sale of a bull, the price also reflects the value of the animal for breeding purposes. It is hard to see that legal barriers based on proprietary rights will significantly increase the costs of breeding stock, which already reflect discounted future benefits. Nevertheless, the scarcity of material affects its cost, and breeders in developing nations as well as others may well pay substantially more for the particular germplasm.

For poultry, which has a much higher replication rate, the marketing pattern is quite different. Here, parental or grandparental lines (which are often inbred) are typically maintained for breeding purposes and are carefully controlled as a form of effective protection. Only the later generations or hybrids are sold. The number of firms is relatively limited, and the genetic base is said to have been reduced significantly (Council for Agricultural Science and Technology, 1984b). Whether it is parental or grandparental material that is maintained in the developed world, the increasingly important large-scale poultry operator in a developing nation must regularly return to a supplier in the developed world to obtain commercial stock (Hill, 1986). Thus, for finished lines, the current terms of international transfer do not favor developing nations. It would take time, investment, and trained animal geneticists to breed new material from the range of germplasm that is available.

The cattle marketing patterns have already responded to the new animal reproductive technologies (such as embryo transfer); it is not clear how or whether they will change further in response to biotechnology and the possible emergence of animal patents. It is not unlikely that a patentable cattle improvement, for example, a cow genetically modified to produce and respond to increased quantities of bovine growth hormone, will be most effectively marketed through the traditional method of selling animals, embryos, or semen (at prices that capitalize the breeding value of the improved animals). Depending on the evolution of the law, however, it might become possible to obtain royalties from further generations of genetically modified animals. In such a case, patent rights might impose significant costs on international transfer of the improved animals. In the absence of such rights over progeny, however, breeders might vertically integrate with the dairy or meat sectors in order to protect their technology.

The effects of possible animal patents are arguably more favorable to developing nations in the case of poultry. Research in this area is oriented toward diagnostics and vaccines as well as increased productivity (Purchase, 1986). If animal patents are actually issued and provide protection for subsequent generations, the poultry firms may be more willing to supply newly patented breeding lines for a long-term royalty. Patenting in the poultry area might also be healthy for the overall diversity of poultry materials. The relatively few lines that provide the basis for current industry might well be expanded.

Microbes

There has been some flow of microbial genetic material from developed to developing nations, for example, as part of the pharmaceutical industry's search for antibiotics. There may be more in the future, particularly in the biological pest control area and because new applications of industrial biotechnology appear to be increasingly appealing (Demain, 1983). At the same time, except for natural fermentations and the like, there have been relatively few applications of industrial biotechnology in the developing world. Moreover, microorganisms are significantly engineerable in the sense that genes for particular functions can be transferred from organism to organism. This is also an area in which there are some explicit legal export restrictions, such as those under the U.S. Export Administration Act (50 U.S.C. App. 2401-24-20) (see also 15 C.F.R. Part 399.1, Suppl. 1, group 7 [including certain culture media and certain chemicals obtained by bioprocessing] and group 9 [including certain viruses and microbes]).

Manufacturers attempt to protect their microorganisms in two ways: physical protection (for example, ensuring that the organism itself is not transferred to other parties) and patenting. There is a trade-off involved in patenting, in that a patent disclosure, as discussed above, must be enabling, which, in the case of a microorganism, typically means that the patent holder must place a sample of the material in a depository so that it is available to other researchers. The strategic cost of providing information to competitors in this way may often lead a firm to use the physical protection method of protection instead of patents.

It can thus be anticipated that in certain sectors, such as those involved in fermentation and pharmaceutical production processes in which the organism can be physically retained, the materials may be closely held rather than patented. Competition among the developed countries in the technology is very strong and information is carefully guarded. It has been said, for example, that the fermentation technology of the pharmaceutical firms is so closely held that university research in the area is well behind. It also follows that the transfer of the fermentation and production processes to firms in the developing world will suffer; these technologies may have to be independently developed in developing nations, possibly relying on local sources of genetic material. Thus, in this sector, the patentability of microorganisms as such is likely, at worst, to make little difference—positive or negative—in providing access to the materials and, at best, to permit some access to materials that would otherwise be held closely.

Patentability, however, will arise as a central issue in those market sectors in which the microorganism itself must be provided to the customer. This is the case, for example, with bakers' yeast and a number of microbial pesticides. In these cases, the opportunity to support substantial research (and to support product effectiveness, safety studies, and licensing costs) is very likely to depend on patentability. The lack of protection in some markets in developing nations may slow application there, depending on the firm's fear of competition and the effort and investment needed to develop the market.

This is an area that risks significant tension. Consider the possibility of a patented gene derived from a microbial source in the developing world and applied in a product that is sold back or used in a process that competes with an export from a developing nation. It is also likely to be very difficult to argue that a purified strain of a natural microorganism (as opposed to an engineered strain) should be protectable or that an individual should be able to protect his or her own cell line, and not to accept an argument that a nation that is the source of such a strain or cell line has a claim as well.

IMPLICATIONS OF THE EXAMPLES

The examples presented above emphasize three points. First, there are major differences among different species: the structures of the international maize and international rice sectors are radically different, as are those of the poultry and cattle sectors. Second, for the species considered here, the impacts of new forms of intellectual property are relatively small. Either there is already some form of proprietary or trade secret protection or there is a significant public sector. Third, significant barriers are imposed by governments in both the developing and the developed world. (It is possible that somewhat different patterns might emerge in such areas as vegetables or legumes.)

For almost all the plant and animal (but not microbial) innovations analyzed here, there is a common pattern, as outlined below.

1. A new development occurs in the developed world (which may or may not depend on use of germplasm from the developing world).

2. This new material costs more than the previously used material, but the incremental cost is less than the immediate or short-term benefit to the farmer.

3. Competitive harm to farmers in developing nations can arise from the possibility that (a) farmers in the developed world will use the product first and be more competitive for some period, (b) foreign exchange costs of importing the advanced material will be prohibitive, or (c) the materials will be more adapted to production in the developed world.

4. As long as public or old material is also available and useful, farmer access to these resources is not actually reduced as a result of the new innovation or its patenting in the developed world. Availability of nonrestricted material is a function of the rate at which such material becomes obsolete because of vulnerability to evolving pathogens, the degree of market competition, the presence of public sector breeding, and demand for products.

5. The increased commercial value of germplasm creates pressures on both the public and the private sectors to control access to their information and genetic materials. To the extent that such access is restricted, whether by institutions in the developed or the developing world, there may be more severe costs for developing nations.

It is this last step that is most important. Extrapolation from the species analyzed here suggests, at least in the short term, that the developing world has relatively little to fear (from a genetic resources perspective) from the developed world's adoption of proprietary rights for life forms. Given the existing private sector restrictions, the more serious issues are deliberate public sector restrictions on access, whether taken as part of a commercialization process or as a political response to proprietary rights. In several cases, moreover, there is a trade-off between disclosure and patentability under which the emergence of patent systems may improve the access of developing nations to materials that would otherwise be retained physically. This is particularly important for hybrid crops, poultry, and microorganisms.

The discussion also suggests several technical legal issues that will be at least as important for developing nations as is the issue of patentability itself.

• To what extent can a patent on a life form be invoked to restrict movement of the products of the patented life form in international commerce? This is more of a commercial than a germplasm issue, but it will be extremely significant.

• To what extent can a patent be used to restrict the use of a life form as one of several sources of genetic material for further breeding purposes? This may be especially important in those systems that allow patents on individual genes, because a claim on the gene itself would reach all offspring that contained the same gene.

• To what extent can reproduction rights in the offspring of a patented organism be retained when the organism is sold? If they can be retained, royalties are likely to be claimed for the lifetime of the patent; if they cannot, some organisms (for example, poultry breeding material) may not be made available at all.

• When deposit of a new strain is required as a condition of patentability, what are the restrictions on access to and use of the deposited strains?

• Are any of the protections (for example, for research purposes or for farmers to use saved seed) inherent in the plant variety protection system built into the new regular patent system as it works in practice? (Note that a seed producer may find it wise from administrative and marketing perspectives to give up any right to prevent farmers from reusing seed and, instead, to price the initial sale in anticipation of such reuse.)

The central issue for developing nations is usually whether or not there is a substantial public sector program or domestic breeding industry in that nation. If there is such an industry, there is likely to be effective fairness in access for farmers in the developing world, and in such a case there is likely ultimately to be a reciprocity that avoids any perception of unfairness deriving from the developed world's access to the developing world's genetic resources.

Finally, although political arrangements for the resolution of disputes over germplasm are discussed in Chapter 14, there may be value in facing some of the issues posed in this chapter through a scientific code of conduct designed to encourage the exchange of germplasm at the precommercial and noncommercial levels. Such a code could usefully be addressed to institutions in the public and private sectors as well as to governments. It might, for example, cover issues of access to materials in depositories, support a common understanding of the line between research and commercial use, and encourage both the public and the private sectors to make available to developing nations research and advanced materials on as liberal a basis as possible.

RECOMMENDATIONS

Although proprietary rights may sometimes constitute barriers to the international flow of germplasm, their effects so far have been much smaller than those of barriers imposed by public institutions, including governments responding to the increased value that biotechnology and growing commercial use have placed on germplasm. Issues of proprietary rights could be wisely addressed through a code of conduct governing the exchange and availability of genetic material.

Intellectual property protection systems should be designed to minimize the potential for restricting the free exchange of germplasm among nations.

Areas of particular concern include (1) terms of availability of material placed in depositories, including depositories acting in support of plant breeders' rights; (2) whether intellectual property rights make provision for protected materials to be usable for research and breeding purposes; and (3) import restrictions intended to prevent the circumvention of process patents.

Public institutions should not respond to the commercialization of germplasm by enacting restrictions that could limit the use of genetic resources by developing nations.

In an era of shrinking research budgets and strong competition to improve the quality of staff and facilities, many public institutions use the royalties from patents to fund research. To the extent that this practice inhibits international exchange of germplasm it should be limited. Certain of the CGIAR research centers may find it necessary to patent their materials to prevent protection being applied by others. If these institutions find it necessary to patent materials, they should provide royalty-free licenses when necessary and possible for the benefit of all countries. This approach is similar to the policy of the Agricultural Research Service of the U.S. Department of Agriculture (Brooks and Murphy, 1989).

Private sector institutions are encouraged to make available their materials in developing nations on as broad a basis as possible.

In particular, this includes material that is not currently the basis of varieties being marketed in developed nations and royalty-free licenses for the use of materials with a low commercial potential.

13

Genetic Resources: Assessing Economic Value

The improvement of plant and animal genetic resources is increasingly an international activity. Improvements in the techniques for breeding and applied genetics (modern biotechnology) enhance the value of genetic resources. The difficulty is in measuring the value. This chapter reviews evidence that provides general support for the proposition that investment in the collection, preservation, and management of genetic resources to support crop improvement is economically sound.

Genetic resources are valued as public consumption goods in the same way that scenic lakes and mountains are. Therefore, their conservation and preservation, usually in in situ states, are also valued. They are also of value in plant and animal breeding programs. This is an indirect or derived value, because it is derived from their contribution to economically improved plants and animals. This value is termed their *producer good* value. (The distinction between the consumption good and producer good value is effectively the same as the nonuse value and use value distinction in the valuation literature.)

VALUING GENETIC RESOURCES

The fundamental difficulty in measuring the value of genetic resources is that genetic resources are seldom traded in markets. Hence, prices cannot be observed except under exceptional circumstances (usually when proprietary rights have been established). Analysts use two basic methods to place value on such nonmarketed goods. The first is the contingent valuation method, in which the analyst seeks to elicit indicators of what people would be willing to pay for a public consumption good such as germplasm. This method is suited to measuring germplasm's value as perceived by the general public, such as in nature reserves. The second is the hedonic pricing method (the productivity method), in which the economic value of the nonmarketed good is estimated or inferred from the value of the marketed good in which it is contained or to which it contributed. This method is suited to germplasm uses in breeding.

Few evaluations have been made for the value of the public consumption good of genetic resources. It appears likely, however, that the demand for conserving genetic resources may be strongly affected by income level. Thus, in general, poor people (as expressed by their organizational leaders) might be less willing to pay for the pure conservation of genetic resources. This would be particularly true in the poorest countries where other concerns rank higher.

By contrast, if the source of value for genetic resources is based on them as producer goods, the picture is much different. Because the proportion of the income of poor people that is devoted to food is high, poorer people generally place a high value on plant and animal improvements that could increase production and hold down prices. Hence, the genetic resources that help to produce such improvements should be highly valued by the poor.

Because plants and animals are generally traded in markets, an economic value can be placed on them. Furthermore, the value of improvements in these plants and animals can be determined directly. Hedonic pricing methods can be used to relate the value of such improvements to the genetic resources and other activities that were used to produce them.

In this chapter, hedonic pricing methods are used to estimate the value of rice germplasm resources in a specific geographic setting as producer goods; their value as consumer goods is not estimated. These estimates are then used to consider the likely similar values of genetic resources in other crops and animals. The extension of the rice analysis to other commodities or other settings is subject to considerable uncertainty.

THE PROBLEM OF LACK OF MARKETS

The fundamental characteristic of germplasm resources that differentiates them from other natural resources or goods is that they are usually replicable at a very low cost. This means that the scarcity of a particular genetic resource may be transient. A genetic source of resistance to a plant disease (for example, grassy stunt virus in rice) may be very valuable, because more rice can be produced when the resistance gene is incorporated into rice varieties. Once the source is discovered (in this example, in *Oryza nivara*, a wild species in the International Rice Research Institute [IRRI] collection), it can be replicated easily and cheaply in rice plants. Thus although initially scarce, it may quickly become abundant.

A private enterprise might not be able to earn a substantial income from selling germplasm resources under these conditions of unrestricted replicability. A firm might, for example, collect landraces and wild species of rice, identify particular traits and characteristics, then seek to sell particular germplasm resources to plant breeders. However, once the genetic resources (via seeds or other reproducible materials) are available to a single breeder, they will be reproduced and distributed. Thus their price will fall. This is the justification for supporting collection and management of genetic resources as well as plant and animal breeding and research in the public sector.

The simple economics of demand for improved seed indicates that as the price declines, more seed is demanded. Two types of costs are associated with supplying improved seed. The first are those of developing the improved genetic combination (the variety). These are the initial costs of locating, developing, testing, storing, and classifying the genetic resource and are fixed, in that they do not depend on the number of units of seed sold. Average fixed costs then decline with growth in sales. Costs of the second type are those of replicating, treating, and packaging the improved genetic combination in seed units. If this replication is easy (for example, if farmers can save their own seed), the marginal variable cost, that is, the cost of replicating the genetic combination in an additional seed unit, is very low. If many suppliers of seed are competitively selling the seed, they will not be able to charge a price that is above this marginal cost. If this price is below average fixed costs, no private firm could capture a return to the plant breeding program through seed sales.

Remedies for Market Failure

There are two remedies for this market failure. The first is to limit sales of the improved genetic material and charge a higher price to buyers. If replicability can be controlled through natural means (for example, seeds produced by hybrid corn plants themselves produce plants of much less value than the original hybrid), the supplier of genetic characteristics, including landraces and advanced lines, could charge a price sufficiently high to cover costs (see Chapter 12). In the absence, or even in the presence, of strong intellectual property rights, the second, and historically more common, remedy to this problem is to establish public sector institutions to engage in plant and animal improvement and in the collection, classification, preservation, and use of germplasm resources associated with this activity.

Naturally occurring germplasm resources are generally not subject to intellectual property rights protection. Even when markets for crop varieties exist, markets for germplasm resources do not exist. Thus, most germplasm resources are collected, catalogued, and maintained in public sector collections. Many private sector collections exist, but these are usually composed of parts of public sector collections and specific advanced breeding materials developed by a private firm.

PRICE EVALUATION METHODS

Improved plants and animals are produced by systematic breeding (genetic recombination) and selection activities. Germplasm resources constitute a type of capital or source material which, when combined with other forms of capital (fields, buildings, and equipment), research labor (crossing and selection), and breeding technology, produces superior plants and animals. The process is subject to uncertain outcomes (stochastic) and has a trial-and-error aspect. (Evenson and Kislev [1975] have presented a stochastic model for agricultural research.)

Nonetheless, these processes are systematic, and the role of germplasm resources is relatively well understood. The relevant germplasm resources can be classified into two groups: (1) the stock of naturally occurring genetic resources (these include landraces, wild species, and related materials) and (2) recombinations or advanced stocks of materials that have been made in past systematic breeding programs. These recombinations may be commercial varieties of different vintages or advanced lines that may or may not be commercially developed. Both types of resources are important. The recombinations may represent years of systematic breeding work and development. Such recombinations are usually based on a small subset of the naturally occurring germplasm resources and usually form the parent materials in breeding programs.

Farmers of the plains of northern Thailand transplant upland rice seedlings. Credit: Food and Agriculture Organization of the United Nations.

Modes for Introducing Germplasm Resources into Advanced Cultivars

There are effectively two modes by which genetic resources are selected for incorporation into advanced breeding stocks. The first mode is through the building of breeder collections, that is, of working collections used for strategic crossing purposes. Most breeders maintain a relatively small set of cultivars for crossing purposes. The local breeding program typically keeps some landrace material and a set of advanced materials. Local breeders rely on information from national and international sources and from the managers of germplasm collections to identify promising new materials. Chang (1976) and Hargrove (1978) have described this process for rice. National breeding programs may maintain larger numbers of germplasm resources (landraces and wild species). They seek to produce new varieties suited to fairly large regions, but they also produce advanced germplasm materials for more local breeding programs. International breeding programs concentrate even more on germplasm development and seed to provide national systems with advanced materials (Chang, 1976).

The second mode is the incorporation of distinctive genetic traits into breeder cores. For example, several serious pest outbreaks of rice such as tungro virus and green leafhoppers (*Nephotettix* spp.), affecting rice variety IR-8 in 1969 and a grassy stunt virus outbreak that occurred in 1977, precipitated collectionwide searches of germplasm resources by IRRI (Chang et al., 1975) and others for genetic traits that conveyed resistance to these diseases.

The Hedonic Pricing Method

The hedonic pricing method used is a statistical procedure in which the value of plant or animal improvements is statistically associated with the germplasm and breeding stocks and other inputs that contributed to the improvement. The value of the improvement is measured in terms of relative amount of product per unit of input (for example, the yield per hectare of a new variety relative to those of older varieties is an index of the value of the new variety) or in terms of the change in the costs of production of a unit of the commodity. The genetic resources can be quantified in terms of the complexity of advanced breeding lines and the number of landraces in the pedigree of the variety. Other inputs into the improvement process include the labor associated with the process and capital equipment.

The usual procedure entails a statistical multiple regression analysis to ascertain the value of the improved plants or animals as a result of the contributing inputs, that is, germplasm stocks, breeding and selection, labor, and other research program resources. From this regression, the marginal contribution of inputs can be estimated in value terms. From these estimates, the value of collecting and maintaining genetic resources stocks can be inferred.

For example, if it was found that improved crop varieties were essentially based on recombinations of existing advanced materials and that no new germplasm resources were included in successful varieties, this would indicate that there would be little value to expanding existing collections of landrace and wild species. On the other hand, if new germplasm material is being incorporated into improved varieties, this would indicate that there is value to maintaining germplasm resource collections. If the genetic resources in question were found in the unusual or fringe materials in the collection, this would indicate that there is value to an expanding collection including such materials.

RICE: AN EXAMPLE OF PRICING METHODS

Pricing methods have been used to establish a value for rice germplasm resources in India (Gollin and Evenson, 1990). This section summarizes that application.

Institutional Background

Cultivated rice falls into two species, *Oryza sativa* and *O. glaberrima*. The former is the common Asian cultigen, whereas the latter accounts for a small fraction of African rice production. In addition to these two cultigens, the genus *Oryza* includes about 20 wild species, although some scholarly disagreement remains on the exact number (Chang, 1985d). It is estimated that about 140,000 cultivars or types of rice exist in the world today (Chang, 1985d). These include cultivated varieties of *O. sativa* and *O. glaberrima* as well as landraces and wild species. About 85,000 of these are in a long-term storage facility at IRRI. Seeds of these germplasm resources are catalogued according to their agronomic and genetic characteristics, and they are kept under conditions of low temperature and humidity (Chang, 1976, 1987).

Unimproved materials from the collection are evaluated by different disciplines at IRRI and are also sent out freely to scientists around the world. In addition, new germplasm sources may be incorporated into the pedigrees of improved lines that are sent out from IRRI through a variety of testing programs (Hargrove, 1978). Other rice germplasm collections are maintained by various national programs and by some regional centers (as in India).

India has three national-level breeding programs that work on rice improvement: the Directorate of Rice Research in Rajendranagar, the Central Rice Research Institute (CRRI) in Cuttack, and the Indian Agricultural Research Institute (IARI) in New Delhi. In addition, there are 21 state programs, some of which have branch stations. The national-level programs generally perform the more technically difficult work of making crosses involving landraces and wild species. The national programs also tend to perform more experimental scientific work and to screen larger quantities of material for certain useful characteristics, such as disease resistance and pest tolerance.

The national stations have the largest collections of germplasm in India. CRRI maintains a collection of about 15,000 accessions of rice. Of an estimated 60,000 to 65,000 accessions held in different collections in India, perhaps 18,000 accessions are held at Raipur, another main germplasm storage location. The National Bureau of Plant Genetic Resources, which has recently been given a mandate to collect and classify the germplasm resources being held in different locations around India, has perhaps 12,000 accessions at present (Gollin and Evenson, 1990).

Analysis of Released Rice Varieties in India

From 1965 to 1986, Indian rice breeders released a total of 306 rice varieties for planting in India. (These included varieties developed in earlier years.) This number also includes 27 high-yielding varieties requiring more intensive chemical inputs (so called green revolution varieties) that were actually developed at IRRI but that were released in India. These 306 varieties were the result of approximately 20,000 crosses made by Indian (or other) breeders since independence from the United Kingdom in 1947.

Pedigree Analysis

A pedigree analysis of each released rice variety was undertaken. It traced lineage back to the original genetic resources in the variety (in most cases landraces, but in some cases wild species). The characteristics emphasized in the development of each variety were recorded. These included disease and insect resistance, stress tolerance, and agronomic and grain quality. Many varieties emphasized more than one characteristic. Disease resistance tended to be the most sought after characteristic; this was followed by insect resistance, but no single characteristic dominated the breeding strategies. The analysis led to a quantitative description of varieties in terms of year of release, releasing institution, characteristics emphasized, parent and grandparent combinations, number of landraces in the pedigree, and the number of generations from crosses of landrace materials.

Varietal Releases

A steadily increasing trend in varietal releases was seen from 1965 to 1975, with approximately constant releases since then (Table 13-1). There were 27 rice varieties developed originally at IRRI and 6 varieties developed in other foreign breeding programs that have been distributed evenly over time.

TABLE 13-1   Varieties of Rice Released, by Year

| Year of Release | Number Released | | | Average Number of | |
	India	IRRI	Other Foreign	Landraces	Generations[a]
1965	2	0	1	1	2
1966	2	1	0	2	1
1967	2	0	0	2	1
1968	6	0	0	2.5	1.5
1969	6	1	0	2	0.9
1970	12	1	1	3.8	2.4
1971	8	1	0	3.6	2.2
1972	17	5	0	4.2	2.6
1973	14	0	0	4.1	2.9
1974	7	0	0	4.0	2.7
1975	14	3	1	5.4	3.4
1976	17	2	1	4.1	2.7
1977	12	0	0	4.2	2.8
1978	11	4	0	4.3	2.9
1979	15	1	0	5.8	3.3
1980	20	0	0	4.3	2.7
1981	11	3	0	6.7	3.9
1982	23	2	1	6.1	4.0
1983	14	0	0	4.8	2.9
1984	7	0	1	5.5	3.6
1985	31	1	0	6.1	3.9
1986	22	2	0	8.7	4.6

NOTE: IRRI, International Rice Research Institute.

[a]Generations refers to the average number of generations since introduction of the oldest landrace in each pedigree.

The average number of landraces in each pedigree and the average number of generations since the oldest landraces in each pedigree by year of varietal release are shown in Table 13-1 (the number of generations is a useful index of genetic complexity). These data show steady growth over time in pedigree complexity. The early, high-yielding rice varieties released before 1970 had relatively simple pedigrees. Recent varietal releases are much more complex, with as many as 27 landraces and as many as 12 generations of crosses going back to original landrace material. Foreign germplasm is found in almost all varieties released in India, even though the country is the source of a large share of the world's rice germplasm resources. (IRRI is the major source of introduced germplasm.)

Institutional Releases

The Central Variety Release Committee releases rice varieties deemed to have broad regional potential. Releases by the states tend to be more location specific. Most major state programs released 10 or more varieties from 1965 to 1986. Approximately 100 of the 306 varieties released since 1966 were planted in 1984. Of these 306 varieties, 118 were released after 1980.

Landrace Appearance

Table 13-2 tabulates landraces by the year of their first appearance in a released variety. These data show that the landrace content for Indian varieties has been expanding (and, by inference, that if

TABLE 13-2   First Appearance of Various Landraces in Indian Rice Pedigrees Between 1965 and 1986

Year	Landraces Appearing in Pedigrees of Released Varieties for the First Time		Total Number, 1965 to 1986
	Number	Percent of Total Released	
1965	1	(0.6)	1
1966	5	(3.0)	6
1967	3	(1.8)	9
1968	6	(3.6)	15
1969	7	(4.2)	22
1970	8	(4.8)	30
1971	7	(4.2)	37
1972	17	(10.1)	54
1973	7	(4.2)	61
1974	2	(1.2)	63
1975	11	(6.5)	74
1976	11	(6.5)	85
1977	6	(3.6)	91
1978	8	(4.8)	99
1979	5	(3.0)	104
1980	13	(7.7)	117
1981	5	(3.0)	122
1982	11	(6.5)	133
1983	5	(3.0)	138
1984	4	(2.4)	142
1985	12	(7.1)	154
1986	14	(8.3)	168

breeders were constrained to work only with the landrace materials being used in breeder collections as of about 1970, they would have been far less productive).

Origins and Inclusion in Varieties

For purposes of further analysis, these landrace materials are classified according to origin (Indian and foreign) and further categorized into pre-1975, post-1975, and specialized groups, with the percentage inclusion in varieties noted (Table 13-3). Pre-1975 and post-1975 landrace categories entered varieties through the building of breeder collections, that is, working collections developed for strategic crossing purposes.

Regression Analysis of Data

The price evaluation analysis used entails a statistical regression relating a measure of varietal improvement in farmers' fields to factors expected to be associated with varietal improvement as described above. District-level measurements of rice yields are available for India. Yields are general productivity indexes and may be influenced by both varietal and nonvarietal factors. Accordingly, nonvarietal research activities and other increasing investments in rural infrastructure must be considered.

For India a two-stage regression analysis was pursued by using data for 240 districts. The first stage was designed to estimate the relative contribution that overall varietal improvement made to pro-

TABLE 13-3 Proportion of Pre- and Post-1975 Materials of Indian or Foreign Origin Included in Released Varieties

Landrace	Proportion of Planted Area
Indian origin	
Pre-1975	0.62
Post-1975	0.11
Specialized	0.12
Foreign origin	
Pre-1975	0.84
Post-1975	0.07
Specialized	0.03

ductivity growth in rice. If it cannot be established that modern high-yielding varieties actually contribute to productivity growth, there is little point in attempting to identify germplasm resource effects, although use of such resources may be essential just to maintain, rather than increase, productivity in the face of new insect or disease pressure. (This maintenance effect is included in the productivity estimate discussed below.) However, if it is shown that varietal improvement does affect productivity, one can proceed to the second stage, where germplasm content variables can be incorporated into the analysis.

The first stage estimates showed that varietal improvement was a significant determinant of rice yields for Indian districts from 1959 to 1984. The dependent variable was the rice yield for the district and year relative to the average rice yield for the district from 1957 to 1960 (Gollin and Evenson, 1990). The following independent explanatory variables were included:

• The proportion of land planted to modern varieties released since 1966. This variable was included to test whether varietal improvement actually affected productivity. By 1986, 60 percent of the rice area in India was planted to modern varieties.
• Indian agricultural research, a stock variable reflecting the contributions of nonvarietal public agricultural research.
• Indian private sector research and development relevant to agriculture.
• Agricultural extension services.
• Literacy of farmers.
• Roads, a road density variable.
• Markets, a measure of regulated market infrastructure.
• Irrigation investments.
• A specialized program providing additional extension and infrastructure to farmers.

The estimates showed that varietal change contributed more than one-third of the rice productivity gains realized over the period after the green revolution (1972 to 1984).

Having shown that rice varietal improvement did contribute to rice productivity, the second-stage analysis can be justified. In this stage, variables measuring the genetic content of varieties actually planted by farmers were substituted in the analysis for the variable representing genetic content of modern varieties. The analysis was undertaken only for the most recent 5-year period (1979 to 1984), because the relevant data for earlier periods were not available. The dependent variable, rice yield, was indexed relative to the average yield from 1972 to 1974. Thus, the analysis focused on yield changes in the post-green-revolution period. It sought to determine whether rice yield gains after the 1972 to 1974 period were systematically related to the germplasm contents of the varieties planted by farmers.

For each district, germplasm content variables were defined for the land actually planted. These germplasm content variables were defined for five clusters of variables: source of breeding materials, varietal characteristics, parental origin, pedigree complexity, and landrace content.

Separate regression analyses were then undertaken for each cluster to estimate the impact of each cluster of genetic variables on yield. From these estimates were derived the percent change in yield as a result of a 1 percent increase in the variable at the expense of its reference group. Details of the analyses are given in Gollin and Evenson (1990).

The analyses of source variables indicated that the varieties released by the Central Variety Release Committee had a higher impact than those released by the states. Varieties of foreign origin that were not released directly by IRRI were also associated with higher productivity. Varieties released directly by IRRI had lower impacts on productivity. Many of these were the early green revolution varieties, or more recent varieties released in response to specific problems thus designed more to maintain than to increase productivity.

The varietal characteristic variables indicate that varieties stressing grain quality, specific agronomic characteristics, or stress tolerance had higher yields than those with no selection strategy for these characteristics. By contrast, varieties with resistance to disease and insects appeared to have lower yields. However, this may reflect pest or disease incidence as these were not monitored.

The impacts of parental origin were particularly complex. In general varieties with mixed foreign and Indian parentage appeared to have better yields than varieties in which both parents were advanced and of foreign origin.

Analysis of pedigree complexity indicated that varieties with higher landrace contents and more generations of development had higher yields. This supports the contention that landrace genetic resources are valuable breeding materials.

Estimates for the impact of landrace content in varieties more directly address this implicit value of genetic resources. The data were compared to material based exclusively on old (pre-1975) national landraces and showed that new landrace resources generally had significant positive impacts on productivity of released varieties.

National landrace materials (which entered pedigrees after 1975) also contributed to increased yields. This suggests that systematic and strategic incorporation of more landrace materials into breeders' active lines has a payoff.

The estimated impacts of landrace materials obtained from specialized national or international searches were quite large. This has implications for genetic resources management. The genes identified in such searches are typically uncommon and found in materials with few or no other valuable traits. The probability of discovering accessions with such uncommon traits is increased if collections are large. The evidence of their value to increasing yield indicates that the general activities of collection, preservation, and maintenance of landraces and related wild species are of value.

Economic Analysis

Forty-one percent of the Indian rice acreage in 1984 was planted to varieties containing pre-1975 international breeding materials. While this acreage includes both dryland and paddy rice, in practice only irrigated and lowland rainfed paddy rice varieties have been genetically improved. These were, effectively, the original green revolution genetic resources. The analysis showed that a 1 percent expansion in the use of these old international materials would still increase average rice yields in India by 2 percent.

The more pertinent question, however, is what the post-1975 materials have contributed to yields. This can be calculated from the 1984 levels for the variables examined. The calculation indicates that yields were higher for all of India by 5.6 percent than they would have been had only the germplasm resources present at the time of the original green revolution been available to breeders. The total increase from modern varieties in the period was 13.4 percent. It can be inferred then that reworking of the original (pre-1975) genetic materials alone would have contributed 7.8 percent to yields (Gollin and Evenson, 1990).

The 5.6 percent added yield increase realized from the newest added germplasm resources up to 1984 is likely to continue to increase as the area planted to the newer varieties expands. If the 5.6 percent is conservatively treated as having been realized at a 0.5 percent rate over the 11 years leading up to 1984, a cost-benefit or rate of return analysis can be undertaken, under the conservative assumption that no further gains after 1989 will be realized. To do this, an estimate of the time lag between incurring costs and realizing the yield gain benefits is required. With that time lag, the present value of the stock of genetic resources can be computed, given a discount rate.

Present Value of Rice Germplasm Stocks

The Indian study indicated that the time period from initial breeding strategy until resultant varieties were fully adapted by farmers was about 9 years. Thus in 1975, the availability of the genetic resources enabled this contribution of 0.5 percent of Indian rice production (worth US$50 million in 1990 dollars) to be realized 20 years later (9 years from initial breeding to varietal release plus 11 years to full adaption). Since this contribution is maintained from 1995 and thereafter, it can be treated as a benefit "stream" and its present discounted value (as of 1975) computed. Using a 10 percent discount rate, this value is US$75 million. (At 5 percent this value is US$377 million.)

These values can then be compared with the costs of maintaining and operating the genetic resources collections. They can also be compared with the costs associated with developing a larger collection. The costs of maintaining the larger world collection of rice material at IRRI are roughly US$700,000 annually (Chang, 1989). The costs of maintaining the Indian collections are roughly US$300,000. Thus, the economic value of genetic resources in India vastly exceeds the costs of maintaining them. If India were to invest, for example, US$20 million over a 10-year period to expand its collections further, this would add to the annual costs of maintaining the collection by the amortization of the US$20 million plus added costs. Even if this raised annual costs by US$3 million per year, the value produced by such additional resources would more than justify the expenditures. Indeed, since much of the estimated value of new materials emanates from the fringe materials, the value of a nearly complete collection relative to its cost is probably higher than the value of the present collection relative to its costs. This presumes, however, that the difficulties of managing very large collections are addressed (see Chapter 5).

Contributions of International Collections

Germplasm collections of rice are international and are exchanged internationally. Collections at IRRI and elsewhere contributed to productivity gains in India. Conversely, Indian germplasm has contributed to productivity growth in other countries. A global calculation for irrigated rice using the Indian estimates shows that a 0.5 percent increase in output per unit of input would be valued at about US$400 million. Using the same 20-year time lag applied to the Indian data, the present discounted value of this benefit's stream is US$594 million (discounted at 5 percent, it is US$3.015 billion). The values can be compared with the current annual costs of rice germplasm maintenance of perhaps US$10 million. If the current rice collections were brought to near-complete status over the next 10 years at a cost of US$10 million per year, this could, when amortized and adjusted for the larger collection sizes, raise the annual cost of these collections to US$30 million per year. This is well below the estimated present annual value of the benefits. Even if the Indian estimates overstate the impact of new genetic resources by a factor of 10, the economics of moving to a near-complete collection justify doing so. However, as the collection approaches completion, the increased effort and cost needed to get the last few samples would probably be unjustified.

IMPLICATIONS FOR OTHER CROP SPECIES

The study of rice in India given above indicates the fundamentals of placing economic value on genetic resources as producer goods. For rice, a distinction was made between a general strategic search by breeders for new genetic materials and a specialized search for genetic materials to address specific problems associated with vulnerability or lack of diversity. This distinction is important, because the general strategic search for new resources provides the case for the economic value of genetic resources already in collections. This value can justify the maintenance and preservation of most materials. The economic case for the collection and preservation of fringe genetic resources, that is, wild species and wild relatives, is based on the second search motive.

For rice, the evidence shows that both general search and specialized vulnerability-based strategies have paid off. It is useful to maintain this distinction in considering other commodities. In fact, the calculations for rice suggested that if there is an economic case for maintaining an ex situ collection, the case for maintaining a near-complete collection is actually stronger than the case for maintaining a partial collection, unless the costs of maintaining collections of fringe materials are extraordinarily high.

The economic calculation for rice showed that the present value of genetic resource impacts 20 years hence was $V \times I \times D$, where V is the value of the commodity, I is the impact estimate (.005 for rice), and D is the discount factor when returns are discounted at 10 percent ($1.486 = 1/6.73 \times 10$).

The estimates for rice may be unrepresentative for other commodities. It appears to be reasonable, if not unduly conservative, however, to consider the value .005 to be an upper bound and the value .002 to be a lower bound (this is approximately the lower bound of statistical significance) on a more general estimate of I. Given this, the estimated genetic research collection and maintenance costs can be compared with the commodity value. The discounted range of values for the ratio of annual costs of maintaining and operating the genetic research system to the value of the commodity for which the system is economically justified is thus .0029 to .0074 (.002 × 1.486 = .0029 and .005 × 1.486 = .0074).

How do actual costs to value ratios compare with this range?

For rice, the ratio of current costs to commodity value is roughly (US\$15 million/US\$100 billion) = .00015, which is well below the lower bound, .0029. Doubling of annual costs to US\$30 million to achieve a near-complete collection raises the ratio to .0003, but this, too, is below the lower bound.

Calculations for wheat and maize appear to be very similar to those for rice. The economic case for moving to a near-complete collection may be strong for these crops as well.

For the less important cereal grains, the ratio in question depends on the size of the potential collection and its replication in different locations. Barley appears to have a large number of potential accessions relative to the crop value. On the other hand, most accessions have already been acquired for barley.

For other crops for which the commodity value is some fraction of that for the big three crops (rice, wheat, and maize), the economic case for near completion of collections is just as strong, as long as the collection's costs are the same fraction of the costs for collection of the more important crops. The case for moving to near-complete collections is strong for most grains of significant economic importance.

The case for roots and tubers is somewhat less certain. Essentially the same arguments apply. Collection costs are higher because the low-cost seed storage options that allow for low-cost grain systems are not available. It is likely, however, that the major root crops have economically viable ratios.

RECOMMENDATIONS

The calculations made here provide general support for the proposition that investment in the collection, preservation, and management of genetic resources as producer goods provides a good return. Rates of returns are likely to be high, probably on the order of those for research in general. This is likely to be the case for all major agricultural commodities. For minor commodities, the case may not hold if costs are high. Yet, for many commodities of low economic importance, conservation costs may also be low. For very minor commodities (annual production costs of US$50 million), the case for conservation cannot be made unless the maintenance costs are very low. There well may be other compelling arguments for conserving genetic resources of minor crops, however. Biotechnology has opened more options for using germplasm and minor crops may well contain some important genes.

National and international investments in research related to collecting, managing, and using genetic resources should be increased.

The analysis suggests that with reasonable collection costs, if it makes sense to maintain a collection it makes even more sense to achieve a near-complete collection in a cost-efficient manner. For most commodities, uncollected materials probably have more value than most materials currently in collections. However, it does not make economic sense to incur extremely high costs in the search of the last few resources. For a well-designed program of collections, however, costs are likely to be reasonable.

Improvements in the techniques for breeding and applied genetics enhance the value of genetic resources. Furthermore, these developments likely enhance the value of fringe materials most. They may also expand the species that can be considered as genetic resources considerably beyond the present state.

The improvement of crops and livestock genetic resources is increasingly an international matter, because most valuable genetic resources cross international borders. International agricultural research centers, such as those of the Consultative Group on International Agricultural Research, and other agencies are critical institutions because of their role in facilitating this transfer. With the growth of private sector collections and intellectual property right protection, free exchange of genetic resources may be impeded. These developments could increase the importance of such public institutions for promoting international exchange of genetic resources.

The study on which this analysis is based is the first of its type. Given the economic impacts of genetic resources, it is important that further economic studies be made to develop a more complete body of evidence on which to base policy.

14

Conflicts Over Ownership, Management, and Use

Outside the relatively small community of breeders, evolutionary biologists, and geneticists, recognition of the importance of germplasm as a crucial resource has been recent. Major factors that led to this change in perspective include the successes of the green revolution in developing countries, which, in part, were based on plant breeders' combining crop plant genes from diverse locations (Plucknett et al., 1987), and the rapid degradation of many natural habitats by development, deforestation, and desertification, raising the specter of the irreversible·loss of countless numbers of plant and animal species (Office of Technology Assessment, 1987a). Awareness has also been heightened by controversies over proprietary rights, control of resources, inadequacies of conservation programs, and concerns about genetic vulnerability and genetic erosion.

POLITICS AND GEOGRAPHY

The conservation, management, and use of crop germplasm have become highly politicized in recent years. The difficulties of conserving food crop germplasm and the political controversies over the distribution and use of genetic resources ultimately have their origin in the vagaries of biology. The ancestors of modern crop species were unevenly distributed. Much of the world's diversity in wild and primitive crop plants is concentrated in the tropics and subtropics, where food crop plants were first domesticated. Only one of the world's 40 major food and industrial crop plants, the sunflower, is native to North America, and only two, oats and rye, are native to northern Europe.

The original genetic material of landraces and high-yielding varieties of crop plants can be traced to several relatively small regions in Asia, Africa, and Latin America. These areas are still characterized by rich genetic diversity. Before the 1960s, this genetic material was collected in developing countries by explorers, military officials, scientists, and agriculture ministry personnel from developed countries.

In the early 1970s, when the green revolution for wheat and rice was less than a decade old, there was a major conference on conserving crop genetic resources sponsored by the Food and Agriculture Organization (FAO) of the United Nations (Frankel, 1985b). O. H. Frankel (1970) and other scientists were beginning to predict that genetic erosion would occur as a result of the deployment of modern varieties (see also Frankel, 1985b,c, 1986, 1987). Genetic erosion inside germplasm banks can also contribute to the loss of traditional germplasm.

The International Board for Plant Genetic Resources

In the early 1970s, the Technical Advisory Committee of the Consultative Group on International Agricultural Research (CGIAR) established a means of coordinating and supporting the conservation of genetic resources. The International Board for Plant Genetic Resources (IBPGR) was established for this purpose in 1974 as one of the international agricultural research centers (IARCs) that now total 17 in the CGIAR network (Baum, 1986).

Awareness of the need to conserve genetic diversity increased in the years following the establishment of IBPGR, and progress was made in organizing a global network of cooperating germplasm banks. However, the IBPGR budget has remained modest (about US$7 million in 1989) (International Board for Plant Genetic Resources, 1990). In addition, national financial resources committed to conserving genetic resources in developing and developed countries have increased only marginally, if at all. Furthermore, it should be stressed that many of the successes of the commodity-based IARCs in germplasm preservation and distribution have been due primarily to their own efforts rather than to those of IBPGR.

Origins of the Conflict over Germplasm

In the late 1970s, some of the basic parameters of the global movements of raw and finished plant germplasm were questioned (Mooney,

1979, 1983). It was argued that there was a pattern of genetic resources being exported from the developing world, with the benefits of these resources accruing disproportionately to the developed world. In some instances, it was argued that finished varieties, consisting primarily of germplasm that originated in developing countries, were being imported by developing countries at high prices (Mooney, 1979, 1983). The assertion was made that the developing world (the South) was being exploited by the industrialized countries (the North).

This chapter examines these concepts, giving principal emphasis to food crop germplasm, with less attention given to industrial crops, such as rubber and oil palm. It should be stressed, however, that the geographic movement of food crop germplasm differs substantially from that of industrial crops. The centers of diversity for food crops are located primarily in developing countries. Accordingly, the principal flow of food crop plant germplasm has been from developing (south) to developed (north) countries. By contrast, industrial crops exchange is largely south to south (Kloppenburg and Kleinman, 1987a), although often mediated by the north.

It is important to recognize that genetic resources became an issue for social activists following the passage of the Plant Variety Protection (PVP) Act (codified at 7 U.S. Code Sections 2321-2583) of 1970

Workers harvest watermelons at a farm in the São Francisco Valley, Brazil. Credit: Food and Agriculture Organization of the United Nations.

in the United States. This also roughly coincided with the development of genetic engineering. Both the controversy over PVP and the rise of genetic engineering increased global interest in genetic resources. The passage of PVP brought the United States broadly in line with the European systems of plant breeders' rights providing limited protection, analogous to a patent, for sexually reproduced crop varieties. Many independent seed companies in the United States and elsewhere were absorbed into large multinational firms during the decade after passage of PVP (Doyle, 1985; Kloppenburg, 1988; Kloppenburg and Kleinman, 1987b; McMullen, 1987).

There are major differences of opinion about the social and conservation implications of restructuring of the seed industry. Some argue that restructuring led to increased investment in private plant breeding, increased the pace of development of improved cultivars, and increased corporate interest in genetic resources conservation (McMullen, 1987). Others see restructuring as less benign (Doyle, 1985; Fowler and Mooney, 1990; Mooney, 1979, 1983, 1985) and have expressed concerns about the distribution of genetic resources. It was feared that large companies would take a short-term, opportunistic view of genetic resources conservation, and would not exchange genetic material (especially proprietary material) with public and private plant breeders. Concern existed that monopolies would be furthered by breeding new crops dependent on the use of agricultural chemicals produced by the multinational corporations that now owned the seed companies.

The emergence of genetic engineering also had a major influence on genetic resources activism. The rise of biotechnology demonstrated how a single gene for tolerance to a broad-spectrum herbicide might be worth millions of dollars. The development of biotechnology raised the perceived value of crop genetic resources and reinforced the perception that developing countries could be increasingly exploited (Fowler and Mooney, 1990; Goldstein, 1988).

INTERNATIONAL DIALOGUE ON PLANT GENETIC RESOURCES

During the twenty-first Biennial Conference of FAO in 1981, Resolution 681 was proposed, which requested the director general to draft an international convention "including legal provisions designed to ensure that global plant genetic resources of agricultural interest will be conserved and used for the benefit of all human beings, in this and future generations, without restrictive practices that limit their availability or exchange, whatever the source of such practices" (Food and Agriculture Organization, 1981). At that conference, a number of developing country governments argued for the creation of an international germplasm bank at FAO and for legally binding commitments by all countries to make germplasm freely available (Mooney, 1985; Witt, 1985). Feasibility studies were requested on each and were considered in advance of the 1983 FAO conference. The fact that the developing countries supported these feasibility studies while most developed countries opposed them was a harbinger of the conflict that was to be manifest at the next FAO convention in 1983.

A draft report presented to the 1983 FAO conference led to the establishment of the FAO Commission on Plant Genetic Resources (Frankel, 1988; Mooney, 1985; Witt, 1985). At that conference, the issue of what were to be considered *genetic resources* covered by the resolution was hotly debated. Several developing country ambassadors actively supported the inclusion of finished varieties, mutant stocks, and breeders' lines, and none of them opposed their inclusion. This position was advanced by developing country FAO ambassadors, because including advanced breeding materials in the resolution would reinforce the free flow of genetic materials, make a wider range of germplasm available to plant breeders in the developing world, and reduce developing country expenditures on imported seed. Developed countries, notably those with private seed industries and plant breeders' rights legislation, argued that only wild relatives, landraces, and varieties not covered by patent or patentlike protection should be included. It was argued that breeders' lines represented combinations of genes unavailable in the landraces and wild relatives and therefore constituted original genetic variation. If private industry was to play a role in plant breeding, a reasonable protection of finished varieties was appropriate, since plant breeding required extensive investments. Furthermore, it was pointed out that use of protected varieties for the purpose of plant breeding was not prevented by plant breeders' rights legislation. Moreover, governments could not require breeding lines held by private industry to be available for exchange.

Nevertheless, the 1983 FAO conference adopted Resolution 8/83, the International Undertaking on Plant Genetic Resources (Frankel, 1988; Mooney, 1985). Article 1 of the annex to the resolution stated that the objective of the undertaking was to ensure that plant genetic resources of economic or social interest, particularly for agriculture, would be explored, preserved, evaluated, and made available for plant breeding and scientific purposes (Food and Agriculture Organization, 1983a). It was based on the presumption that plant genetic resources are a heritage of humankind that should be available without restriction. The resources would include native cultivars, landraces, wild relatives, newly developed varieties, and special genetic stocks (including elite and current breeders' lines and mutants). Thus, the undertaking asserted that proprietary germplasm was part of that available through the principle of common heritage. After acrimonious debate, the undertaking was passed by a vote in which the developing country majority voted for and the developed countries voted against. Nonetheless, it is significant that the issue required a vote, because FAO work is generally accomplished through consensus.

Continuing International Controversy

In the aftermath of the passage of the undertaking there has been continued geopolitical controversy. Some advocates for developing countries have continued to contend that current germplasm arrangements are unsatisfactory (Fowler and Mooney, 1990; Mooney, 1985). They argue that the concerted efforts of the governments of developed countries to preserve crop genetic resources demonstrates the value of these resources, even in their unfinished forms (landraces, wild relatives). Because finished varieties and elite breeding lines are in part derived from germplasm of developing country origin that was obtained without a fee, they should be freely available to developing country researchers.

The position of developed countries generally has been that, although most agricultural crop germplasm originated in the developing world and that the developed world has benefited from this germplasm, these arrangements are not inequitable (Brown, 1988). Primitive forms of germplasm have little economic value without evaluation and enhancement, the lengthy process of incorporating genes through breeding into advanced lines to develop improved cultivars. Moreover, it is argued that developing countries lose nothing, because only a small number of seeds is gathered by collecting and duplicates are placed in their seed banks whenever possible. Advanced breeding lines and proprietary varieties from developed countries are very seldom directly useful in developing countries. Thus, it is maintained that although unimproved germplasm is the common heritage of humankind, improved germplasm is not. Those who support farmers' rights take the position that landraces are, in fact, improved germplasm and could legitimately be regarded as no more the common heritage of humankind than the advanced cultivars of the developed world.

It was demonstrated earlier (see Chapter 13) that there is a potential economic benefit in obtaining the landraces and wild species that are not in present collections. However, this benefit cannot be realized without investment in the research, testing, breeding, and selection that are integral parts of germplasm enhancement.

Despite objections over what was to be included as common heritage, FAO member countries were allowed to sign the undertaking and, by stating their reservations on specific articles, indicate the extent to which they were prepared to concur with its principles. No industrial country with a well-established private seed company sector agreed to support the undertaking without reservations (although many have ratified the undertaking with reservations) (Table 14-1).

The central point of the geopolitical debate over crop genetic resources is, as Kloppenburg and Kleinman (1987a:196) have noted, that

> in a world economic system based on private property, each side in the debate wants to define the other side's possessions as common heritage. The advanced industrial nations wish to retain free access to the developing world's storehouse of genetic diversity, while the developing nations would like to have the proprietary varieties of advanced nations' seed industry declared a similarly public good.

It can be argued that neither side in the debate has a fully defensible or realistic position. As noted earlier, some representatives to FAO from developing countries have pointed out that governments of developed countries cannot logically advocate that unfinished genetic resources in other sovereign states are common heritage when some of these developed country governments have declined to recognize the resources of the ocean floors or Antarctica as common heritage (even when these resources do not even lie within the boundaries of another sovereign state).

Likewise, it makes little sense for developing countries to place so much emphasis on access to elite breeding lines and finished varieties developed by firms in developed countries for use by a developed country when little of this material would be of use in the agroecological contexts of developing countries (Brown, 1988). In a world economy dominated by market relations one may ask if it is likely that the major industrial powers would relinquish a resource as valuable as proprietary germplasm.

Indeed, these inconsistencies in each side's claims in the debate strongly suggest that the debate, at least to some degree, reflects deeper political antagonisms and differences of economic interest. The north-south genetic resources debate has been based at least as much on premises that apply more to larger issues of the fairness of the global economy, monetary system, and foreign aid apparatus, as on the genetic resources issues that are the major concern of this report.

TABLE 14-1   Countries That Joined the Commission on Plant
Genetic Resources of the Food and Agriculture Organization (FAO)
or Agreed to Support the International Undertaking on Plant
Genetic Resources

Regions and Country	Member of FAO Commission	Supports Undertaking
Africa		
Angola	•	—
Benin	•	•
Botswana	•	—
Burkina Faso	•	•
Cameroon	•	•
Cape Verde	•	•
Central African Republic	•	•
Chad	•	•
Congo	•	•
Côte D'Ivoire	—	•
Equatorial Guinea	•	•
Ethiopia	•	•
Gabon	—	•
Gambia	•	—
Ghana	•	•
Guinea	•	•
Guinea-Bissau	•	—
Kenya	•	•
Liberia	•	•
Madagascar	•	•
Malawi	—	•
Mali	•	•
Mauritania	•	•
Mauritius	•	•
Morocco	•	•
Mozambique	—	•
Niger	•	•
Rwanda	•	•
Senegal	•	•
Sierra Leone	•	•
South Africa	—	•
Sudan	•	•
Tanzania	•	•
Togo	•	•
Uganda	•	—
Zaire	•	—
Zambia	•	•
Zimbabwe	•	•

TABLE 14-1   *Continued*

Regions and Country	Member of FAO Commission	Supports Undertaking
Asia and the Southwest Pacific		
Australia	•	•
Bangladesh	•	•
Fiji	—	•
India	•	•
Indonesia	•	—
Japan	•	—
Korea, Democratic People's Republic of	•	•
Korea, Republic of	•	•
Malaysia	•	—
Myanmar	•	—
Nepal	—	•
New Zealand	•	•
Pakistan	•	—
Philippines	•	•
Samoa	•	•
Solomon Islands	—	•
Sri Lanka	•	•
Thailand	•	—
Tonga	—	•
Vanuatu	•	—
Europe		
Austria	•	•
Belgium	•	•
Bulgaria	•	•
Cyprus	•	•
Czechoslovakia[a]	•	•
Denmark	•	•
Estonia	•	—
Finland	•	•
France	•	•
Germany	•	•
Greece	•	•
Hungary	•	•
Iceland	•	•
Ireland	•	•
Italy	•	•
Liechtenstein	—	•
Lithuania	•	—
Netherlands	•	•
Norway	•	•
Poland	•	•
Portugal	•	•

continued

TABLE 14-1   *Continued*

Regions and Country	Member of FAO Commission	Supports Undertaking
Europe—*continued*		
Romania	•	•
Russia	—	•
Spain	•	•
Sweden	•	•
Switzerland	•	•
United Kingdom	•	•
Yugoslavia[a]	•	•
Latin America and the Caribbean		
Antigua and Barbuda	—	•
Argentina	•	•
Barbados	•	•
Belize	•	•
Bolivia	•	•
Brazil	•	—
Chile	•	•
Colombia	•	•
Costa Rica	•	•
Cuba	•	•
Dominica	•	•
Dominican Republic	•	•
Ecuador	•	•
El Salvador	•	•
Grenada	•	•
Guatemala	•	—
Guyana	•	—
Haiti	•	•
Honduras	•	•
Jamaica	—	•
Mexico	•	•
Nicaragua	•	•
Panama	•	•
Paraguay	—	•
Peru	•	•
Saint Christopher and Nevis	•	—
Saint Lucia	•	—
Saint Vincent and the Grenadines	•	—
Suriname	•	—
Trinidad and Tobago	•	•
Uruguay	•	—
Venezuela	•	—

TABLE 14-1 *Continued*

Regions and Country	Member of FAO Commission	Supports Undertaking
Near East		
Afghanistan	•	—
Bahrain	—	•
Egypt	•	•
Iran	•	•
Iraq	•	•
Israel	•	•
Jordan	•	—
Kuwait	—	•
Lebanon	—	•
Libya	•	•
Oman	—	•
Syria	•	•
Tunisia	•	•
Turkey	•	•
Yemen	•	•
North America		
Canada	•	—
United States of America	•	—

NOTE: Of the 135 countries listed, 118 are members of the commission and 107 have adhered to the International Undertaking as of April 18, 1993.

[a]Information pertaining to the republics formed from the dissolution of Czechoslovakia and Yugoslavia was not available.

SOURCE: Unpublished information from the Food and Agriculture Organization of the United Nations.

Nonetheless, although the rhetorical debate is rooted in fundamental differences of interest and opinion, there is room for common ground on the genetic resources question. More specifically, there is room for developed countries to accede to developing countries on common heritage in a modified form, and for countries of the south to do likewise with respect to plant breeders' rights. This would enhance the interests of both sides. Krasner (1985) has provided a useful examination of the fundamental geopolitical differences of interest between developed and developing countries. There are several clear respects in which developed and developing countries' interests depart substantially and with little hope of resolution. Unfortunately, these divergent views have become manifest in genetic resources issues, an area in which there is room for compromise for mutual benefit. An important forum for reaching that compromise has been the Keystone International Dialogue on Plant Genetic Resources (Keystone Center, 1988, 1990, 1991).

Evolution of the North-South Dialogue on Genetic Resources

At its meeting in 1987, the FAO Commission on Genetic Resources discussed a number of reports prepared to clarify certain issues (Food and Agriculture Organization, 1987b,c,d,e). These included studies of the legal status of base and active collections of plant genetic resources, including seed legislation and plant breeders' rights (Food and Agriculture Organization, 1987d); the legal arrangements, with a view to the possible establishment of an international network of base collection germplasm banks, under the auspices or jurisdiction of FAO (Food and Agriculture Organization, 1987e); the status of in situ conservation of plant genetic resources (Food and Agriculture Organization, 1987c); and the feasibility of establishing an international fund for plant genetic resources (Food and Agriculture Organization, 1987b).

These reports contributed significantly to a clearer perspective on the underlying issues of genetic conservation and diffused some of the more political and confrontational views expressed in previous meetings on ownership of genetic resources. This understanding was further strengthened during the 1989 meeting of the FAO Commission on Plant Genetic Resources (Food and Agriculture Organization, 1989b; Keystone Center, 1990). Some of this moderation resulted from the first of a series of dialogues organized by the Keystone Center, which brought together a mix of social activists, public and private plant breeders, conservationists, curators of germplasm banks, and government officials with the goal of finding areas of consensus and compromise in the debate (Keystone Center, 1988).

It is now generally accepted that plant breeders' rights legislation represents a legitimate interest and does not necessarily constitute an impediment to access to protected varieties for the purpose of research and the creation of new varieties, although no legal guarantees exist at the international level for the free exchange of such resources. It was suggested that a network of base collections be established for conserving genetic resources and that unrestricted access be ensured through an intergovernmental authority such as FAO (Food and Agriculture Organization, 1987d,f). Various models were proposed, ranging from complete control over base collections being exercised by FAO to a much looser arrangement in which a government or institution would formally agree to carry out certain fundamental obligations toward FAO regarding the base collections (Food and Agriculture Organization, 1987d,f).

The commission also discussed the need to establish an international fund to ensure adequate financial resources for genetic conservation (Food and Agriculture Organization, 1987b,f). Such a fund would seem essential to allow developing countries, particularly those situated in major areas of genetic diversity, to play an appropriate role in safeguarding the genetic variation threatened by erosion (Keystone Center, 1988).

Linked to the issue of plant breeders rights was the concept of farmers' rights to landraces (Food and Agriculture Organization, 1987f). This was introduced to express the recognition that local landraces typically are the result of centuries of work by farmers, including those of the present day, that has resulted in the development of enormous variety in plant types. These landraces constitute an important source of the genetic diversity used in breeding programs.

The commission further suggested that farmers' rights to landraces are, to a point, comparable to breeders' rights to modern varieties. It argued that farmers' rights should be interpreted as an obligation of the world community at large, to ensure adequate safeguards for conserving landraces (Food and Agriculture Organization, 1987f). However, communal forms of ownership are not easily translated into practical arrangements of financial compensation. It is difficult to decide who should be compensated and for what. As presented in the 1989 agreed interpretations of the 1983 Undertaking, the farmers' rights concept amounted to a general obligation of the developed nations to help the developing nations in their genetic resources conservation and use activities.

Ownership and Stewardship

Much of the recent debate in the FAO conference on genetic resources deals with ownership and stewardship of genetic resources. Mooney (1985) argued that over the past decade a massive gene drain has taken place from farmers' fields in the south to the seed banks in the north. He estimated that more than 90 percent of the samples collected by IBPGR came from developing countries. About 40 percent of that group went to the seed banks of Europe and North America, an equivalent portion was deposited with the IARCs, and only 15 percent ended up in the seed banks of the developing countries.

At FAO, questions were raised as to access and ownership of such materials, specifically those materials in the north. It was argued that even though the IARCs are located in developing countries, control over the IARCs and the IBPGR rested with its donors, who are largely developed countries and institutions from them. It was suggested that to safeguard the interests of developing countries, proper international control over collections and access to these collections are required.

IBPGR and the scientists involved in genetic resources conservation disagreed with the often highly political arguments questioning their objectives (Plucknett, 1985; Plucknett et al., 1987). Although developing countries were limited by inadequate facilities and financial resources for safeguarding genetic resources, the specter of what appeared to be a worldwide conspiracy to rob them of genetic resources for the benefit of commercial interests in the north was considered a more immediate cause for concern (Mooney, 1979, 1983).

The debate, initially highly politicized, is now more technical and pragmatic. It is accepted that the existing global network of base and active collections provides a good basis for conservation, but there is renewed attention to the longstanding view that the role of national seed banks in centers of diversity should be strengthened (Keystone Center, 1988, 1990). This strengthening of the role of national collections will require long-term international funding (as well as national commitment), because many countries in the centers of diversity lack sufficient resources.

Restructuring of the Seed Industry

One of the issues brought up in the genetic resources debate is the ongoing restructuring of agroindustry in the north and its possible influence on agriculture in the south (Doyle, 1985; Fowler and Mooney, 1990; McMullen, 1987). A notable feature of this is that many seed firms in Europe and North America, previously in the hands of a large number of mostly small family enterprises, are now subsidiaries of relatively few large transnational corporations (Doyle, 1985; McMullen, 1987). Many of these corporations have major interests in the production and marketing of pesticides and fertilizers and are leaders in biotechnology research and development. It is argued that this industrial complex represents enormous interests that should be analyzed in historical perspective and in the context of genetic resources.

Over the past few decades, plant breeding and seed production in Europe and North America have been dominated by private industry. Early in this century, small nurseries and seed merchants started selecting their own varieties, often for limited regional markets, using regionally available landraces and imported materials as their base material. As plant breeding evolved, companies became more specialized and started to use scientific methods of plant breeding; this work was supported by government through research institutes and universities.

Private investment in plant breeding has benefited by specialized legislation, such as the Plant Patent Act of 1930 in the United States, developed in a number of industrial countries. It provides plant breeders legal ownership of their varieties and regulates the marketing of such varieties (but usually without restricting their use for further breeding). In 1961, the plant breeders' rights systems of several countries were harmonized in the Union for the Protection of New Varieties (UPOV; see Chapter 12). Plant breeders' rights legislation, in general, required that protected varieties satisfy specific requirements of distinctiveness, uniformity, and stability. It has led to additional seed legislation regulating the nature and quality of seeds brought to the market (essentially a form of consumer protection). This system has worked reasonably well in industrial countries.

Only a handful of developing countries, however, have so far joined the UPOV convention. There are several reasons for this that mainly pertain to the historical and structural aspects of their varietal development and seed distribution systems. Plant breeding and seed production in developing countries have largely followed a different path of development from that in the developed countries. In developing countries plant breeding began mainly in government institutions. A major development in plant breeding in food crops was set in motion in the 1960s with the establishment of four IARCs by the Rockefeller and Ford foundations; they cooperated closely with national research systems in developing countries. The primary objectives are to increase food production and, more recently, to improve the livelihood of the majority of often small and poor farmers in developing countries. This has led to a nonproprietary system of plant breeding on an international scale, with largely free exchange of finished varieties, advanced breeding lines, and unimproved materials such as landraces and wild species. Local multiplication of varieties and distribution of seeds by farmers are encouraged.

The key issue is whether the current system in developing countries will evolve into one similar to that of the UPOV countries, involving private industry. The present system has several clear advantages in the developing world. First, the free exchange of breeding lines widens the scope of material available to local plant breeders. Second, it protects the interests of poorer farmers who cannot afford expensive inputs (since private breeding companies will, for economic reasons, concentrate on the larger and more capital-intensive sector of the farming community). Third, this system involves less need for the genetic uniformity necessitated by the criteria that are generally necessary to receive proprietary protection. Finally, it is argued that the current system avoids an economic linkage between crop varieties and chemical inputs (pesticides, herbicides, fertilizers). Furthermore, heavy dependence of national seed supplies on transnational corporations may not be in the best interests of many countries.

Typically, the major disadvantage of public-dominated plant breeding systems in the developing world is in the area of seed production and distribution, particularly their tendency toward a lack of efficiency and effectiveness compared with that of private industry. However, intermediary forms of public and private institutional arrangements may well be possible in this sphere. It is also argued that some form of proprietary protection will be necessary if private industry is to play a role in plant breeding in these countries, as may well be considered desirable at some point in their development. Protagonists of such protection argue that public sector breeding programs may benefit from such arrangements by receiving additional funding.

How, then, do these different systems affect the conservation of genetic diversity? Private industry in Europe and North America, with some notable exceptions, has so far contributed little to the conservation of genetic resources. Germplasm collections in Europe and North America generally were, with only a few exceptions, limited to material essential for ongoing breeding programs.

Little is known about the germplasm collections of private industry, but it is doubtful that they constitute much beyond what is available in publicly supported collections. The notion that international commercial plant breeding interests try to monopolize genetic resources thus far has little foundation. The notion that private industry has an interest in the loss of crop genetic diversity also has no basis. However, commercial plant breeding largely has short-term objectives and greatly values the use of materials that already satisfy most of the requirements being sought. Thus, private firms have little incentive to devote resources to genetic resources conservation and, with few exceptions, have little direct concern for this activity. The use of genetic resources such as landraces and wild relatives takes place primarily in university and government research programs, which release improved breeding materials for practical plant breeding in later stages of development.

RESPONSES TO THE CONFLICT OVER GENETIC DIVERSITY

Much that has been written on the basic conflict over genetic diversity deals with a conflict of principles. On one side is the principle of free flow of genetic resources, while on the other, albeit with a somewhat greater internal diversity of opinion, is the principle of establishing and enforcing property rights as incentive to innovation. Under the principle of free flow of genetic resources, many genetic materials move freely from nation to nation. During the first three decades after World War II, this free flow of genetic resources was primarily from the developing to the developed world. More recently, however, there have been greater movements of germplasm in the opposite direction. Under intellectual and private property principles, some germplasm materials may not, because of fees or proprietary restrictions, be readily exchanged.

The conflict is also one of outcome and process, however. The two sides have different judgments about the national and international institutions involved. One side is more optimistic about the international political process; the other side is more optimistic about informal interchange among scientists. Moreover, the two sides have different interpretations of the impact of international intellectual property rights as they relate to the developing world. In large part, these rights can be legally enforced only in nations in which the relevant patents or plant breeders' rights are issued. Thus, the direct effect of patents and plant breeders' rights on developing nations may be minor, although there may be indirect effects associated with the confidentiality and the commercialization of research.

A Basis for Resolution

The disagreements are real, and the issues are already complex and difficult to resolve. The issues will become more complex as questions of plant genetic material and plant breeders' rights give way to questions involving all forms of genetic material and the proprietary rights associated with biotechnology. They will also become more difficult as the developed countries become less willing to support public sector agricultural research both at home and in the developing countries and as such research becomes more and more centered in the private sector (de Janvry and Dethier, 1985; Ruttan, 1982). Nonetheless the following facts show promise for progress:

• Developing countries need increased foreign technical assistance to enhance their ability to conserve genetic resources and to do the research necessary to incorporate genes into new, more productive cultivars rather than to acquire advanced foreign breeding lines, proprietary genetic material, and varieties developed by private firms for the agronomic environments of developed countries. There is a growing realization that the locally adapted landraces are usually the most important base materials for breeding, specifically for areas with less favorable production environments and where low levels of external inputs are used.

• The basis for expecting assistance from the developed world are (1) developed countries and some of their firms have benefited greatly, at minimal cost, from transfers of genetic material from developing countries, and (2) it is in the long-term interest of the developed countries to encourage agricultural and overall economic development in the developing world.

• The bulk of the crop genetic resources material derived from developing countries consists of landraces that have been developed and improved by generations of farmers. Thus, it may be argued that these resources have value and their use may merit either compensation or contributions to an international fund.

• Developed countries are extremely unlikely to negotiate over any treaty or other international agreement that would have the direct or indirect effect of making improved genetic material a common property resource.

• Although it is in the scientific interest of both developed and developing countries that access to genetic resources remains open, it is in the interest of developing nations that access be available only for a fee—and such a fee may provide some incentive for conserving genetic resources.

• Ideally, global rules governing plant genetic resources should have the effect of encouraging conservation and utilization activities in the future, rather than relying largely or entirely on realizing rents from previous activities and investments. The royalty rights of intellectual property are normally justified because they provide forward-looking incentives to encourage invention and innovation, although it should be recognized that, in practice, proprietary rights in plant genetic materials may serve to protect marketing investments and augment the value of previous breeding investments. The need for a forward-looking incentive system was recognized as well at the Keystone dialogues. A consensus emerged that a desirable compensation system would be one that would provide incentives to developing countries to conserve and use plant genetic resources in the future.

Avenues for Resolution

It is therefore essential to look for solutions. The committee envisaged three possibilities. The first is to negotiate a treaty that defines a compromise position on intellectual property and free flow of crop germplasm. The second is to create an international payment mechanism, most likely a system linked to the value of seed sales, whose proceeds would support genetic diversity conservation programs. The third is to focus attention on strengthening plant breeding and building biotechnology research capacity in the developing world rather than on legal arrangements. This would restore reciprocity between the industries of the developed and developing countries and would remove some of the difficulties that now make the problem so intractable.

Treaty Solution

It might be possible to achieve a compromise between developed and developing countries to create an intellectual property and free-flow regime. This would require that a set of principles, which make scientific and institutional sense, be found; and it would almost certainly involve an agreed boundary between the following two regimes: (1) a free-flow regime for some materials, including those traditionally handled through free scientific exchange, as well as, presumably, access to genetic materials that are found in wild form or purchased in markets; and (2) an intellectual property or property rights agreement (at least one that nations may optionally choose if they so desire). This compromise would permit private industry to appropriate a portion of the benefits of innovation under some circumstances. Such appropriation might be accomplished by patent or breeders' rights or by recognition of private property.

A similar compromise was achieved at the United Nations Conference on the Environment and Development (UNCED) at Rio de Janeiro in June 1992. Under its terms, genetic resources already in the international germplasm banks or international research system would continue to be freely available. Looking to the future, however, genetic resources will be subject to the control of national governments. Access, "where granted," is to be on "mutually agreed terms" (United Nations, 1992). Thus, this agreement breaks with the pattern of the International Undertaking on Plant Genetic Resources— genetic resources will no longer be freely available and developing nations may attempt to charge for access to their resources.

The corresponding intellectual property issues are dealt with in separate articles on technology transfer (Article 16) and biotechnology (Article 19). In general, both create an obligation on the part of the developed world to make the relevant technology available to developing nations. In addition, the technology transfer article requires respect for intellectual property rights and also requires cooperation to ensure that those rights be supportive of the objectives of the convention.

Thus the end of free access is accepted, together with what appears to be a relatively superficial resolution of the intellectual property issues. Although the convention was rejected by the United States at that time, it was accepted by most nations. Developing nations may restrict access to their genetic resources, regardless of the U.S. position, in which case, future access would depend on compensation.

International Payment

If there is broad agreement that nations should be paid in some fashion for germplasm that derives in part from within their boundaries, the compromise solution may encounter difficulty because the mechanisms of payment may become the central issue.

The design of a payment system is not an easy task (Barton and Christensen, 1988). Part of the question is what the payment should be for. It could be an equitable compensation for all the material provided in the past or for costs imposed by the plant breeders' rights system. Alternatively, it could be an incentive compensation designed to encourage future collection and management (including evaluation) of genetic material. Systems developed on the basis of these two patterns would look very different and would distribute revenues quite differently.

The total benefit derived from germplasm that originated in developing countries is beyond any form of practical calculation. Likewise, the actual cost of plant breeders' rights to the developing world is also difficult to estimate, but an order-of-magnitude approximation is feasible. At one extreme, there may have been net benefits to the developing world because plant breeders' rights in the developed world may have led to breeding innovations that have been shared with the developing world. At the other extreme, there may have been a net cost to the developing world.

Limited data from the U.S. seed industry was used by Barton and Christensen (1988) to develop an "extremely approximate estimate" of the cost impact of proprietary rights. There is anecdotal evidence that, in one area, protected varieties cost about twice as much as competing unprotected varieties (Barton and Christensen, 1988; Butler and Marion, 1985). By aggregating available data on protected and unprotected varieties (including hybrid maize as effectively protected), it can be estimated that of the almost $4 billion in U.S. seed usage in 1979 about $1 billion was in protected varieties. If the protected seeds did cost twice as much, proprietary protection led to a 12 percent increase in overall seed costs in 1979. The impact of this on the developing world would be indirect, because the rights are not enforceable beyond national borders.

Assuming that this indirect effect is perhaps one-tenth of the direct effect, or an effect of about 1 percent on sales, and that total annual seed sales in the developing world are about US$10 billion, the greatest plausible annual cost of plant breeders' rights to developing nations (remembering that there may actually be net benefit rather than a cost) would be about US$100 million. This sum is larger than total global expenditures on seed bank maintenance in 1989 (about US$50 million), but somewhat smaller than the roughly US$220 million spent on public international agricultural research in that year (J. Barton, Stanford Law School, personal communication, September 1990).

Although approximate, these numbers suggest that providing an incentive is a potentially more successful mechanism. Discussions of compensation approaches during the late 1980s generally focused on devising a formula that would reimburse developing countries for their past activities or for past acquisitions of their genetic resources by developed countries. This type of system would have been essentially retrospective and difficult to apply. As noted earlier, the Keystone dialogues recognized the difficulties of implementing a retrospective compensation system and developed a consensus that compensation should be forward looking and provide incentives to the developing countries to conserve and use their genetic resources in the future.

Incentives could bring benefits in the form of increased conservation of genetic resources. These new conservation incentives would balance the innovation incentives provided by intellectual property protection. Thus, it is reasonable to create a fund on the order of US$300 million or more to increase efforts to conserve genetic resources.

In the absence of any global financing agreement, nations will probably exercise their control over genetic resources by refusing to permit access unless the collector accepts a material transfer agreement. Most likely, such an agreement will require the collector to agree to discuss compensation and pay a royalty in the event that a commercial product derives from the material—and will of course prohibit the collector from transferring the material to any third party without exacting a similar agreement. Those negotiated agreements will roughly reflect a portion of the economic value of the material, just as patent royalties reflect a portion of the economic value of the innovation. Alternatively, the international community could create a fund to be used to support the conservation of genetic diversity. This fund, perhaps gained through national contributions linked to seed sales, would be disbursed to developing nations to help them conserve genetic materials.

Although most economists prefer a solution in which the market defines the size of the payment, the tax on seed sales is probably the more feasible. Even though developing nations could carry out the first approach unilaterally or by agreement among themselves, they are likely to gain more effective enforcement through an international convention. However, a convention of this kind will need to recognize intellectual property rights as well as supplier's rights in genetic material, but these are the issues that make the treaty solution difficult. The allocation of payments for a cultivar that contains genes from many different sources will be difficult. Comparable problems with copyright payments for musical recordings played on the radio are sometimes handled by creating a fund and dividing the proceeds according to formulas that reasonably approximate what the complicated and expensive negotiations might produce. Note, also, that existing public sector germplasm banks may be placed in a very awkward position with respect to accepting material under the agreements likely to be imposed by developing nations just as they would be by accepting patented materials (including genes) from the private sector.

In contrast, a fund can be created and used to support the conservation of genetic diversity. The most logical source for a fund is the Global Environment Facility (GEF), created by the United Nations Environment Program, United Nations Development Program, and World Bank and recognized as the primary source for funding genetic resources in developing nations. It ran US$1.3 billion before the UNCED meeting and was both expanded and restructured at that negotiation (Bureau of National Affairs, 1992). The GEF's basic concept is to provide international funds for those situations in which a project cannot be justified in term of domestic costs and benefits but can be justified in terms of global costs and benefits (Anonymous, 1991).

Should this not be practical, one might consider a seed tax or its equivalent. Assuming that US$300 million is the minimum amount for such a fund, a global tax on seed sales of about 1 percent would be sufficient. This approach would provide secure long-term funding for the conservation of genetic diversity. It would also facilitate participation by developing nations without asking them to undertake the primary expense. The expense is borne, ultimately, for the good of all, not just of developing countries.

The disadvantages are those inherent to an earmarked tax. The fund will develop its own constituencies and survive at a predetermined level, whether or not that level remains rational. There will also be political conflict over whether control should be through a technical body like the IBPGR, or through one reflecting greater political input, such as FAO's International Fund for Plant Genetic Resources, or under the control of a higher commission of the United Nations. These will not be simple conflicts between interested parties, but will reflect fundamental differences over the form of international review essential to ensure that the funds are used well. Such differences were a central part of the funding diplomacy at UNCED.

Scientific Balance

The third approach changes the terms in which the problem is defined. To the extent that the concerns of developing countries about genetic resources are a reflection of their concerns about the status of their plant breeding and biotechnology sectors, it may be better to work directly to improve the positions of developing countries in these sectors by providing them with the means of using their own resources more effectively.

An immediate answer is that the developing world does not have a chance against the large multinational corporations, but there are reasons to believe that this is not the case. First, as demonstrated by the existence of the Seminario Panamericano de Semillas (Pan American Seeds Seminar), there is already an industry in Latin America. The same is true of some areas of Asia: The needs for local multiplication and testing and for adapting varieties to meet local needs provide a foundation for developing nations in specific sectors of the seed industry. As they build on that, some of them could begin to export their seeds. Second, as new biotechnologies are adopted, there is a good chance for some new developing world firms to be successful globally. In the United States, small firms may take the leadership in new technologies and, in turn, supply them to the larger multinational corporations. Some of these new small firms could be from developing countries.

To make this happen, a public sector scientific, regulatory, and economic infrastructure is needed. A scientific infrastructure that provides education, basic research, and travel to ensure the integration of scientists and technologists from the developing world into the global scientific community is crucial. To be successful, it must be built with national or international funds. The public sector must also orient research toward subsistence farmers, who are otherwise likely to be ignored by the private sector. Finally, the business infrastructure is equally important and requires access to management skills, venture capital, avoidance of overly restrictive regulations, and assurance that domestic and international markets are open. These considerations may not be generally relevant to all types of economies, however.

Many firms in developing countries have a good understanding of their own markets but need to know what technologies may be useful and when to apply them. First, they will need formal ties to technology suppliers, such as national agricultural research programs or firms in developed countries. They may then require ties to foreign markets, either through marketing arrangements similar to those of small biotechnology firms in developed countries, or through new marketing organizations in developing countries.

Public programs in developing countries will require support for education and will need access to advanced scientific information. It is also crucial to ensure that access to technology and markets is not closed off by reluctance on the part of developed countries that worry about providing technology to create new competition. It may be necessary to negotiate agreements to ensure continued access to technology and markets.

This business infrastructure approach deals directly with the underlying problem and provides an answer in the real world rather than in the legal world. It expends resources on the development of a global scientific and technological community rather than on new bodies of law. It takes advantage of the opportunities available from biotechnology, avoids complicated intellectual property negotiations, and builds on the rapid evolution of the seed industries of developing countries.

However, this solution does not have a clear endpoint. It is focused on science and technology rather than on genetic resources and has no single agreement or step to which political leaders can point. It could benefit some countries much more than others because, for example, not all countries will build a biotechnology industry. In the poorer nations, international public sector programs may have to be the dominant source of technology for the foreseeable future.

RECOMMENDATIONS

What earlier seemed to be an increasingly contentious debate has, since 1988, moved significantly in the direction of compromise and consensus. The questions raised over the accountability of current germplasm conservation activities at both the international and national levels and the need for adequate safeguards to ensure free access to genetic diversity are generally accepted as important and relevant. The present challenge is to achieve both aims in a worldwide system of genetic conservation that is practical and effective, and that does justice to the aspirations of the world community.

There is a growing awareness among scientists, policymakers, and international organizations that germplasm is an important natural resource. In a polarizing international environment with conflicting economic systems and a skewed distribution of wealth and control over agricultural production and markets, the problems of germplasm conservation and management extend beyond purely technical issues to include the broader social and economic ramifications of research.

The conservation of genetic resources must receive a greater level of national and international effort and financial support.

There is a need for legal, institutional, and other arrangements that guarantee use and accessibility of genetic diversity for the benefit of society as a whole. Global responsibility must find its expression in truly international funding of genetic resources activities. The source of such finances, however, should not, as is often the case now, derive from unstable funding such as international aid programs. A more structured form of financing may be necessary to facilitate both in situ and ex situ conservation in the most appropriate locations (notably, in present geographic centers of diversity). International control over this fund is logical and desirable. The fund could be derived from national contributions that would be linked to the value of commercial seed sales. Its governance would be at the international level through an existing agency or specially designated commission. The proposal of the Keystone International Dialogue for a fund of US$300 million per year to support global and national efforts is an important first step (Keystone Center, 1991).

Disagreements over the ownership, control, and use of genetic resources must be resolved.

If left unresolved these issues will seriously hamper international cooperation and the unrestricted, free flow of genetic resources between countries. As necessary as political will and cooperative relations among scientists are to germplasm exchange, the funding levels and technical standards of genetic resources collection and preservation activities are major influences on distribution. Major barriers to the free exchange of genetic resources remain because of the lack of funding and staff to maintain and distribute materials upon request. There is blame to share and responsibility to be accepted on the part of developing and developed countries alike. The governments of developed countries that are hosts to major collections should maintain them and make materials available for free distribution, since many of the collected samples will not be reliably maintained at the national centers.

Government officials, agricultural scientists, and germplasm conservation personnel in developed and developing countries must recognize the necessity of cooperation in genetic resources conservation, exchange, and use. Cooperation is needed between the developing countries, which serve as centers of diversity, locations for rejuvenation of seed, and sites for in situ conservation, and the developed countries, which serve as sources of funding, preservation and distribution sites, and training programs (Chang, 1992). International cooperation and exchanges of germplasm can be substantially strengthened by international agencies such as FAO, IBPGR, and the IARCs. Cooperation cannot be secured by statutory means alone. It must begin with working scientists who share a common interest in the management and use of the world's genetic resources.

The capacities for plant breeding and biotechnology in developing countries should be strengthened.

The governments of developing nations could help their plant breeding and biotechnology sectors by establishing appropriate economic incentives and institutions. Public sector and extension activities should be structured to ensure that subsistence farmers participate in the benefits provided by these technologies. Accomplishing these goals will require assistance from developed countries in the form of training, facilities, and operating support.

Until recently, it has been relatively easy politically to conduct a large-scale, international, public sector, agricultural technology development program. That type of program, which was responsible for the green revolution, was quite successful. Today, multilateral programs may be the objects of political criticism. Agricultural technology is also increasingly being developed in the private sector rather than in the public sector. These transitions underlie the germplasm controversy; they pose the challenge of ensuring the capacity for private sector agricultural biotechnology research and expanded opportunities in developing countries.

15

National and International Programs

Crop genetic resources have national and global importance. They are a valuable part of the cultural agricultural heritage of all nations, and their effective management and use are of strategic importance to national food security. Genetic resources are important to the agriculture of nations far beyond the geographic centers of origin or diversity. Access to and control of germplasm and information are important international issues, which are extensively debated at the international level. This chapter addresses the role of national programs in managing genetic resources and the need to develop an equitable and stable international system to guide the collection, management, conservation, and use of crop genetic resources for the benefit of all nations.

NATIONAL PROGRAMS

The primary purpose of a national system is to ensure that the genetic resources needed in agriculture, forestry, and conservation programs are available, evaluated, conserved, and used. This is a long-term requirement to safeguard national food production now and in the future, and it should be a responsibility of an official public agency. A national agency should assume the responsibility for carrying out a national policy on germplasm use and conservation and interact with other parts of the government structure that have responsibility for quarantine and research.

The germplasm activity in a country may be centralized or diffused, public or private, large or small, or various combinations of these. When several organizations are involved in germplasm management, the centralized coordination of information and documentation of the holdings in various collections is desirable. A national plant germplasm system must have the capacity and a policy to assemble germplasm (through active collecting or exchanging); process and store materials; grow, test, and evaluate samples; regenerate and distribute samples; and maintain appropriate data.

There is considerable variation among countries in the level of genetic resources activity, ranging from virtually none to rather large programs, such as that in the United States (National Research Council, 1991a), usually as part of national agricultural research systems. Only a limited number of countries, however, have clearly defined national objectives and policies and with an adequate infrastructure to preserve the germplasm resources necessary beyond their immediate needs.

Why Are National Programs Necessary?

Access to diverse breeding materials essential to crop improvement is a major benefit provided by national programs. While the genetic diversity of major crop species has been extensively collected, much of the genetic diversity of others has not. This is especially true of many indigenous species that are poorly known outside the country or region where they are used. National programs are most appropriate for conserving these resources, since these minor species (in a global sense) may not command a high priority for international action.

The Importance of Exchange

Collections frequently include accessions obtained from other collections, and germplasm exchange operates in both directions. Plant germplasm exchange has played an important role in broadening the base of plant breeding, even in those countries that have a wide range of indigenous genetic resources. It has been shown that nations around the world benefit from the introduction of crops from other regions (Kloppenburg and Kleinman, 1988) (Table 15-1).

Participants in National Programs

Most existing collections have originated through the efforts of individuals involved in plant breeding, botanical and evolutionary studies, or other research. For example, botanical gardens and arboretums have historically played a role in the collection, conservation, and exchange of certain types of germplasm (Plucknett et al., 1987); although intraspecific variation maintained is generally quite limited. In most countries, active collections are held in public agricultural research institutes. In some countries, private institutions and universities are the principal repositories.

Status of National and Regional Programs

National plant genetic resources conservation programs vary considerably in organizational structure, in the nature of materials conserved, and how those are used. Although many nations have some level of plant germplasm activity related to agriculturally important plants, there is no nation, to the committee's knowledge, that has a comprehensive approach to the identification and conservation (in situ or ex situ) of all, or even the most common, plant species extant in the country. Collections of agricultural crop species are most frequently handled by a branch of the crops research unit of the agriculture ministry. It could be argued that food and fiber crop species (including forestry species) should receive added priority attention in conservation programs, and they generally do when these types of programs exist. However, the conservation of economically important species as well as the protection of ecosystems would often be better if national policies and strategies dealt comprehensively with biological diversity.

The committee examined many national and regional efforts to manage crop genetic resources. Although not exhaustive, this activity has provided an overview of the present general status and needs of conservation, management, and use at the national level.

The Commonwealth of Independent States and Eastern Europe

Genetic resources work in Eastern Europe has a long tradition going back to the pioneering research and discoveries of Russian academician N. I. Vavilov. In the former socialistic system, genetic resources programs were guided by the Scientific-Technical Board for Wild Species and Cultivated Agricultural Plant Collection as part of the Council for Mutual Economic Assistance encompassing the Soviet Union and all East European countries. They were coordinated by the N. I. Vavilov All-Union Scientific Research Institute of Plant Industry (VIR). As a result, the collection and conservation of genetic resources were well established in that region. Local landraces

TABLE 15-1 Percentages of Regional Food Crop Production and Industrial Crop Area Accounted for by Crops Associated with Different Regions of Diversity

Regions of Production	Regions of Diversity										Total Dependence
	Chino-Japanese	Indo-chinese	Austra-lian	Hindu-stanean	West Central Asiatic	Mediter-ranean	African	Euro-Siberian	Latin American	North American	
Food crops (percent production)											
Chino-Japanese	37.2	0.0	0.0	0.0	16.4	2.3	3.1	0.3	40.7	0.0	62.8
Indochinese	0.9	66.8	0.0	0.0	0.0	0.0	0.2	0.0	31.9	0.0	33.2
Australian	1.7	0.9	0.0	0.5	82.1	0.3	2.9	7.0	4.6	0.0	100.0
Hindustanean	0.8	4.5	0.0	51.4	18.8	0.2	12.8	0.0	11.5	0.0	48.6
West Central Asiatic	4.9	3.2	0.0	3.0	69.2	0.7	1.2	0.8	17.0	0.0	30.8
Mediterranean	8.5	1.4	0.0	0.9	46.4	1.8	0.7	1.2	39.0	0.0	98.2
African	2.4	22.3	0.0	1.5	4.9	0.3	12.3	0.1	56.3	0.0	87.7
Euro-Siberian	0.4	0.1	0.0	0.1	51.7	2.6	0.4	9.2	35.5	0.0	90.8
Latin American	18.7	12.5	0.0	2.3	13.3	0.4	7.8	0.5	44.4	0.0	55.6
North American	15.8	0.4	0.0	0.4	36.1	0.5	3.6	2.8	40.3	0.0	100.0
World	12.9	7.5	0.0	5.7	30.0	1.4	4.0	2.9	35.6	0.0	

Industrial crops
(percent
area)

Chino-Japanese	8.3	4.7	0.0	1.4	7.4	27.5	0.1	0.0	45.4	5.1	91.6
Indochinese	5.0	43.5	0.0	7.1	2.9	0.0	22.6	0.0	18.8	0.0	56.4
Australian	0.0	51.2	0.0	0.0	1.8	3.3	0.0	0.0	15.4	28.3	100.0
Hindustanean	2.6	14.2	0.0	7.2	20.5	17.2	0.9	0.0	35.2	2.1	92.7
West Central Asiatic	1.5	14.7	0.0	0.0	4.5	14.2	0.1	0.0	56.6	8.4	95.5
Mediterranean	0.0	3.9	0.0	0.2	2.4	25.3	0.0	0.0	31.8	36.5	74.9
African	1.3	16.3	0.0	0.1	10.6	0.4	22.4	0.0	46.0	3.0	77.7
Euro-Siberian	0.4	0.0	0.0	0.1	12.8	41.3	0.0	0.0	17.5	27.9	100.0
Latin American	0.2	30.4	0.0	0.4	5.9	0.4	25.7	0.0	28.0	9.1	72.1
North American	0.0	3.7	0.0	0.0	8.3	33.1	0.0	0.0	39.6	15.3	84.7
World	2.1	13.7	0.0	2.0	10.8	18.2	8.3	0.0	34.4	10.5	

NOTE: Reading the table horizontally along rows, the figures can be interpreted as measures of the extent to which a given region of production depends on each of the regions of diversity. The column labeled "total dependence" shows the percentage of production for a given region of production that is accounted for by crops associated with nonindigenous regions of diversity. Due to rounding error the figures in each row do not always sum to 100.

SOURCE: Kloppenburg, J. R., Jr., and D. L. Kleinman. 1988. Seeds of controversy: National property versus common heritage. Pp. 175–203 in Seeds and Sovereignty: The Use and Control of Plant Genetic Resources, J. R. Kloppenburg, Jr., ed. Durham, N.C.: Duke University Press. Reprinted with permission, ©1988 by Duke University Press.

and old cultivars have been well collected and programs begun notably in Bulgaria, the Czech Republic, (formerly East) Germany, Hungary, Poland, the Republics of Montenegro and Serbia, and the Slovak Republic (Table 15-2). VIR also holds one of the largest worldwide collections of cultivated species.

However, today it seems unlikely that there will be adequate central and local support to sustain work on genetic resources at these centers. For example, letters written by the director of the program in Bulgaria point out the extreme lack of financial resources that is forcing this center to cut back its work drastically, or even cease operating altogether, with a major loss to the world community. Similar problems exist at VIR.

In general, genetic resources have had a higher priority in agricultural research in Eastern Europe than in Western Europe. Although the principle of free exchange of genetic resources has been supported, the accessibility of the collections to the international community has been hampered by a backlog in the development and availability of adequate computerized information systems. A major step forward has been realized through the establishment of a number of European data-base systems, as part of the European Cooperative Program for Crop Genetic Resources Networks (ECP/GR), for specific crops. However, these systems are under serious threat in the wake of political developments in the former Soviet Union and in Eastern Europe, and the enormous economic problems that have followed. Shortage of funds and changed national priorities in a transfer to a market economy are taking their toll. While it is still difficult to assess the actual situation, it is obvious that international support is needed to avert the wholesale loss of valuable collections in most of these countries.

Western Europe

Since the 1950s, conservation of crop genetic resources has received increasing attention in western Europe. Previously, genetic variation was primarily assembled as part of breeding programs or in the context of taxonomic and ecogeographical studies at universities and other specialized institutions. However, concern about genetic erosion has become widespread, and many countries have established a variety of activities specifically aimed at conservation of genetic variation (Table 15-2). Since 1959, many of these activities in Europe have been stimulated by the European Association for Research on Plant Breeding (EUCARPIA) through the Genetic Resources Section.

TABLE 15-2 European Data Bases That Serve Important Roles in Coordinating Working Group Activities in the European Program

Crop or Species	Institute
Allium spp.	HRI, Wellesbourne, United Kingdom
Avena spp.	FAL, Germany (West)
Barley	ZIGuK, Germany (East)
Beta spp.	CGN, Netherlands
Prunus spp.	NGB, Sweden
Rye	PBAI, Poland
Cultivated *Brassica* spp.	PBAI, Poland
Pisum spp.	Wiatrowo Experimental Station, Poland
Sunflower	
Cultivated species	CRI, Szeged, Hungary
Wild species	IFVC, Republics of Montenegro and Serbia
Forages	
Tristeum flavescens and	Research Station of Grasses, Rožňava,
Arrhenatherum elatius	Slovak Republic
Poa spp.	FAL, Germany (West)
Lathyrus latifolius, L.	Institut de Biocénotique Experimentale
sylvestris, L. heterophyllus,	des Agrosystèmes, Université de Pau et
and *L. tuberosus*	des Pays de l'Adour, France
Medicago, perennial species	Groupe d'Etude et de Contrôle des Varietés
	et des Semences, INRA, La Minière,
	Guyancourt, France
Bromus spp.	RCA, Hungary
Trifolium alexandrinum,	Field Crops Department, Faculty of
T. resupinatum, and	Agriculture, Hebrew University of
wild related taxa	Jeurusalem, Rehovot, Israel
Lolium spp.	Instituto del Germoplasma, CNR, Bari,
Annual species,	Italy
Phalaris spp., *Vicia* spp.,	
and *Hedysarum*	
Dactylis spp. and *Festuca* spp.	PBAI, Poland
Trifolium subterraneum,	INIA, Spain
annual species	
Phleum spp.	NGB, Sweden
Trifolium pratense	Federal Agricultural Research Station,
	Changins, Switzerland
Lolium multiflorum, L. perenne,	WPBS, Aberystwyth, Wales
and *Trifolium repens*	

NOTE: The following acronyms are listed: CGN, Center for Genetic Resources; CNR, Consiglio Nazionale delle Ricerche; CRI, Cereal Research Institute; FAL, Institut für Pflanzenbau und Pflanzenzüchtung der Bundesforschungsanstalt für Landwirtschaft; IFVC, Fruit and Viticulture Research Institute; HRI, Horticultural Research Institute; INIA, Instituto Nacional de Investigaciones Agrarias; INRA, Institut National de la Recherche Agronomique; NGB, Nordic Gene Bank; PBAI, Plant Breeding and Acclimatization Institute; RCA, Research Center for Agrobotany, Institute for Plant Production and Qualification; WPBS, Welsh Plant Breeding Station; ZIGuK, Zentralinstitut für Genetik und Kulturpflanzenforschung.

SOURCE: Adapted from International Board for Plant Genetic Resources. 1990. Annual Report 1989. Rome: International Board for Plant Genetic Resources.

National genetic resources conservation is organized in a variety of ways, ranging from national germplasm banks, such as those of Germany [West], Greece, Italy, The Netherlands, Spain, Turkey, and a consortium of the Nordic countries, to coordinated institutional activities, such as those in Austria, Belgium, Cyprus, France, and Portugal. Mandates and objectives vary from conserving what is still available nationally to occasionally more basic approaches, for example, sampling overall genetic variation of certain crop species.

Programs tend to be largely opportunistic, with a few exceptions for individual crops, emphasizing exploitation rather than systematic long-term conservation of overall biodiversity. However, there are signs of change. The European Parliament decided in 1991 to establish a new and separate budget line for a European Community Program on the Conservation of Plant Genetic Resources to be executed through the commission's Directorate-General for Agriculture. This program is presently under consideration.

Eastern, Southern, and Southeastern Asia

Among the major germplasm banks of the world are those of the People's Republic of China which hold 400,000 samples, with 200,000 accessions in long-term storage. Japan and India, each with under 100,000 accessions in long-term storage, are also among the most recent in establishing centralized modern long-term storage facilities. The Japanese seed storage facility at Tsukuba is, perhaps, the most mechanically sophisticated seed storage of its kind in the world. It went into operation in 1979 and expanded in 1986. In Beijing, a modern seed storage facility has recently been put into operation, and in India, a new central seed storage facility and laboratories are being developed with bilateral financial assistance.

As in other regions of the world, there is variety in the organizational pattern of genetic resources conservation programs in the various countries of Asia. Most are nationally coordinated programs, with a centralized institutional germplasm bank for cultivars, for example, in Japan, Bangladesh, Pakistan, the Philippines, Malaysia, Sri Lanka, India, Nepal, Indonesia, the People's Republic of China, Thailand, and the Republic of Korea. Some countries have only special collections while others (for example, Myanmar) have no germplasm programs. Many have central coordinating councils or committees, but these need to be strengthened. Important to the region is the germplasm collection of the Asian Vegetable Research and Development Center in Taiwan.

In addition to seed collections, clonal materials are maintained in

Researchers selectively pollinate an experimental crop of hybrid onions growing in breeding cages. Several successful varieties released by the Indian Institute of Horticulture Research have provided higher yields and longer storage duration. Credit: Food and Agriculture Organization of the United Nations.

the People's Republic of China, India, Indonesia, the Philippines, and several other Asian countries. Botanic gardens, arboreta, and parks play an important role in conservation programs in Sri Lanka, India, and Indonesia. In Indonesia, education of the public on the importance of conserving genetic resources is an integral part of the national effort, and public institutions, as well as the general public, are encouraged to participate in maintaining many endangered species (Sastrapradja et al., 1985).

Latin America and the Caribbean

Argentina, Brazil, Colombia, Cuba, Guatemala, and Mexico have established coordinated plant genetic resources conservation programs organized within the framework of the agricultural research institutions of the ministries of agriculture. Similar programs are emerging in Peru, Bolivia, Ecuador, and Chile. The Centro Agronómico Tropical de Investigación y Enseñanza (CATIE, Tropical Agriculture Research and Training Center) in Costa Rica serves a coordinating role in the Central American region. In other countries, modest collections are held by public or private institutions. For the most part, there is little coordination among them. Considering the importance of the wide array of plant species throughout Latin America and the Caribbean, the present level of organization and support for genetic resources conservation is alarmingly inadequate.

Many gaps exist in national collections, especially collections of locally important plant species. The largest national program and ex situ collection of material is that of the Centro Nacional de Recursos Genéticos e Biotecnologia (CENARGEN, National Center for Genetic Resources and Biotechnology) in Brazil. CENARGEN and independent institutions that are coordinated through CENARGEN and Empresa Brasileira de Pesquisa Agropecuária (EMBRAPA, Brazilian Enterprise for Agricultural Research) conduct extensive collection efforts. Some collection efforts are made by international centers and germplasm banks outside Brazil, but in the absence of the necessary infrastructure (staff, laboratories, and storage facilities), the level of collection activities is relatively low.

In situ conservation is important in both the lowland and highland tropics and subtropics and, thus, so is the need to conserve natural ecosystems. Although national authorities in Latin America recognize the urgency to build the capacity to acquire, conserve, and use germplasm, economic constraints and personnel shortages frequently place this activity at a low priority.

One model of a cooperative interregional program in Latin America is the Andean crop network of Bolivia, Colombia, Chile, Peru, and Ecuador that focuses on indigenous crop resources. For the past 25 years these countries have been collecting, documenting, maintaining, and characterizing more than 30 crops, including grains, tubers, and fruits that have been and are major staples for the highland populations of these countries (National Research Council, 1989b).

The CATIE germplasm bank in Costa Rica serves as a regional research and training center designed primarily to serve the Central American nations. In practice, its activities and influence extend beyond that region. CATIE initiated the Genetic Resources Project in 1976 with the help of international funding. It has an active evaluation program and maintains 5,000 accessions, representing 335 species, that are being grown out. There are about 6,000 accessions of seed in long-term, cold storage facilities that are capable of holding seed materials at –17°C and 5 to 7 percent relative humidity.

Africa

The national genetic resources conservation programs in Africa have recently been described in considerable detail (Attere et al., 1991;

Ng et al., 1991). In general, most activities have occurred over the past decade, although several germplasm collections originated from earlier breeding programs (for example, sorghum in eastern Africa and plantation crops in western and central Africa). Africa is endowed with great genetic diversity, but this wealth has not been adequately conserved and utilized (Thitai, 1991). There is, however, a growing awareness of the need to avert the loss of these important resources.

Currently, there is activity on genetic resources in about 30 countries that ranges from collection work to more coordinated comprehensive national programs (Attere et al., 1991; Thitai, 1991). The increased efforts that have occurred over the past decade are not widely recognized. Although these changes are paralleled in other regions of the developing world, they are more impressive in Africa because of the paucity of prior activities.

Africa covers a number of traditional centers of diversity. In recent years, a degree of regionality has emerged, particularly in the Southern African Development and Coordination Conference countries (Angola, Botswana, Lesotho, Malawi, Mozambique, Swaziland, Tanzania, Zambia, and Zimbabwe). Consultations have led to the proposal for a regional program for southern Africa and a regional germplasm bank (Kyomo, 1991). A program of collection, conservation, and distribution has been established for the Great Lakes countries of Burundi, Rwanda, and Zaire (ne Nsaku, 1991). The program in Ethiopia, which was established in 1976, presently holds more than 45,000 accessions in a national center (Thitai, 1991).

Because of the interrelationships between Mediterranean countries and earlier regional programs, most countries of northern Africa have national programs, all with some form of seed storage and a national coordinator. More than 25 countries comprise eastern and southern Africa and the island states. Ethiopia, Kenya, Sudan, Tanzania, and Uganda (eastern Africa); Burundi, Rwanda, and Zaire (central Africa); Malawi, South Africa, Zambia, and Zimbabwe (southern Africa); and Mauritius have defined national programs (Attere et al., 1991). National genetic resources programs are at various stages of development in Ethiopia, Kenya, and Sudan (Thitai, 1991). Genetic resources activities are ongoing in many other countries.

The 22 countries of western Africa show the widest range of development of national programs. Twelve have no national program at all, whereas Cameroon, Côte d'Ivoire, Ghana, Nigeria, and Senegal have more advanced national programs.

Australia, New Zealand, and Papua New Guinea

Until the early 1970s, little attention was paid to the conservation of Australia's own native genetic resources or to the extensive materials introduced from overseas. The prevailing view was that additional material could easily be collected if it were required. The cost of recollection was considered lower than the cost of long-term conservation. This view gradually began to change with the increasing threats of genetic erosion in both Australia and overseas. International developments prompted Australian scientists to propose a national network of genetic resources centers that would conserve germplasm of Australian's most important crop and pasture species and tie this network in with the international network (Francis, 1986) sponsored by the Food and Agriculture Organization (FAO) of the United Nations. Although this plan was accepted in principle by the state and federal governments in the mid-1970s, it took two major committees of enquiry and almost a decade to develop firm proposals for an Australian network of genetic resources centers. The complete network came into being in the late 1980s.

Eight centers, each with a professionally qualified curator, cover temperate crops and pasture plants of greatest economic importance in Australia. This includes a small center in the Commonwealth Scientific and Industrial Research Organization's Division of Plant Industry, Canberra, Australia, concerned with conserving relatives of crop plants indigenous to Australia that are often of great interest to other countries and that provide a valuable asset in reciprocal exchange programs. Although the situation with respect to the conservation of genetic resources is improved, there are still substantial gaps in the established network. For example, important groups of plants, such as horticultural and forest species, have yet to be included in the Australian national system. Furthermore, there is an urgent need to develop national crop data bases on genetic resources that can be accessed through any or all of the centers.

New Zealand does not have a genetic resources program at present. It does, however, participate in international exchange activities and generally supports the concepts of conservation of biodiversity.

Germplasm activities in Papua New Guinea are carried out in the Crops Research Division of the Department of Agriculture and Livestock. The Laloki Station has responsibility for overall coordination of the plant genetic resources program. Vegetatively propagated materials are maintained in field germplasm banks or in active collections at various locations around the country. The main collections include sweet potato, banana, taro, yam, cassava, sugarcane, and sago.

Traditional vegetable, fruit and nut tree species, as well as cash crop species are maintained among various institutes and experiment stations. Seed collections, consisting mainly of introduced species are held in short- to medium-term storage. Except for winged bean and certain other indigenous food plants, the seed collections are mainly used for experimental purposes and not as germplasm stores. With the help of IBPGR, collecting trips made during 1986 and 1987 added samples to the collections of taro, yams, cassava, bananas, and other plants. Additional collecting activities have broadened the materials held in the respective field germplasm banks. Of particular significance are the banana and root and tuber crops.

North America

The United States and Canada have what they consider to be centrally coordinated national systems; however, many other public and private institutions participate in each system.

The U.S. National Plant Germplasm System (NPGS) was recently reviewed by the National Research Council (1991a). It contains more than 350,000 individual accessions, with more than 240,000 of these being in long-term storage at the National Seed Storage Laboratory. On average, NPGS distributes about 125,000 samples to more than 100 countries each year.

Although germplasm activities began much earlier (National Research Council, 1991a), the Office of Seed and Plant Introduction was not established until 1898. It was expanded in 1946 with the establishment of regional plant introduction stations. The present NPGS emerged in 1974 following a major reorganization of the Agricultural Research Service of the U.S. Department of Agriculture (National Research Council, 1991a).

Plant Gene Resources of Canada operates as a unit under the Plant Research Center within the Research Branch of the Ministry of Agriculture. The system began in 1970 following a series of consultations. The system holds about 82,000 accessions. Both mid- and long-term storage facilities are maintained at the Central Experimental Farm, Ottawa, Ontario. Materials are freely exchanged as long as they are available and plant quarantine procedures are met.

INTERNATIONAL PROGRAMS

Although the primary responsibility for the management of plant genetic resources rests with each nation, there are important reasons why nations should act in concert to protect, preserve, and use these resources. First, the task is enormous, so the involvement of all nations would more likely ensure proper protection of the world's plant genetic resources. Second, it is in the self-interest of each nation to participate in a global system. Plant distributions are often transnational, and different plant species occur in different regions of the world. The alleles that are needed or that are useful in one region are often found in another. Finally, there are disparities among nations with regard to the location and holding of plant germplasm, the financial resources available for conservation and use, and the technical expertise and other vital components of germplasm management that only a global system of collaboration can help to structure for the equitable benefit of all concerned.

Despite the recognition by scientists since the early 1930s of the importance of genetic diversity and the accelerating pace at which landraces and natural populations of wild species related to crops were being lost (Frankel and Bennett, 1970; Frankel and Hawkes, 1975b; Harlan, 1975; National Research Council, 1972), there was no organized international effort until the 1960s to ensure the conservation and preservation of genetic diversity of even the most important food crops. In the 1940s and 1950s, many countries began to assemble their own collections based on collections developed by plant specialists or crop breeders. Concepts such as the conservation of the total spectrum of diversity was not a high priority in the early stages, until international organizations lent a vastly expanded dimension to these efforts (Chang, 1985e; Hawkes, 1985).

Food and Agriculture Organization

FAO convened the first international meeting on plant genetic resources in 1961 and later established the Panel of Experts on Plant Exploration and Introduction in 1965 (Frankel and Hawkes, 1975c; Williams, 1984). A parallel panel of experts on forest gene resources was established by FAO in 1968 (National Research Council, 1991b). FAO also established a crop ecology and genetic resources unit in 1968 to deal with activities related to the collection, conservation, and documentation of genetic resources (Williams, 1984). Two major technical conferences, in 1967 and 1973, specifically on genetic resources were cosponsored by the International Biological Programs and FAO.

Consultative Group on International Agricultural Research

The greatest impetus to the internationalization of germplasm was provided with the establishment of the international agricultural research centers (IARCs) in the Consultative Group on International Agricultural Research (CGIAR) (Baum, 1986). The CGIAR system was established in 1971, at which time there were four operational IARCs (Baum, 1986). These and subsequent CGIAR centers amassed large germplasm collections as a consequence of their work on particular crops. Today they represent a major element of an emerging global germplasm system.

International Board for Plant Genetic Resources

The technical conferences on genetic resources and the FAO Panel of Experts consistently recommended that a global network of crop genetic resources centers be established. The United Nations Conference on Human Environment held in Stockholm, Sweden, in 1972, gave FAO the responsibility to assist in the establishment of an international genetic resources program (Frankel, 1988). Subsequently, FAO requested that the Technical Advisory Committee (TAC) of the newly formed Consultative Group on International Agricultural Research (CGIAR) establish a mechanism to encourage, coordinate, and support conservation of genetic resources and to make them available for use by all nations. Following a series of modifications of recommendations by TAC, CGIAR approved the formation of the International Board for Plant Genetic Resources (IBPGR) (Baum, 1986). In 1974, IBPGR was organized to be associated with FAO in Rome, Italy. FAO provided logistical support to IBPGR and continued publishing its documents related to germplasm activities including its genetic resources newsletter.

In October 1991, a previously ratified agreement was signed by board members from China, Denmark, Kenya, and Switzerland, which established the International Plant Genetic Resources Institute (IPGRI). After ratification by the Italian government is obtained, IPGRI will assume the duties of IBPGR and the latter will cease to exist.

European Association for Research on Plant Breeding

In 1959 the European Association for Research on Plant Breeding (EUCARPIA) established the Wild Species and Primitive Forms Section (currently, the Genetic Resources Section). It highlighted the importance of using wild species related to cultivated crops as sources of genetic traits for disease and pest resistance, and as tools for understanding the evolution of crop gene pools. Since 1960, genetic conservation and the danger of genetic erosion in wild species and primitive forms have become major concerns of the association.

SELECTED INTERNATIONAL CROP DATA BASES

To address documentation, the IBPGR encourages the establishment of international data bases for various crops. These would logically be handled by base collection holders. Several international data bases are being developed. Through the European Cooperative Program for Conservation and Exchange of Crop Genetic Resources, several regional European data bases have been set up (see table below). Some of these regional data bases have spread beyond the confines of Europe and may eventually become part of international data bases. The extensive documentation system of the International Rice Research Institute may serve as an example of other international crop data bases under consideration.

Selected International Crop Data Bases

Data Base/Location	Description
Centro Internacional de Mejoramiento de Maíz y Trigo (CIMMYT), Mexico	Maize germplasm in major Latin American germplasm banks. Software, ACCESSION EDITOR, developed to facilitate compilation of standardized maize passport data. Later, query software will be produced to facilitate the management of this large data base.
North Carolina State University, United States of America	Inventory of samples of wild *Arachis* spp. in major collections in Argentina, Brazil, Colombia, India (International Crops Research Institute for the Semi-Arid Tropics [ICRISAT]), and the United States. Project initiated after recommendations of the joint meeting of the Centro Internacional de Agricultura Tropical (CIAT), International Board for Plant Genetic Resources (IBPGR), and ICRISAT held in February 1989 at CIAT.
International Center for Agricultural Research in the Dry Areas (ICARDA), Syria	Mediterranean forages. Project started in 1987 at IBPGR headquarters and continued in 1989 at ICARDA. Particular emphasis on filling gaps in records received earlier from germplasm banks in 15 countries. Data base on wild species closely related to wheat. Project was initially carried out at IBPGR headquarters (in 1988) and then moved to ICARDA.
University of Southampton, United Kingdom	Ecogeographical data base for *Vicia faba, Vicia sativa,* and their relatives in *Vicia* subgenus *Vicia.* Data base will complement the *Vicia* section of the Mediterranean forages system.

Selected International Crop Data Bases—*continued*

Data Base/Location	Description
University of Malaya, Malaysia	Records of *Citrus* spp. herbaria data necessary for studying the distribution of *Citrus* and its wild relatives in southeastern Asia. In 1989 IBPGR assisted the university in acquiring a microcomputer.
University of Reading, United Kingdom	First version of data base on primitive cacao germplasm (begun in 1988 with support of IBPGR and the Biscuit, Cake, Chocolate, and Confectionery Alliance of the United Kingdom) completed and distributed in 1989. Further development of the system includes information on morphology and disease resistance.
Escuela Técnica Superior de Ingenieros Agronómos (ETSIA), Spain	Data base on *Cucumis* germplasm. After completion of documentation of comprehensive collection at Instituto Nacional de Investigaciones Agrícolas, Spain, the work in 1989 concentrated on collating passport data from other major *Cucumis* collections.
Universidad Politecnica, Spain	Data base for cultivated *Brassica* species in the Mediterranean; follow-up to collecting missions sponsored by IBPGR.
Plant Breeding and Acclimatization Institute (PBAI), Poland	Data base for cultivated *Brassica* germplasm maintained in European collections.
International Network for the Improvement of Banana and Plantain (INIBAP), France	International *Musa* Conservation Network. Development of a *Musa* data base to serve the network.
International Rice Research Institute (IRRI), Philippines	Worldwide rice germplasm conservation and dissemination services with special emphasis on tropical cultivars and wild species of Asia. Also serves as coordinator of the International Rice Testing Program. Data base includes germplasm, international nurseries, and breeding records, all inter-linked and freely distributed. Germplasm data base includes passport, characterization, and evaluation data. Seed storage, rejuvenation, viability monitoring, and distribution are computer managed. Also develops and supplies software for personal computers for national germplasm use.

SOURCES: International Board for Plant Genetic Resources. 1990. Annual Report 1989. Rome: International Board for Plant Genetic Resources; International Rice Research Institute. 1991. Rice Germplasm: Collecting, Preservation, Use. Los Baños, Philippines: International Rice Research Institute.

EUCARPIA was instrumental in recommending a series of germplasm banks that spanned Europe. The national germplasm banks of Italy and Germany (then, the Federal Republic of Germany) and the Nordic germplasm bank subsequently emerged. A germplasm bank committee of EUCARPIA continued to meet and has been closely related with IBPGR and the ECP/GR, which forged cooperation between eastern and western Europe largely through the development of European crop data bases.

THE ROLE OF INTERNATIONAL PROGRAMS

An original concept of the FAO Panel of Experts on Plant Exploration and Introduction was to establish regional exploration centers within the major regions of diversity of major crops. A regional approach makes botanical sense and was thought to be more effective because many centers of diversity are located in the less developed regions of the world that could not devote major resources to germplasm banks. Attempts to develop regional centers in the areas of major crop diversity were not successful (Baum, 1986). It became clear that the essential operating unit was the national program.

The basic function of IBPGR was to promote an international network of genetic resources centers in cooperation with national programs, donor nations and institutions, and individual scientists (International Board for Plant Genetic Resources, 1988c). It was to act as a catalyst for stimulating the action needed to sustain a viable network of institutions for conserving plant genetic resources and to provide leadership and advice on establishing and managing germplasm collections. IBPGR activities included exploring and collecting important plant species; encouraging development of facilities for preserving and using collections; training germplasm specialists and technicians; fostering establishment of base collections of important crops; establishing operational standards for germplasm banks; creating a data and information center; publishing an annual status report on germplasm conservation activities; and assisting in the development of networks to facilitate the documentation and exchange of germplasm materials. The ultimate goal is a globally integrated genetic resources system.

Between 1978 and 1988, an estimated 230,000 samples of various crops were added to germplasm collections through multi-institutional collaborations, and field collection activities were carried out in more than 100 countries. The criteria for selecting crops and regions included the immediate or potential danger of genetic erosion, the economic and social importance of the crops, the requirements of plant breeding, and the availability of existing collections.

IBPGR policies today stress the need to analyze the available information on species distribution and collection holdings and has moved to more systematic programs based on effective sampling of gaps in collections or rescuing threatened germplasm (van Sloten, 1990b). IBPGR has given increased priority to strategic research on genetic resources and to raising the scientific standards applied in germplasm work worldwide.

In 1983, the Twenty-Second Biennial FAO Conference adopted Resolution 8/83, the International Undertaking on Plant Genetic Resources. Under Article 9.2 of the annex to the resolution, the FAO council established the Commission on Plant Genetic Resources (see Chapter 14). The FAO commission was to

• Monitor the international arrangements referred to in Article 7 that relate to the strengthening of international activities and development of an integrated global system;

• Recommend measures that would ensure that the global system would be comprehensive and have an efficient operation; and

• Review matters relating to the policy, programs, and activities of FAO in the field of plant genetic resources and to give advice to FAO.

The FAO is developing plans for a Fourth International Technical Conference on Plant Genetic Resources to be held in 1995. Preparations include development of a report on the state of the world's plant genetic resources. This will be derived from individual country studies, regional reports, and a survey of national programs that is being conducted and analyzed in cooperation with IBPGR. From this, FAO hopes a plan of action will emerge to guide future plant genetic resources activities.

THE GLOBAL GENETIC RESOURCES NETWORK

Most available genetic resources are part of active collections. These have been assembled to serve the ongoing breeding and introduction programs of several hundred institutions throughout the world. Of the estimated 2.6 million accessions held worldwide, there are only about 1.3 million unique samples and about 35 percent of these are held by the crop-oriented IARCs (van Sloten, 1990a). To safeguard collections held outside the developed world and outside the IARCs, and to formalize worldwide responsibility, IBPGR proposed and helped to set up a network of institutions holding base collections of particular crops. As of 1990 about 50 institutions had voluntarily entered into an agreement to accept such responsibilities (International Board for Plant Genetic Resources, 1991). These include the IARCs within the CGIAR system, regional germplasm banks, and larger national programs in both developed and developing countries (Table 15-3). The intent is to duplicate each accession in at least two base collections. A good start has been made, but many minor crops are not yet stored in base collections; and many duplicate base collections have yet to be identified or stored. Several institutes and programs have accepted responsibility for maintaining selected vegetatively propagated crops as field or greenhouse collections (Table 15-4).

Regeneration is a major problem for seed collections, and few base collection holders have the capacity to do it. Another problem is that environmental specificity often requires multiplication under conditions not available in the country of the base collection holder. Furthermore, there are problems of limited support, inadequately trained staff, and poor management. Since the important materials are in long-term storage, however, there is still an opportunity to organize regeneration networks. (Chapter 5 contains a more detailed discussion of regeneration.)

Documentation is another problem. Accessions entered in a base collection should be accompanied by adequate passport data and, if possible, characterization data (see Chapter 8). This is often not the case, limiting the usefulness of such accessions to breeders.

IBPGR has spent much effort toward standardizing data collection and developing descriptor lists for various crops. The development of appropriate information systems specially designed for use in genetic collections and for facilitating the exchange of information between germplasm banks and with users and collection managers is only slowly improving.

OTHER INTERNATIONAL CENTERS

The IARCs form the backbone of the international network. They are responsible for collecting, distributing, and conserving the genetic resources of many specific food crops. An effective lead was provided by the program embarked on by IRRI at Los Baños, Philippines, which began operations in 1962. It now holds the major collection of the world's landraces of Indica and Javanica rices and is still expanding, and it has given added emphasis to wild rice species. Other IARCs are following suit, notably Centro Internacional de la Papa (potatoes and sweet potato), International Crops Research Institute for the Semi-Arid Tropics (sorghum, millet, chickpea, pigeonpea, and groundnut), Centro Internacional de Agricultura Tropical (landraces

TABLE 15-3   Base Collections of Seed Crops That Accepted
Responsibility for Long-Term Conservation

Crop and Species Covered	Scope of Collection		Institute
	Global	Regional	
Cereals			
Barley	•		PGR, Ottawa, Canada
		European	NGB, Lund, Sweden
		African	PGRC\E, Addis Ababa, Ethiopia
		Asian	NIAR, Tsukuba, Japan
	•		ICARDA, Syria
Maize		New World	NPGS, United States
		Asian	NIAR, Tsukuba, Japan
		Asian	TISTR, Bangkok, Thailand
		European	VIR, Leningrad, Russia
		Mediterranean	Portuguese Genebank, Braga, Portugal
Millets			
Pennisetum spp.	•		NPGS, United States
	•		PGR, Ottawa, Canada
	•		ICRISAT, India
Eleusine spp.	•		PGRC/E, Addis Ababa, Ethiopia
	•		ICRISAT, India
Minor Indian millets		Indian	NBPGR, New Delhi, India
Eragrostis	•		PGRC/E, Addis Ababa, Ethiopia
Panicum miliaceum	•		ICRISAT, India
Setaria italica	•		ICRISAT, India
Oats (*Avena* spp.)	•		PGR, Ottawa, Canada
	•		NGB, Lund, Sweden
	•		FAL, Braunschweig, Germany (West)
Rice			
Oryza sativa			
O. indica	•		IRRI, Philippines
O. javanica	•		IRRI, Philippines
O. japonica	•		NIAR, Tsukuba, Japan
		African	IITA, Ibadan, Nigeria
	•		NPGS, United States
Wild species	•		IRRI, Philippines
Rye	•		Polish Genebank, Radzikow, Poland
	•		NGB, Lund, Sweden
Sorghum	•		NPGS, United States
	•		ICRISAT, India

continued

TABLE 15-3 *Continued*

Crop and Species Covered	Scope of Collection		Institute
	Global	Regional	
Wheat			
Cultivated species	•		VIR, Leningrad, Russia
	•		CNR, Bari, Italy
	•		NPGS, United States
		Asian	CAAS, Beijing, People's Republic of China
	•		ICARDA, Syria
Wild species	•		Plant Germplasm Institute,
(*Triticum* and			University of Kyoto, Japan
Aegilops spp.)	•		ICARDA, Syria
Food legumes			
Chickpea	•		ICRISAT, India
	•		ICARDA, Syria
Cicer spp.	•		ICARDA, Syria
Faba bean	•		CNR, Bari, Italy
	•		ICARDA, Syria
Groundnut	•		ICRISAT, India
		South American	INTA, Pergamino, Argentina
Lentil	•		ICARDA, Syria
Lens spp.	•		ICARDA, Syria
Lupin	•		ZIGuK, Gatersleben, Germany (East)
		European	INIA, Madrid, Spain
Pea	•		NGB, Lund, Sweden
		Mediterranean	CNR, Bari, Italy
		Central and East European	Polish Genebank, Radzikow, Poland
Phaseolus			
Wild species	•		JBNB, Bruxelles, Belgium
Cultivated species	•		CIAT, Colombia
	•		NPGS, United States
		European	FAL, Braunschweig, Germany (West)
Pigeonpea	•		ICRISAT, India
	•		NBPGR, New Delhi, India
Soybean	•		NIAR, Tsukuba, Japan
	•		NPGS, United States
Wild perennial	•		CSIRO, Canberra, Australia
Vigna spp.			
Wild species	•		JBNB, Bruxelles, Belgium
V. mungo	•		NBPGR, New Delhi, India
V. radiata	•		IPB, Los Baños, Philippines
	•		AVRDC, Taiwan
V. umbellata	•		NBPGR, New Delhi, India
V. unguiculata	•		IITA, Nigeria
	•		NPGS, United States

TABLE 15-3 *Continued*

Crop and Species Covered	Scope of Collection		Institute
	Global	Regional	
Winged bean	•		IPB, Los Baños, Philippines
	•		TISTR, Bangkok, Thailand
Root crops			
Carrot	•		IHR, Wellesbourne, United Kingdom
Cassava (seed)	•		CIAT, Colombia
Solanum spp.	•		CIP, Peru
Sweet potato (seed)	•		NPGS, United States
		Asian	AVRDC, Taiwan
	•		NIAR, Tsukuba, Japan
Vegetables			
Allium	•		CGN, Wageningen, Netherlands
	•		IHR, Wellesbourne, United Kingdom
	•		NPGS, United States
		South and East European	RCA, Tápiószele, Hungary
		Asian	NIAR, Tsukuba, Japan
Amaranthus	•		NPGS, United States
		Asian	NBPGR, New Delhi, India
Capsicum	•		CATIE, Turrialba, Costa Rica
	•		AVRDC, Taiwan
		Asian	NBPGR, New Delhi, India
Cruciferae			
Brassica carinata	•		FAL, Braunschweig, Germany (West)
	•		PGRC/E, Addis Ababa, Ethiopia
B. oleracae	•		CAAS, Beijing, People's Republic of China
	•		IHR, Wellesbourne, United Kingdom
	•		CGN, Wageningen, Netherlands
Raphanus	•		CAAS, Beijing, People's Republic of China
	•		IHR, Wellesbourne, United Kingdom
		Asian	NBPGR, New Delhi, India
Wild species	•		Universidad Politécnica, Madrid, Spain
	•		Tohoku University, Sendai, Japan

continued

TABLE 15-3 *Continued*

Crop and Species Covered	Scope of Collection		Institute
	Global	Regional	
Oilseeds and green manures			
B. campestris,	•		PGR, Ottawa, Canada
B. juncea		Asian	NBPGR, New Delhi, India
B. napus, Sinapis alba	•		FAL, Braunschweig, Germany (West)
Vegetables and fodders			
B. campestris, B. juncea, B. napus	•		IHR, Wellesbourne, United Kingdom
B. napus	•		FAL, Braunschweig, Germany (West)
All Cruciferae crops		East Asian	NIAR, Tsukuba, Japan
Lactuca spp.	•		IHR, Wellesbourne, United Kingdom
	•		CGN, Wageningen, Netherlands
Okra	•		NPGS, United States
	•		NBPGR, New Delhi, India
Safflower	•		NBPGR, New Delhi, India
Tomato	•		CATIE, Turrialba, Costa Rica
	•		ZIGuK, Gatersleben, Germany (East)
	•		NPGS, United States
		Asian	IPB, Los Baños, Philippines
Southeastern Asian vegetables		Southeast Asian	IPB, Los Baños, Philippines
Cucurbitaceae			
Benincasa, Luffa, Momordica, Trichosanthes spp.	•		IPB, Los Baños, Philippines
Cucumis, Citrullus, Cucurbita spp.	•		NPGS, United States
Citrullus, Cucurbita spp.	•		VIR, Leningrad, Russia
Cucumis, Citrullus spp.	•		INIA, Madrid, Spain
Eggplant	•		CGN, Wageningen, Netherlands
	•		NPGS, United States
	•		NBPGR, New Delhi, India
Industrial crops			
Beet	•		FAL, Braunschweig, Germany (West)
	•		NGB, Lund, Sweden
		Mediterranean	Greek Gene Bank, Thessaloniki

TABLE 15-3 *Continued*

Crop and Species Covered	Scope of Collection		Institute
	Global	Regional	
Cotton		Mediterranean	Greek Gene Bank, Thessaloniki
Sugarcane (seed)	•		NIAR, Tsukuba, Japan
	•		NPGS, United States
Tobacco		Mediterranean	Greek Gene Bank, Thessaloniki
Jute and kenaf	•		BJRI, Dhaka, Bangladesh
Forages			
Legumes			
Centrosema spp.	•		CIAT, Colombia
	•		CENARGEN, Brazil
	•		CSIRO, Brisbane, Australia
Desmodium spp.	•		CIAT, Colombia
	•		CSIRO, Brisbane, Australia
Desmanthus spp.	•		CSIRO, Brisbane, Australia
Stylosanthes spp.	•		CIAT, Colombia
	•		CSIRO, Brisbane, Australia
Leucaena spp.	•		NPGS, United States
Lotononis spp.	•		ILCA, Ethiopia
	•		Seed Bank, RBG, Kew, United Kingdom
Macroptilium spp.	•		CENARGEN, Brazil
	•		CSIRO, Brisbane, Australia
Neonotonia		African	ILCA, Ethiopia
	•		Seed Bank, RBG, Kew, United Kingdom
Zornia spp.	•		NPGS, United States
	•		CIAT, Colombia
Trifolium spp.		African	ILCA, Ethiopia
	•		Seed Bank, RBG, Kew, United Kingdom
Grasses			
Cynodon spp.	•		NPGS, United States
Cenchrus spp.	•		Seed Bank, RBG, Kew, United Kingdom
	•		ILCA, Ethiopia
	•		CSIRO, Brisbane, Australia
Digitaria spp.	•		ILCA, Ethiopia
	•		CSIRO, Brisbane, Australia
	•		Seed Bank, RBG, Kew, United Kingdom
Pennisetum spp.	•		NPGS, United States
Paspalum spp.	•		NPGS, United States
Urochloa spp.	•		CSIRO, Brisbane, Australia

continued

TABLE 15-3 *Continued*

Crop and Species Covered	Scope of Collection		Institute
	Global	Regional	
Others			
Tree species	•	(Fuel and environmental stabilization in arid areas)	Seed Bank, RBG, Kew, United Kingdom
Sesame	•		KARI, Nairobi, Kenya
	•		RDA, Republic of Korea
Sunflower		European, Mediterranean	Plant Production Research Institute, Slovak Republic

NOTE: The following acronyms are listed: AVRDC, Asian Vegetable Research and Development Center; BJRI, Bangladesh Jute Research Institute; CAAS, Chinese Academy of Agricultural Sciences; CATIE, Centro Agronómico Tropical de Investigación y Enseñanza; CENARGEN, Centro Nacional de Recursos Genéticos; CGN, Center for Genetic Resources; CIAT, Centro Internacional de Agricultura Tropical; CIP, Centro Internacional de la Papa; CNR, Consiglio Nazionale delle Ricerche; CSIRO, Commonwealth Scientific and Industrial Research Organization; FAL, Institut für Pflanzenbau und Pflanzenzüchtung der Bundesforschungsanstalt für Landwirtschaft,; ICARDA, International Center for Agricultural Research in the Dry Areas; ICRISAT, International Crops Research Institute for the Semi-Arid Tropics; IHR, Institute of Horticulture Research; IITA, International Institute of Tropical Agriculture; ILCA, International Livestock Center for Africa; INIA, Instituto Nacional de Investigaciones Agrarias; INTA, Instituto Nacional de Tecnologia Agropecuaira; IPB, Institute of Plant Breeding; IRRI, International Rice Research Institute; JBNB, Jardin Botanique National de Belgique; KARI, Kenyan Agricultural Research Institute; NBPGR, National Bureau of Plant Genetic Resources; NGB, Nordic Gene Bank; NIAR, National Institute of Agrobiological Research; NPGS, National Plant Germplasm System; PGR, Plant Gene Resources of Canada; PGRC/E, Plant Genetic Resources Center; RBG, Royal Botanic Gardens; RCA, Research Center for Agrobotany, Institute for Plant Production and Qualification; RDA, Rural Development Administration; TISTR, Thailand Institute of Scientific and Technological Research; VIR, N.I. Vavilov Institute of Plant Industry; ZIGuK, Zentralinstitut für Genetik und Kulturpflanzenforschung.

SOURCE: Updated from International Board for Plant Genetic Resources. 1990. Annual Report 1989. Rome: International Board for Plant Genetic Resources. Reprinted with permission, ©1990 by International Board for Plant Genetic Resources.

TABLE 15-4   Field Germplasm Banks That Accepted Responsibility for Conservation (Active Collections for Vegetative Material), 1990

Crop and Species Covered	Geographical Representation		Institute
	Global	Regional	
Roots and tubers			
Cassava	•		CIAT, Colombia
		Central American	INIA, Mexico
		African	IITA, Nigeria
Sweet potato		Asian and Pacific	AVRDC, Taiwan
	•		IITA, Nigeria
Fruits			
Banana	•		Banana Board, Jamaica
		Southeast Asian	PCARRD, Philippines
		African	DGRST, Cameroon
Citrus		East Asian	Fruit Tree Research Station, Tsukuba, Japan
		Mediterranean	INIA, Valencia, Spain
		Mediterranean and African	IRFA, Corsica, France
		North American	USDA, United States
		Latin American	CENARGEN, Brazil
		South Asian	IIHR, India[a]
Subfamily Aurantioideae	•		University of Malaya, Kuala Lumpur, Malaysia
Industrial crops			
Cacao	•		University of the West Indies, Trinidad and Tobago
	•		CATIE, Costa Rica
Sugarcane	•		Sugarcane Breeding Institute, Coimbatore, India
	•		USDA, Florida, United States
Perennial species			
Allium spp.			
Short-day spp.	•		Hebrew University of Jerusalem, Israel[a]
Long-day spp.	•		Research Institute for Vegetable Growing and Breeding, Olomouc, Czech Republic
Arachis spp. (wild)		Latin American	CENARGEN, Brazil
Glycine spp. (wild)	•		CSIRO, Australia

continued

TABLE 15-4 *Continued*

NOTE: The following acronyms are listed: AVRDC, Asian Vegetable Research and Development Center; CATIE, Centro Agronómico Tropical de Investigación y Enseñanza; CENARGEN, Centro Nacional de Recursos Genéticos; CIAT, Centro Internacional de Agricultura Tropical; CSIRO, Commonwealth Scientific and Industrial Research Organizaton; DGRST, Delegation Generale a La Recherche Scientifique; IIHR, Indian Institute of Horticultural Research; IITA, International Institute of Tropical Agriculture; INIA, Instituto Nacional de Investigaciones Agrarias; IRFA, Institut de Recherches sur les Fruits et Agrumes; PCARRD, Philippine Council for Agricultural and Resources Research and Development; and USDA, U.S. Department of Agriculture.

*a*Under discussion or awaiting formal agreement.

SOURCE: Updated from International Board for Plant Genetic Resources. 1990. Annual Report 1989. Rome: International Board for Plant Genetic Resources. Reprinted with permission, ©1990 by International Board for Plant Genetic Resources.

of *Phaseolus* species and cassava), International Institute for Tropical Agriculture (cowpea, yams, and rice), International Center for Research in Agroforestry (multipurpose trees), and others. These institutes have direct linkages, through their plant breeding programs and germplasm collections, to national agricultural research systems, and they serve as effective centers of germplasm exchange.

There is no systemwide genetic resources program formally coordinated within the CGIAR that addresses the complete range of plant genetic resources issues. However, the IARCs have organized an intercenter plant genetic resources working group to address issues of common interest.

CONSTRAINTS TO EFFECTIVE INTERNATIONAL PROGRAMS

The institutional problems largely stem from asymmetry in the availability and, more important, the use of genetic resources in plant breeding and biotechnology in developed and developing countries. Although, with few exceptions, the free exchange of genetic resources is taking place, there is an underlying uneasiness that increased involvement of the private sector in plant breeding may affect this free exchange, if not now, possibly in the future (see Chapter 14). The most important part of the solution is to improve the position of developing countries in plant breeding and seed production. Much of the earlier scholarship and fellowship support to train plant breeders and other agricultural scientists is no longer available. Some progress is being made through the activities of the CGIAR in cooperation with national programs. Much more needs to be done, however, especially in improving the quality of collections. It is generally becoming appreciated that no country is self-sufficient in genetic resources and that all countries can benefit from the free flow of germplasm, especially if there is a strong capability in place to use it (Keystone Center, 1988, 1990, 1991).

The standards of management and the facilities that house the collections are also of concern. In response to criticism that genetic erosion was taking place in numerous germplasm banks because of inadequate conservation practices, IBPGR initiated a review of standards. Those germplasm banks that did not meet these standards were encouraged to make the necessary improvements, and some have done so (International Board for Plant Genetic Resources, 1989a).

Much has been achieved in collection efforts, training, the establishment of germplasm banks (especially in developing countries), documentation, and research support. However, the funding available for genetic resources activities at both the national and international level does not reflect the substantially increased public and political awareness of the importance of genetic resources (Keystone Center, 1988, 1991). The CGIAR has consistently been a major provider of funds since the early 1970s. Funds from most other international sources have been smaller and, the debates of recent years notwithstanding, have not provided the substantial support that is needed. Clearly enhancement of the CGIAR's role would be the logical means of improving these programs. However, even within the CGIAR system, there is much variability in the emphasis on and quality of germplasm resource endeavors.

THE FUTURE

The United Nations Conference on Environment and Development, held in Rio de Janeiro in 1992, gave the biodiversity issue greater visibility and a higher place on the international political agenda. A draft Convention on Biological Diversity has received wide support. While vague and open ended in its general recommendations, the convention's message is clear: people are concerned about the general decline of biodiversity and recognize the need for international action to realize effective programs of plant genetic resources conservation, notably in centers of diversity.

The need for substantial additional and sustainable funding by the international community is acknowledged. The convention makes no specific recommendations about the level of funding and its governance or about how policies, strategies, program priorities, and eligibility criteria are to be established. These issues will be subject to further negotiations in future conferences of parties to the convention and during the designation of a secretariat at an existing international organization.

Plant genetic resources relevant to agriculture are only part of overall biodiversity. In dealing with plant genetic resources, it is to be expected that existing structures and organizations will be taken into account. While still inadequate for reasons indicated in this chapter, a basic global structure is in place for plant genetic resources. It includes the FAO Commission on Plant Genetic Resources, which provides an intergovernmental platform for policies and oversight; the IBPGR, which coordinates and provides technical support at the program level; various regional programs; and ultimately international and national plant genetic resources programs. International cooperation in this structure is well established and accepted as the norm by most participating institutions, with the apparent emphasis in the draft convention on national sovereignty over indigenous plant genetic resources as a basic principle.

Various scenarios can be envisaged for the further development of international collaboration in germplasm conservation and use. These include a continued emphasis on national programs, internationalization through CGIAR, internationalization through FAO, and internationalization through a new consortium of institutions.

Continued Emphasis on National Programs

Considerable progress has been made at many levels of germplasm conservation and use since 1970. Global activities take place in a coherent framework provided by IBPGR, the IARCs, and many regional programs. An increased awareness of environmental issues and the need to conserve nature in general benefit genetic resources conservation. The primary responsibilities, however, rest with national germplasm banks and the willingness of national governments to make available the resources necessary for crop germplasm conservation or a national genetic resources conservation strategy. A major problem with this approach is that a significant burden rests on developing countries in regions with major plant diversity. The need for development and the scarcity of resources make it unlikely that such countries will be able to assume this burden.

Genetic resources issues are attracting considerable political and public attention (see Chapter 14). However, genetic erosion continues to take place inside and outside germplasm banks, and genetic resources cannot be saved by political rhetoric. Only a few developing countries have sound genetic resources programs, other than the

IARC programs. This situation is unlikely to change significantly unless substantial and permanent international funding is made available to assist developing countries in financing their germplasm banks, which were largely built with foreign assistance.

In the long term, a global system that is able to coordinate and foster activities from in situ conservation to collection, evaluation, storage, distribution, and use would be in the interest of all nations. As the principal users and beneficiaries of conserved germplasm, national programs should have a significant role in this global effort.

Internationalization Through CGIAR

The CGIAR centers so far provide the strongest programs and links in genetic resources conservation. Although much is being achieved through CGIAR, its programs cover only a limited number of food crops and forages. There is limited coordination among the various commodity-oriented centers and in their relationship to IBPGR. CGIAR could strengthen the genetic resources work of those centers, inasmuch as they already have an international working base, essential facilities, and trained staff; however, this will require additional funding and improved coordination.

Most of the commodity-oriented IARCs are located in or near the major centers of species diversity for a large number of the world's food crops. Their present mandate for genetic resources covering specific crops could be extended to include wider regions. Although national germplasm banks currently cooperate with the genetic resources programs of the IARCs for individual crops, in an expanded program, the IARCs could actively promote and support both regional and national conservation programs. This would, however, broaden their responsibilities beyond breeding for specific crops. It would require a systemwide program that would set objectives and priorities independent from those of the individual crop research programs of the various IARCs. Such a program could be strengthened further by establishing within the overall budget of CGIAR a separate item for the genetic resources programs distributed over the IARCs. Regional committees and an overall advisory group could be devised to satisfy technical and national interests. This would have consequences for the role of the IBPGR, which would require consideration and proper linkages to such activities at other centers, and for large areas of the world not served by IARCs. The IARCs are not staffed to assume these tasks at present, however, and some may have little interest in pursuing them, even if funding were made available (Hawkes, 1985).

Internationalization Through FAO

The FAO Commission on Plant Genetic Resources and its International Undertaking on Plant Genetic Resources are based on the notion that genetic resources are the common heritage of humankind (see Chapter 14). This concept implies international responsibility for genetic resources, which is further heightened by the fact that important centers of genetic diversity are located in the less developed and poorer regions of the world. The framework of the undertaking includes the following:

• A basic legal structure (the undertaking itself) approved by members of the Twenty-Second Session of the FAO Conference;
• An international forum (the Commission on Plant Genetic Resources) established at the 1983 FAO conference that offers an opportunity for countries to discuss issues related to plant genetic resources; and
• A financial mechanism (the International Fund for Plant Genetic Resources) through which contributions are received and applied toward conserving and using plant genetic resources.

The activities of the Commission on Plant Genetic Resources are directed toward the further refinement and implementation of the intent of the undertaking to strengthen the preservation, use, and availability of germplasm. It also addresses issues related to, for example, the establishment of a legal network of base collections, an international fund, and in situ conservation. There is an overlap between the aims of IBPGR and the stated objectives of the FAO commission, as there are among many agencies working in the area of genetic resources.

A major proposal of the commission has been the establishment of an international fund to finance genetic resources activities as well as activities in plant breeding and seed production in developing countries. The fund's specific objectives and the types of activities to be supported have yet to be defined.

Internationalization Through a New Consortium of Institutions

Many efforts related to the conservation of genetic resources, biodiversity, and the environment in general are under way. Many of them could be merged into a new structure, but efforts to do so have not progressed substantially. The programs of FAO and the IARCs, and many national programs, are focused primarily on food crops. Other plant conservation activities are usually undertaken by different organizations including national governments and private organizations, yet it is increasingly evident that these diverse efforts are inextricably linked, or should be. Food crop and related species are one part of the total biological diversity. The species covered under the current efforts of FAO, the IARCs, and national programs must be expanded to perennial tree crops and other woody species of the tropics, many of which are of great value in sustainable agriculture systems and for improving the quality of life of poor people in rural areas.

Furthermore, although nature conservation efforts have generally focused on in situ methods and genetic resources programs have focused on ex situ methods, in situ and ex situ technologies need to be more broadly applied in all genetic resources conservation programs. A properly structured and adequately funded organization could eliminate some of the institutional isolation and help to bring about a more holistic approach to the conservation of germplasm resources.

RECOMMENDATIONS

The world scientific and technical community has put in place the beginning of a functional network of both national and international genetic resources programs that are capable of safeguarding germplasm for the future. Present knowledge and rapid developments in a number of fields related to genetic conservation, including modern biotechnologies, provide a good basis for rational and selective conservation. Adequate operational funding remains the bottleneck.

Other problems still exist at the geopolitical level, however. Although the debate on the legal and effective ownership continues, genetic erosion also continues to take place, notably in developing countries, where most of the world's untapped genetic resources exist. It is in the interest of world agriculture that countries in regions with major genetic diversity be provided with the means to participate more fully in genetic resources conservation and use of biological resources.

International responsibility for conserving, managing, and using genetic resources must be translated into a workable form of funding within a coherent framework that satisfies the basic principles of the FAO International Undertaking on Plant Genetic Resources and the Convention on Biological Diversity.

The options described are complementary. Any strategy should build on the existing framework and activities, stressing national involvement and the cooperation of FAO, IBPGR, and the other CGIAR centers. A major criticism of the CGIAR has been that it has no formal legal basis for action among governments.

An adequate and appropriate funding mechanism must be established to support national and international conservation, management, and use of genetic resources.

Various funding mechanisms have been suggested, ranging from individual donations by interested countries to donations determined by specific criteria (see Chapter 13). International dialogue on the issue has yet to lead to a fully acceptable system. However, a consensus has recently emerged on establishing a global plant genetic resources initiative with a call for a trust fund that would be used to foster growth of local, national, regional, and global programs (Keystone Center, 1991).

Developed countries can afford, if they so desire, to maintain their own national germplasm banks in association with active breeding programs. However, external funding is needed to support the national germplasm banks, active collections, and international programs that serve developing countries. Support to national programs should be selective and based on such criteria as the availability of important genetic diversity and the specific interest and long-term commitment of the concerned government. On that assumption, only about 30 to 40 base collection germplasm banks worldwide would be needed to safeguard genetic resources. An annual contribution of US$5 million to each germplasm bank would probably allow a reasonable genetic resources program to be carried out (see Chapter 13). An additional US$40 million annually will be needed for general applied research, evaluation, and documentation programs. This adds up to a total of about US$240 million annually. Although this is a sizable sum, global commercial seed sales were estimated to be about US$30 billion (Barton and Christensen, 1988; Office of Technology Assessment, 1984), most of which were in developed countries.

When a germplasm bank has an efficient design and cost-effective equipment, the operating cost, excluding regeneration, is modest. For instance, the power consumption at IRRI costs under US$1 per accession each year (T. T. Chang, personal communication, June 1992). Increased collaboration and pooling of existing resources are essential to implementing new international efforts in the face of dwindling resources, both financial and biological.

References

Acree, J. A. 1989. Accessibility of world germplasm resources: Animal health constraints. Pp. 445–458 in Biotic Diversity and Germplasm Preservation, Global Imperatives, L. Knutson and A. K. Stoner, eds. Dordrecht, Netherlands: Kluwer Academic Publishers.

Adams, M. W. 1977. An estimation of homogeneity in crop plants, with special reference to genetic vulnerability in the dry bean, *Phaseolus vulgaris* L. Euphytica 26:665–679.

Adams, R. P. 1993. The conservation and utilization of genes from endangered and extinct plants: DNA Bank-Net. Pp. 43–64 in Gene Conservation and Exploitation, J. P. Gustafson, ed. Stadler Genetics Symposium 20. New York: Plenum.

Adams, R. P., Nhan Do, and Chu Ge-Lin. 1992. The preservation of DNA in plant specimens from tropical species by desiccation. Pp. 135–152 in Conservation of Plant Genes: DNA Banking and *In Vitro* Biotechnology, R. P. Adams and J. E. Adams, eds. San Diego, Calif.: Academic Press.

Alcorn, J. B. 1981. Huastec noncrop resource management. Human Ecol. 9:395–417.

Aleva, J. F., G. W. M. Barendse, J. Tolstra, and G. J. C. M. van Vliet. 1986. Computerization in the Botanical Gardens of The Netherlands and Belgium (Dutch), 2d ed. Wageningen, Netherlands: Agricultural University Wageningen.

Allard, R. W. 1988. Genetic changes associated with the evolution of adaptedness in cultivated plants and their wild progenitors. J. Hered. 79:225–238.

Allkin, R., and F. A. Bisby, eds. 1984. The Systematics Association Special Volume No. 26: Databases in Systematics. New York: Academic Press.

Allsopp, D., D. L. Hawksworth, and R. Platt. 1989. The CAB International Mycological Culture Collection Database, Microbial Culture Information Service (MiCIS) and Microbial Information Network Europe (MINE). Int. Biodeterior. 25:169–174.

Altieri, M., and L. C. Merrick. 1987. In situ conservation of crop genetic resources through maintenance of traditional farming systems. Econ. Bot. 41(1):86–96.

Altieri, M., M. K. Anderson, and L. C. Merrick. 1987. Peasant agriculture and the conservation of crops and wild plant resources. Conserv. Biol. 1:49–58.

Anagnostakis, S. C. 1982. Biological control of chestnut blight. Science 215:466–471.

Anderson, W. R., and D. J. Fairbanks. 1990. Molecular markers: An important tool for plant genetic resources conservation. Diversity 6:51–53.

Anderson, J. R., R. W. Herdt, and G. M. Scobie. 1988. Science and Food: The CGIAR and Its Partners. Washington, D.C.: World Bank.

Anishetty, N. M., J. Toll, W. G. Ayad, and J. R. Witcombe. 1982. Directory of Germplasm Collections. 3. Cereals. 4. Barley. AGP/IBPGR/81/98. Rome: International Board for Plant Genetic Resources.

Anonymous. 1982. Country Abbreviations for Use with IBPGR Descriptors. Plant Genet. Resources Newsl. 49:45–49.

Anonymous. 1989. Directory of Online Databases. New York: Cuadra/Elsevier.

Anonymous. 1991. The Global Environment Facility. Finance Dev. 28(March):24.

Apinis, A. E. 1972. Facts and problems. Mycopathol. Mycol. Appl. 48:93–109.

Apple Crop Advisory Committee. 1987. Apple (Malus) germplasm needs: Collection, evaluation, enhancement, preservation. Paper prepared for the Agricultural Research Service, U.S. Department of Agriculture, Washington, D.C.

Aronson, A. I., W. Backman, and P. Dunn. 1986. Bacillus thuringiensis and related insect pathogens. Microbiol. Rev. 50:1–24.

Ashri, A. 1971. Evaluation of the world collection of safflower, Carthamustinetorius L. 1. Reactions to several diseases and associations with morphological characters in Israel. Crop Sci. 11:253–257.

Asian Vegetable Research and Development Center. 1988. Progress Report Summaries 1987. Tainan, Taiwan: Asian Vegetable Research and Development Center.

Asian Vegetable Research and Development Center. 1992. 1991 Progress Report. Shanhua, Taiwan: Asian Vegetable Research and Development Center.

Attere, F., H. Zedan, N. Q. Ng, and P. Perrino, eds. 1991. Crop Genetic Resources of Africa, Volume 1. Rome: International Board for Plant Genetic Resources, United Nations Environment Program, and International Institute for Tropical Agriculture.

Awoderu, V. A., M. S. Alam, G. Thottappilly, and K. Alluri. 1987. Rice yellow mottle virus in upland rice. FAO Plant Prot. Bull. 35(1):32–33.

Ayala, F. J., and J. A. Kiger, Jr. 1980. Modern Genetics. Menlo Park, Calif.: Benjamin/Cummings.

Bachmann, B. J. 1988. The Escherichia coli. Genetic Stock Center, U.S. Fed. Cult. Collec. Newsl. 18:7.

Ball, C. R. 1930. History of American wheat improvement. Agric. Hist. 4:48–71.

Banno, T., and T. Sakane. 1979. Viability of various bacteria after L-drying. Inst. Ferm. Osaka Res. Commun. 9:35–45.

Barr, P. J., M. D. Power, L.-N. Chung Tee, H. L. Gibson, and P. A. Luciw. 1987. Expression of active human immunodeficiency virus reverse transcriptase in Saccharomyces cerevisiae. Bio/Technology 5:486–489.

Barton, J. 1982. The international breeder's rights system and crop plant innovation. Science 216:1071–1075.

Barton, J. H., and E. Christensen. 1988. Diversity compensation systems: Ways to compensate developing nations for providing genetic material. Pp. 338–355 in Seeds and Sovereignty: The Use and Control of Plant Genetic Resources, Jack Kloppenburg, Jr., ed. Durham, N.C.: Duke University Press.

Barton, J. H., C. O. Qualset, D. N. Duvick, and R. F. Barnes, eds. 1989. Intellectual Property Rights Associated with Plants. ASA Special Publication 52. Madison, Wis.: American Society for Agronomy.

Baum, W. C. 1986. Partners Against Hunger. The Consultative Group on International Agricultural Research. Washington, D.C.: World Bank.

Beier, F., R. Crespi, and J. Straus. 1985. BioTechnology and Patent Protection: An International Review. Paris: Organization for Economic Cooperation and Development.

Benda, I. 1982. Wine and brandy. Pp. 293–402 in Prescott & Dunn's Industrial Microbiology, 4th ed., G. Reed, ed. Westport, Conn.: AVI Publishing.

Bennett, J. W., and L. L. Lasure, eds. 1985. Gene Manipulations in Fungi. Orlando, Fla.: Academic Press.

Bent, S. A. 1987. Intellectual Property Rights in Biotechnology Worldwide. New York: Stockton.

Berdy, J. 1974. Recent developments of antibiotic research and classification of antibiotics according to chemical structure. Advances Appl. Microbiol. 18:309–406.

Berg, G. 1977. Postentry and intermediate quarantine. Pp. 315–326 in Plant Health and Quarantine in International Transfer of Genetic Resources, W. B. Hewitt and L. Chiarappa, eds. Cleveland: CRC Press.

Berland, J.-P., and R. Lewontin. 1986. Breeders' rights and patenting life forms. Nature 322:785–788.

Bettencourt, E., and J. Konopka. 1988. Directory of Germplasm Collections. 5. II. Industrial Crops. Rome: International Board for Plant Genetic Resources.

Birch, G. G., K. J. Parker, and J. T. Worgan, eds. 1976. Food from Waste. London: Applied Science Publishers.

Blanch, H. W., S. Drew, and D. I. C. Wang, eds. 1985. Comprehensive Biotechnology, Vol. 3. Oxford: Pergamon.

Blixt, S. 1984. Applications of Computer to Genebanks and Breeding Programmes. Pp. 293–310 in Crop Breeding; A Contemporary Basis, P. B. Vosse and S. G. Blixt, eds. New York: Pergamon.

Bofu, S., Q. D. Yu, and P. Vander-Zaag. 1987. True potato seed in China: Past, present, and future. Am. Potato J. 64:321–327.

Böing, J. T. P. 1982. Enzyme production. Pp. 634–708 in Prescott & Dunn's Industrial Microbiology, 4th ed., G. Reed, ed. Westport, Conn.: AVI Publishing.

Borlaug, N. E. 1958. The use of multiline or composite varieties to control air-borne epidemic disease of self-pollinated crop plants. Pp. 12–16 in Proceedings of the First International Wheat Genetic Symposium. Winnipeg, Manitoba, Canada: University of Manitoba.

Bos, L. 1977. Seed-borne viruses. Pp. 39–69 in Plant Health and Quarantine in International Transfer of Genetic Resources, W. B. Hewitt and L. Chiarappa, eds. Cleveland: CRC Press.

Bousefield, I. J. 1988. Patent protection for biotechnological inventions. Pp. 115–161 in Living Resources for Biotechnology: Filamentous Fungi, D. L. Hawksworth and B. E. Kirsop, eds. New York: Cambridge University Press.

Brady, B. L. 1981. Fungi as parasites of insects and mites. Biocon. News Inform. 4:281–296.

Brandt, D. A. 1982. Distilled beverage alcohol. Pp. 468–491 in Prescott & Dunn's Industrial Microbiology, 4th ed., G. Reed, ed. Westport, Conn.: AVI Publishing.

Bray, R. A. 1983. Strategies for gene maintenance. In Genetic Resources of Forage Plants, J. G. McIvor and R. A. Bray, eds. Melbourne, Australia: Commonwealth Scientific and Industrial Research Organization.

Breese, E. L. 1989. Regeneration and multiplication of germplasm resources in seed gene banks: The scientific background. Rome: International Board for Plant Genetic Resources.

Bretting, P. K., M. M. Goodman, and C. W. Stuber. 1987. Karyological and isozyme variation in West Indian and allied American mainland races of maize. Am. J. Bot. 74:1601–1613.

Bretting, P. K., M. M. Goodman, and C. W. Stuber. 1990. Isozymatic variation in Guatemalan races of maize. Am. J. Bot. 77:211–225.

Bridge, P. D., and J. W. May. 1988. Changes from homothallism to heterothallism in *Schizosaccharomyces malidevorans:* Biochemical and taxonomic consequences. Trans. Br. Mycol. Soc. 91:439–442.

Bridge, P. D., D. L. Hawksworth, Z. Kozakiewicz, A. H. S. Onions, and R. R. M. Paterson. 1986. Morphological and biochemical variation in single isolates of Penicillium. Trans. Br. Mycol. Soc. 87:389–396.

Brim, C. A., and W. M. Schutz. 1968. Inter-genotypic competition in soybeans. 2. Predicted and observed performance of multiline mixtures. Crop Sci. 8:735–739.

Brockway, L. H. 1988. Plant science and colonial expansion: The botanical chess game. Pp. 49–66 in Seeds and Sovereignty: The Use and Control of Plant Genetic Resources, J. R. Kloppenburg, Jr., ed. Durham, N.C.: Duke University Press.

Brooks, H. J., and C. F. Murphy. 1989. Ownership of plant genetic material. Pp. 493–497 in Biotic Diversity and Germplasm Preservation, Global Imperatives, L. Knutson and A. K. Stoner, eds. Boston, Mass.: Kluwer Academic Publishers.

Brown, W. L. 1988. Plant genetic resources: A view from the seed industry. Pp. 218–230 in Seeds and Sovereignty: The Use and Control of Plant Genetic Resources, J. R. Kloppenburg, Jr., ed. Durham, N.C.: Duke University Press.

Brown, A. D. H. 1989a. The case for core collections. Pp. 136–156 in The Use of Plant Genetic Resources, A. D. H. Brown, O. H. Frankel, D. R. Marshall, and J. T. Williams, eds. New York: Cambridge University Press.

Brown, A. D. H. 1989b. Core collections: A practical approach to genetic resources management. Genome 31(2):818–824.

Brown, A. H. D., and G. F. Moran. 1981. Isozymes and the genetic resources of forest trees. Pp. 1–10 in Isozymes of North American Forest Trees and Forest Insects, M. T. Conkle, coordinating ed. Berkeley, Calif.: Pacific Southwest Forest and Range Experiment Station, U.S. Department of Agriculture.

Brown, A. H. D., O. H. Frankel, D. R. Marshall, and J. T. Williams, eds. 1989. The Use of Plant Genetic Resources. New York: Cambridge University Press.

Browning, J. A. 1972. Corn, wheat, rice, man: Endangered species. J. Environ. Quality 1:209–210.

Browning, J. A. 1988. Current thinking on the use of diversity to buffer small grains against highly epidemic and variable foliar pathogens: Problems and future prospects. Pp. 76–90 in Breeding Strategies for Resistance to the Rusts of Wheat, N. W. Simmonds, and S. Rajaram, eds. Mexico City: Centro Internacional de Mejoramiento de Maíz y Trigo.

Browning, J. A., and K. J. Frey. 1969. Multiline cultivars as a means of disease control. Annu. Rev. Phytopathol. 7:355–382.

Brush, S. 1977. Farming the edge of the Andes. Nat. Hist. 86:32–41.

Buddenhagen, I. W. 1977. Resistance and vulnerability of tropical crops in relation to their evolution and breeding. Ann. N.Y. Acad. Sci. 287:309–326.

Buddenhagen, I. W. 1985. Maize disease in relation to maize improvement in the tropics. Pp. 243–275 in Monografie Agrarie Subtropicali e Tropicali, A. Brandolini and F. Salamini, eds. Florence, Italy: Food and Agriculture Organization of the United Nations.

Bureau of National Affairs. 1992. Compromise reached on financing; Developing nations dismayed with accord. BNA Int. Environ. Reporter (June 17):395–397.

Butler, L. J., and B. W. Marion. 1985. The Impacts of Patent Protection on the U.S.

Seed Industry and Public Plant Breeding. North Central Regional Research Publication 304. Madison, Wis.: Research Division, College of Agricultural and Life Sciences, University of Wisconsin.

Byrne, N. 1985. Plants, animals, and industrial patents. Int. Rev. Ind. Property Copyright Law 16:1–18.

Byrne, N. 1986. Patents for plants, seeds and tissue cultures. Int. Rev. Ind. Property Copyright Law 17:324–330.

Cadman, D. 1985. The protection of microorganisms under European patent law. Int. Rev. Ind. Property Copyright Law 16:311–317.

Canhos, V. P. 1989. World Federation of Culture Collections. Minutes of the general meeting, 2 to 4 November 1988, College Park, Md. Int. J. Syst. Bacteriol. 39:501.

Canhos, V. P., F. Yokoya, and D. A. L. Canhos, eds. 1986. Catalogo Nacional de Linhagens, 2d ed. Campinas, Brazil: Fundacao Tropical de Pesquisas e Tecnologia "Andre Tose."

Cantrell, R. P. 1989. Recent developments in the CIMMYT maize program. Pp. 2–4 in Toward Insect Resistant Maize for the Third World: Proceedings of an International Symposium on Methodologies for Developing Host Plant Resistance to Maize Insects. Mexico City: Centro Internacional de Mejoramiento de Maíz y Trigo.

Castillo-Gonzalez, F., and M. M. Goodman. 1989. Agronomic evaluation of Latin American maize accessions. Crop Sci. 29:853–861.

Centro Internacional de Agricultura Tropical (CIAT). 1987. CIAT Report 1987. Cali, Colombia: Centro Internacional de Agricultura Tropical.

CIAT. 1989. CIAT Report 1989. Cali, Colombia: Centro Internacional de Agricultura Tropical.

Centro Internacional de Mejoramiento de Maíz y Trigo (CIMMYT). 1985a. CIMMYT World Wheat Facts and Trends. Report Three: A Discussion of Selected Wheat Marketing and Pricing Issues in Developing Countries. Mexico City: Centro Internacional de Mejoramiento de Maíz y Trigo.

CIMMYT. 1985b. Yield stability of HYVs. Pp. 51–66 in CIMMYT Research Highlights 1984. Mexico City: Centro Internacional de Mejoramiento de Maíz y Trigo.

CIMMYT. 1986. CIMMYT Research Highlights 1985. Mexico City: Centro Internacional de Mejoramiento de Maíz y Trigo.

CIMMYT. 1987a. The *Septoria* Diseases of Wheat: Concepts and Methods of Disease Management. Mexico City: Centro Internacional de Mejoramiento de Maíz y Trigo.

CIMMYT. 1987b. 1986 CIMMYT World Maize Facts and Trends: The Economics of Commercial Maize Seed Production in Developing Countries. Mexico City: Centro Internacional de Mejoramiento de Maíz y Trigo.

CIMMYT. 1988. Maize Germplasm Bank Inquiry System. Mexico City: Centro Internacional de Mejoramiento de Maíz y Trigo.

CIMMYT. 1989. CIMMYT 1988 Annual Report: Delivering Diversity. Mexico City: Centro Internacional de Mejoramiento de Maíz y Trigo.

Chandan, R. C. 1982. Other fermented dairy products: Milk, cream, yogurt. Pp. 113–184 in Prescott & Dunn's Industrial Microbiology, 4th ed., G. Reed, ed. Westport, Conn.: AVI Publishing.

Chang, T. T. 1976. Manual on Genetic Conservation of Rice Germ Plasm for Evaluation and Utilization. Los Baños, Philippines: International Rice Research Institute.

Chang, T. T. 1984a. Conservation of rice genetic resources: Luxury or necessity? Science 224:251–256.

Chang, T. T. 1984b. The role and experience of an international crop-specific genetic resources center. Pp. 35–45 in Conservation of Crop Germplasm—An International Perspective, W. L. Brown, ed. Madison, Wis.: Crop Science Society of America.

Chang, T. T. 1985a. Evaluation and documentation of crop germplasm. Iowa State J. Res. 59:379–397.

Chang, T. T. 1985b. Germplasm enhancement and utilization. Iowa State J. Res. 59:399–424.

Chang, T. T. 1985c. Evaluation of exotic germplasm for crop improvement. Pp. 95–103 in Proceedings of the International Symposium on South East Asian Plant Genetic Resources, K. L. Mehra and S. Sastrapradja, eds. Bogor, Indonesia: Lembaga Biologi Nasional.

Chang, T. T. 1985d. Crop history and genetic conservation: Rice—A case study. Iowa State J. Res. 59(4):425–456.

Chang, T. T. 1985e. Principles of genetic conservation. Iowa State J. Res. 59(4):325–348.

Chang, T. T. 1987. The impact of rice on human civilization and population expansion. Interdiscipl. Sci. Rev. 21(1):63–69.

Chang, T. T. 1988. Sharing the benefits of bioproduction with the developing world. Paper presented at the XXIII International Union of Biological Scientists' General Assembly, Canberra, Australia. Unpublished.

Chang, T. T. 1989. The case for large collections. Pp. 123–135 in The Use of Plant Genetic Resources, A. H. D. Brown, O. H. Frankel, D. R. Marshall, and J. T. Williams, eds. New York: Cambridge University Press.

Chang, T. T. 1992. Availability of plant germplasm for crop use in improvement. Pp. 17–35 in Plant Breeding in the 1990s, H. T. Stalker and J. P. Murphy, eds. Wallingford, U.K.: CAB International.

Chang, S. T., and W. A. Hayes, eds. 1978. The Biology and Cultivation of Edible Mushrooms. London: Academic Press.

Chang, T. T., and C. C. Li. 1991. Genetics and breeding. Pp. 23–101 in Rice. I. Production, B. S. Luh, ed. New York: Van Nostrand Reinhold.

Chang, T. T., S. H. Ou, M. D. Pathak, K. C. Ling, and H. E. Kauffman. 1975. The search for disease and insect resistance in rice germplasm. Pp. 183–200 in Crop Genetic Resources for Today and Tomorrow, O. H. Frankel and J. G. Hawkes, eds. Cambridge: Cambridge University Press.

Chang, T. T., C. Zuno, A. Marciano-Romena, and G. C. Loresto. 1985. Semidwarfs in rice germplasm collection and their potentials in rice improvement. Phytobreeding 1:1–9.

Chater, K. F. 1980. Genetic instability, with special reference to industrial microorganisms. Pp. 7–10 in The Stability of Industrial Microorganisms, B. E. Kirsop, ed. Kew, U.K.: Commonwealth Mycological Institute.

Chelliah, S., and E. A. Heinrichs. 1980. Factors affecting insecticide-induced resurgence of the brown planthopper, *Nilaparvata lugens* in rice. Environ. Entomol. 9(6):773–777.

Chiarappa, L., and J. F. Karpati. 1984. Plant quarantine and genetic resources. Chapter 11 in Crop Genetic Resources: Conservation and Evaluation, J. H. W. Holden and J. T. Williams, eds. London: George Allen & Unwin.

Clegg, M. T., and R. W. Allard. 1972. Patterns of genetic differentiation in the slender wild oat species *Avena barbata*. Proc. Natl. Acad. Sci. USA 69:1820–1824.

Cohen, J. I., and R. Bertram. 1989. Plant genetic resource initiatives in international development. Pp. 459–476 in Biotic Diversity and Germplasm Preservation, Global Imperatives, L. Knutson and A. K. Stoner, eds. Dordrecht, Netherlands: Kluwer Academic Publishers.

Commission on Plant Genetic Resources. 1987. Legal Status of Base and Active Collections of Plant Genetic Resources. Report CPGR/87/5. Rome: Food and Agriculture Organization of the United Nations.

Committee on Varietal Standardization. 1920. Report of the Committee on Varietal Standardization. J. Am. Soc. Agron. 12:234–235.

Constantine, D. R. 1986. Micropropagation in the commercial environment. Pp. 175–186 in Plant Tissue Culture and its Agricultural Applications, L. A. Withers and P. G. Alderson, eds. London: Butterworths.

Consultative Group on International Agricultural Research. 1985. Report of the Second External Program and Management Review of the International Board for Plant Genetic Resources. Washington, D.C.: Technical Advisory Committee, Consultative Group on International Agricultural Research, World Bank.

Council for Agricultural Science and Technology (CAST). 1984a. Plant Germplasm Preservation and Utilization in U.S. Agriculture. Report. No. 106. Ames, Iowa: Council for Agricultural Science and Technology.

CAST. 1984b. Animal Germplasm Preservation and Utilization in Agriculture. Report No. 101. Ames, Iowa: Council for Agricultural Science and Technology.

CAST. 1985. Plant Germplasm Preservation and Utilization in U.S. Agriculture. Report No. 106. Ames, Iowa: Council for Agricultural Science and Technology.

CAST. 1987. Pests of Plants and Animals: Their Introduction and Spread. Report No. 112. Ames, Iowa: Council for Agricultural Science and Technology.

Cox, T. S., J. P. Murphy, and D. M. Rodgers. 1986. Changes in genetic diversity in the red winter wheat regions of the United States. Proc. Natl. Acad. Sci. USA 83:5583–5586.

Cox, T. S., J. P. Murphy, and M. M. Goodman. 1988a. The contribution of exotic germplasm to American agriculture. Pp. 114–144 in Seeds and Sovereignty: The Use and Control of Plant Genetic Resources, J. R. Kloppenburg, Jr., ed. Durham, N.C.: Duke University Press.

Cox, T. S., J. P. Shroyer, L. Ben-Hui, R. G. Sears, and T. J. Martin. 1988b. Genetic improvement in agronomic traits of hard red winter wheat cultivars from 1919 to 1987. Crop Sci. 28:756–760.

Crespi, A. 1985. Microbiological inventions and the patent law—The international dimension. Biotech. Genet. Eng. Rev. 3:1–37.

Cronk, Q., V. H. Heywood, and H. Synge. 1988. Biodiversity: The key role of plants. Kew, U.K.: International Union for the Conservation of Nature and Natural Resources and World Wildlife Fund.

Crop Science Society of America, Varietal Registration Committee. 1968. Registration of field-crop cultivars, parental lines, and elite germplasm. Crop Sci. 8:261–262.

Dahal, G., H. Hibino, H. Cabunagan, E. R. Tiongco, Z. M. Flores, and V. M. Aguiero. 1990. Changes in cultivar reactions to tungro due to changes in virulence of the leafhopper vector. Phytopathology 80(7):659–665.

Dalrymple, D. G. 1980. Development and Spread of Semi-Dwarf Varieties of Wheat and Rice in the United States: An International Perspective. USDA Agricultural Economic Report 455. Washington, D.C.: U.S. Department of Agriculture.

Dalrymple, D. G. 1985. The development and adoption of high-yielding varieties of wheat and rice in developing countries. Am. J. Agric. Econ. 67(5):1067–1073.

Dalrymple, D. G. 1986a. Development and Spread of High-Yielding Wheat Varieties in Developing Countries, 7th ed. Washington, D.C.: U.S. Agency for International Development.

Dalrymple, D. G. 1986b. Development and Spread of High-Yielding Rice Varieties in Developing Countries, 7th ed. Washington, D.C.: U.S. Agency for International Development.

Dalrymple, D. 1988. Changes in wheat varieties and yields in the United States, 1919–1984. Agric. Hist. 62:20–36.

Darrah, D. L., and M. S. Zuber. 1986. 1985 United States farm maize germplasm base and commercial breeding strategies. Crop Sci. 26:1109–1113.

Darwin, C. 1859. The Origin of Species. New York: Random House.

DaSilva, E. J., A. C. J. Burgers, and R. J. Olembo. 1977. UNEP and the international community of culture collections. Pp. 107–120 in Proceedings of the Third International Conference on Culture Collections, March 14–19, 1977, F. Fernandes and R. C. Pereira, eds. Bombay, India: University of Bombay.

Day, P. R. 1993. Exploitation of genetic resources. In Gene Conservation and Exploitation, J. P. Gustafson, ed. Stadler Genetics Symposium 20. New York: Plenum.

de Janvry, A., and J.-J. Dethier. 1985. Technological Innovation in Agriculture. CGIAR Study Paper No. 1. Washington, D.C.: World Bank.

Demain, A. L. 1983. New applications of microbial products. Science 219:709–714.

Dereuddre, J., C. Scottez, Y. Arnaud, and M. Duron. 1990. Résistance d'apex caulinaires de vitroplants de Poirier (*Pyrus communis* L. cv Beurré Hardy), enrobés dans l'alginate, à une déshydration puis à une congélation dans l'azote liquide: Effet d'un endurcissement préalable au froid. Compt. Rend. Acad. Sci. 310(Série III):371–323.

Dilday, R. H. 1990. Contribution of ancestral lines in the development of new cultivars of rice. Crop Sci. 30(4):905–911.

Dobzhansky, T. 1970. Genetics of the Evolutionary Process. New York: Columbia University Press.

Doebley, J. F., M. M. Goodman, and C. W. Stuber. 1985. Isozyme variation in races of maize from Mexico. Am. J. Bot. 72:629–639.

Douglas, J. 1980. Successful Seed Programs: A Planning and Management Guide. Boulder, Colo.: Westview.

Doyle, J. 1985. Altered Harvest: Agriculture, Genetics, and the Fate of the World's Food Supply. New York: Viking Penguin.

Drake, V. C. 1989. The IUCN: Its role in biological diversity. Part I. The organization for global conservation action. Diversity 5(1):14–15.

Duvick, D. N. 1984a. Genetic diversity in major farm crops on the farm and in reserve. Econ. Bot. 38:161–178.

Duvick, D. N. 1984b. Progress in conventional plant breeding. Pp. 17–31 in Gene Manipulation in Plant Improvement, 16th Stadler Genetics Symposium, J. P. Gustafson, ed. New York: Plenum.

Duvick, D. N. 1987. Sources of genetic advance for the future. In Proceedings of the Ninth Annual Seed Technology Conference, J. S. Burris, ed. Ames, Iowa: Seed Science Center.

Duvick, D. N. 1992. Genetic contributions to advances in yield of U.S. maize. Maydica 37:69–79.

Dyck, V. A., and B. Thomas. 1979. The brown planthopper problem *Hilaparvata lugens*, rice plant, Asia. Pp. 3–17 in Brown Planthopper, Threat to Rice Production in Asia. Los Baños, Philippines: International Rice Research Institute.

Ebner, H. 1982. Vinegar. Pp. 802–834 in Prescott & Dunn's Industrial Microbiology, 4th ed., G. Reed, ed. Westport, Conn.: AVI Publishing.

Elgin, J. H., Jr., and P. A. Miller. 1989. Enhancement of plant germplasm. Pp. 311–322 in Biotic Diversity and Germplasm Preservation, Global Imperatives, L. Knutson and A. K. Stoner, eds. Dordrecht, Netherlands: Kluwer Academic Publishers.

Ellis, R. H. 1985. Information required within genetic resources centres to maintain and distribute seed accessions. Pp. 11–20 in Documentation of Genetic Resources: Information Handling Systems for Genebank Management, J. Konopka and J. Hanson, eds. Rome: International Board for Plant Genetic Resources.

Entwistle, P. F. 1983. Control of insects by virus diseases. Biocon. News Inform. 4:203–235.

Esquinas-Alcazar, J. T. 1989. The FAO global system on plant genetic resources. Diversity 5(2,3):7–9.

European Commission. 1989. Proposal for a council directive on the legal protection of biotechnological inventions. P. C10/3 in Official Journal of the European Communities, January 13. Brussels: European Commission.

Evenson, R. E., and Y. Kislev. 1975. Agricultural Research and Productivity. New Haven, Conn.: Yale University Press.

Falk, D. 1993. Potential genetic resources in rare native species. In Gene Conservation and Exploitation, J. P. Gustafson, ed. Stadler Genetics Symposium 20. New York: Plenum.

Food and Agriculture Organization (FAO). 1981. Resolution on Plant Genetic Resources. C6/81. Rome: Food and Agriculture Organization of the United Nations.

FAO. 1983a. Twenty-Second Session, Sixteenth Meeting. C83/II/PV/16. Rome: Food and Agriculture Organization of the United Nations.

FAO. 1983b. Plant Genetic Resources. Report of the Director General. C83/25. Rome: Food and Agriculture Organization of the United Nations.

FAO. 1987a. ACCIS Guide to United Nations Information Sources on Food and Agriculture. Rome: Food and Agriculture Organization of the United Nations.

FAO. 1987b. Feasibility Study on the Establishment of an International Fund for Plant Genetic Resources. CPGR 87/10. Rome: Food and Agriculture Organization of the United Nations.

FAO. 1987c. Status of In-situ Conservation of Plant Genetic Resources. CPGR 87/7. Rome: Food and Agriculture Organization of the United Nations.

FAO. 1987d. Study on Legal Arrangements with a View to the Possible Establishment of an International Network of Base Collections in Gene Banks under the Auspices or Jurisdiction of FAO. CPGR 87/6. Rome: Food and Agriculture Organization of the United Nations.

FAO. 1987e. A Study on the Legal Status of Base and Active Collections of Plant Genetic Resources (Including Seed Legislation and Breeders' Rights). CPGR 87/5. Rome: Food and Agriculture Organization of the United Nations.

FAO. 1987f. Report of the Second Session of the Commission on Plant Genetic Resources. CPGR/87/REP. Rome: Food and Agriculture Organization of the United Nations.

FAO. 1989a. Plant Genetic Resources. Rome: Food and Agriculture Organization of the United Nations.

FAO. 1989b. Agreed interpretation of the international undertaking. Resolutions 4/89 and 5/89 of the Twenty-Fifth Session of the Food and Agriculture Organization of the United Nations Conference, Rome, Italy, November 11–29, 1989.

Fehr, W. R. 1984. Genetic Contributions to Yield Grains of Five Major Crop Plants. Special Publication No. 7. Madison, Wis.: Crop Science Society of America.

Fehr, W. R. 1989. The impact of biotechnology on public and private plant breeding. Diversity 5(4):35–37.

Fillatti, J. J., J. Kiser, R. Rose, and L. Comai. 1987. Efficient transfer of a glyphosate tolerance gene into tomato using a binary *Agrobacterium tumefaciens* vector. Bio/ Technology 5:726–730.

Ford-Lloyd, B., and M. Jackson. 1986. Plant Genetic Resources: An Introduction to Their Conservation and Use. London: Edward Arnold.

Fowler, C., and P. Mooney. 1990. Shattering: Food, Politics, and the Loss of Genetic Diversity. Tucson, Ariz.: University of Arizona Press.

Francis, C. M. 1986. Conservation of plant genetic resources—A continuing saga. A. Aust. Inst. Agric. Sci. 52(1):3–110.

Frankel, O. H. 1970. Genetic dangers in the green revolution. World Agric. 19(3):9–13.

Frankel, O. H. 1975a. Genetic resources centres—A co-operative global network. Pp. 473–481 in Crop Genetic Resources for Today and Tomorrow, O. H. Frankel and J. G. Hawkes, eds. New York: Cambridge University Press.

Frankel, O. H., and J. G. Hawkes, eds. 1975b. Crop Genetic Resources for Today and Tomorrow. New York: Cambridge University Press.

Frankel, O. H., and J. G. Hawkes. 1975c. Genetic resources—The past ten years and the next. Pp. 1–14 in Crop Genetic Resources for Today and Tomorrow, O. H. Frankel and J. G. Hawkes, eds. London: Cambridge University Press.

Frankel, O. H. 1985a. Into the second decade: Genetic resources and the plant breeder. Pp. 26–31 in Proceedings of the International Symposium on South East Asian Plant Genetic Resources, K. L. Mehra and S. Sastrapradja, eds. Bogor, Indonesia: Lembaga Biologi Nasional.

Frankel, O. H. 1985b. Genetic resources: The founding years. 2. The movement's constituent assembly. Diversity 8:30–32.

Frankel, O. H. 1985c. Genetic resources: The founding years. 1. Early beginnings, 1961–1966. Diversity 7:26–29.

Frankel, O. H. 1986. Genetic resources: The founding years. 3. The long road to the international board. Diversity 9:30–33.

Frankel, O. H. 1987. Genetic resources: The founding years. 4. After twenty years. Diversity 11:25–27.

Frankel, O. H. 1988. Genetic resources: Evolutionary and social responsibility. Pp. 19–46 in Seeds and Sovereignty: The Use and Control of Plant Genetic Resources, J. R. Kloppenburg, Jr., ed. Durham, N.C.: Duke University Press.

Frankel, O. H. 1989a. The Keystone international dialogue on plant genetic resources: A scientist's evaluation. Diversity 5(2,3):59–60.

Frankel, O. H. 1989b. Principles and strategies of evaluation. Pp. 245–262 in The Use of Plant Genetic Resources, A. D. H. Brown, O. H. Frankel, D. R. Marshall, and J. T. Williams, eds. New York: Cambridge University Press.

Frankel, O. H., and E. Bennett, eds. 1970. Genetic Resources in Plants—Their Exploration and Conservation. International Biological Program Handbook No. 11. Oxford: Blackwell.

Frankel, O. H., and A. H. D. Brown. 1984. Plant genetic resources today: A critical appraisal. Pp. 249–257 in Crop Genetic Resources: Conservation and Evaluation, J. H. W. Holden and J. T. Williams, eds. London: George Allen & Unwin.

Frankel, O. H., and J. G. Hawkes, eds. 1975. Crop Genetic Resources for Today and Tomorrow. New York: Cambridge University Press.

Frankel, O. H., and M. E. Soulé. 1981. Conservation and Evolution. New York: Cambridge University Press.

Frese, L., and T. J. L. van Hintum. 1989. The international data base for *Beta*. Pp. 17–

35 in Report of an International *Beta* Genetic Resources Workshop. Rome: International Board for Plant Genetic Resources.

Fungal Genetics Stock Center. 1988. Catalogue of Strains, 2d ed. Kansas City, Mo.: University of Kansas Medical Center.

Gale, J. S., and M. J. Lawrence. 1984. The decay of variability. Pp. 77–101 in Crop Genetic Resources: Conservation and Evaluation, J. H. W. Holden and J. T. Williams, eds. London: George Allen & Unwin.

Gallun, R. L., and G. S. Khush. 1980. Genetic factors affecting expression and stability of resistance. Pp. 64–85 in Breeding Plants Resistant to Insects, F. G. Maxwell and P. R. Jennings, eds. New York: John Wiley & Sons.

Gams, W., G. L. Hennebert, J. A. Stalpers, D. Janssens, M. A. A. Schipper, J. Smith, D. Yarrow, and D. L. Hawksworth. 1988. Structuring strain data for storage and retrieval of information on fungi and yeasts in MINE, the Microbial Information Network Europe. J. Gen. Microbiol. 134:1667–1689.

Garcia, P., F. J. Vences, M. Peréz de la Vega, and R. W. Allard. 1989. Allelic and genotypic composition of ancestral Spanish and colonial Californian gene pools of *Avena barbata:* Evolutionary implications. Genetics 122(3):687–694.

General Accounting Office. 1981. Better Collection and Maintenance Procedures Needed to Help Protect Agriculture's Germplasm Resources. Washington, D.C.: U.S. Government Printing Office.

Genetic Resources Conservation Program. 1988. Evaluation of the University of California Tomato Genetics Stock Center: Recommendations for Its Long-Term Management, Funding, and Facilities. Report No. 2. Oakland: University of California Genetic Resources Conservation Program.

Gherna, R. L. 1981. Preservation. Pp. 208–217 in Manual of Methods for General Bacteriology, P. Gerhardt, R. G. E. Murrary, R. N. Costilow, E. W. Nester, W. A. Wood, N. R. Krieg, and G. B. Phillips, eds. Washington, D.C.: American Society for Microbiology.

Giacometti, D. C., and C. O. Goedert. 1989. Brazil's National Genetic Resources and Biotechnology Center preserves and develops valuable germplasm. Diversity 5(4):8–11.

Goldstein, D. J. 1988. Molecular biology and the protection of germplasm: A matter of national security. Pp. 315–337 in Seeds and Sovereignty: The Use and Control of Plant Genetic Resources, J. R. Kloppenburg, Jr., ed. Durham, N.C.: Duke University Press.

Gollin, D., and R. E. Evenson. 1990. Genetic Resources and Rice Varietal Improvement in India. Economic Growth Center, Yale University, New Haven, Conn. Unpublished manuscript.

Goodfellow, M., S. T. Williams, and M. Mordarski, eds. 1988. Actinomycetes in Biotechnology. London: Academic Press.

Goodman, M. M. 1984. An evaluation and critique of current germplasm programs. Pp. 195–249 in the Report of the 1983 Plant Breeding Research Forum. Des Moines, Iowa: Pioneer Hi-Bred International.

Goodman, M. M. 1985. Exotic maize germplasm: Status, prospects, and remedies. Iowa State J. Res. 59:497–528.

Goodman, M. M., and F. Castillo-Gonzalez. 1991. Germplasm: Politics and realities. Forum Appl. Res. Public Policy 6(3):74–85.

Goodman, M. M., and J. M. Hernandez. 1991. Latin American maize collections: A case for urgent action. Diversity 7(1):87–88.

Goodman, M. M., and C. W. Stuber. 1983. Races of maize. VI. Isozyme variation among races of maize in Bolivia. Maydica 28:169–187.

Gould, F. 1983. Genetics of plant-herbivore systems: Interaction between applied and basic study. Pp. 599–653 in Variable Plants and Herbivores in Natural and Managed Systems. New York: Academic Press.

Gould, F. 1986. Simulation models for predicting durability of insect-resistant germplasm: Hessian fly (Diptera: Cecidomyiidae)-resistant winter wheat. Environ. Entomol. 15:11–23.

Granett, J., J. A. DeBenedictus, J. A. Wolpert, E. Weber, and A. C. Goheen. 1991. Phylloxera on rise: Deadly insect pest poses increased risk to north coast vineyards. Calif. Agric. 45(2):30–32.

Grape Commodity Advisory Committee. 1987. Recommendations for U.S. Grape Germplasm. Paper prepared for the Agricultural Research Service, U.S. Department of Agriculture, Washington, D.C.

Hardman, D. J. 1987. Microbial control of environmental pollution: The use of genetic techniques to engineer organisms with novel catalytic capabilities. Pp. 295–317 in Environmental Biotechnology, C. F. Forster and D. A. J. Wase, eds. Chichester, U.K.: Ellis Horwood.

Hargrove, T. R. 1978. Diffusion and Adoption of Genetic Materials Among Rice Breeding Programs in Asia. IRRI Research Paper Series No. 18. Los Baños, Philippines: International Rice Research Institute.

Hargrove, T. R., W. R. Coffman, and V. L. Cabanilla. 1980. Ancestry of improved cultivars of Asian rice. Crop Sci. 20:721–727.

Hargrove, T. R., V. L. Cabanilla, and W. R. Coffman. 1985. Changes in Rice Breeding in 10 Asian Countries, 1965–84; Diffusion of Genetic Materials, Breeding Objectives, and Cytoplasm. IRRI Research Paper Series No. 111. Manila, Philippines: International Rice Research Institute.

Harlan, J. 1975. Our vanishing genetic resources. Science 188:618–621.

Harlan, J. R. 1992. Crops and Man, Second Edition. Madison, Wis.: American Society of Agronomy and Crop Science Society of America.

Harlan, H. V., and M. L. Martini. 1929. A composite hybrid mixture. J. Am. Soc. Agron. 21:487–490.

Harlan, H. V., and M. L. Martini. 1936. Problems and results in barley breeding. In Yearbook of Agriculture 1936. Washington, D.C.: U.S. Department of Agriculture.

Harley, J. L., and S. E. Smith. 1983. Mycorrhizal Symbiosis. London: Academic Press.

Harris, E. H. 1984. The Chlamydomonas Genetics Center at Duke University. Plant Mol. Biol. Rep. 2:29–41.

Harris, E. H., J. E. Boynton, and N. W. Gillham. 1987. Chlamydomonas reinhardtii. Genet. Maps 4:257–277.

Hawkes, J. G. 1985. Plant Genetic Resources: The Impact of the International Agricultural Research Centers. Consultative Group on International Agricultural Research, Study Paper No. 3. Washington, D.C.: World Bank.

Hawksworth, D. L. 1985. Fungus culture collections as a biotechnological resource. Biotech. Genet. Eng. Rev. 3:417–453.

Hawksworth, D. L. 1991. The fungal dimension of biodiversity: Magnitude, significance and conservation. Mycol. Rev. 95:641–655.

Hawksworth, D. L., and W. Greuter. 1989. Report of the first meeting of a working group on Lists of Names in Current Use. Taxon 38:142–148.

Hawksworth, D. L., and B. E. Kirsop, eds. 1988. Living Resources for Biotechnology: Filamentous Fungi. New York: Cambridge University Press.

Hawksworth, D. L., B. C. Sutton, and G. C. Ainsworth. 1983. Ainsworth & Bisby's

Dictionary of the Fungi, 7th ed. Kew, U.K.: Commonwealth Mycological Institute.

Haymon, L. W. 1982. Fermented sausage. Pp. 237–245 in Prescott & Dunn's Industrial Microbiology, 4th ed., G. Reed, ed. Westport, Conn.: AVI Publishing.

Heinrichs, E. A., F. G. Medrano, and H. R. Rapusas. 1985. Genetic Evaluation for Insect Resistance in Rice. Los Baños, Philippines: International Rice Research Institute.

Heiser, C. B. 1975. The Sunflower. Norman: University of Oklahoma Press.

Heiser, C. B. 1981. Seeds to Civilization, The Story of Food. San Francisco: W. H. Freeman.

Helbert, J. R. 1982. Beer. Pp. 403–467 in Prescott & Dunn's Industrial Microbiology, 4th ed., G. Reed, ed. Westport, Conn.: AVI Publishing.

Henshaw, G. G. 1975. Technical aspects for tissue culture storage for genetic conservation. Pp. 349–358 in Crop Genetic Resources for Today and Tomorrow, O. H. Frankel and J. G. Hawkes, eds. New York: Cambridge University Press.

Hernandez, E., and G. Alanís. 1970. Estudio morfológico de cinco nuevas razas de maíz de la Sierra Madre Occidental de Mexico: Implicaciones filogenéticas y fitogeográficas. Agrociencia 5:3.

Heywood, V. H. 1989. The IUCN: Its role in biological diversity. Part II. Plant conservation. Diversity 5(2,3):11–14.

Heywood, V. H. 1993. Broadening the basis of plant germplasm conservation. In Gene Conservation and Exploitation, J. P. Gustafson, ed. Stadler Genetics Symposium 20. New York: Plenum.

Hill, W. 1986. Breeding for bioindustrial ecosystems. Pp. 73–87 Bioindustrial Ecosystems, D. J. A. Cole, and G. C. Brander, eds. New York: Elsevier.

Holdeman, Q., ed. 1986. Plant Pests of Phytosanitary Significance to Importing Countries and States, 5th ed. Sacramento, Calif.: Department of Food and Agriculture.

Holden, J. H. W. 1984. The second ten years. Pp. 277–282 in Crop Genetic Resources: Conservation and Evaluation, J. H. W. Holden and J. T. Williams, eds. London: George Allen & Unwin.

Holden, J. H. W., and J. T. Williams, eds. 1984. Crop Genetic Resources: Conservation and Evaluation. London: George Allen & Unwin.

Holley, R. N., and M. M. Goodman. 1988. Yield potential of hybrid maize derivatives. Crop Sci. 28:213–218.

Holm, L. G., D. L. Plucknett, J. V. Pancho, and J. P. Herberger. 1977. The World's Worst Weeds, Distribution and Biology. Honolulu: University Press of Hawaii.

Holm, L., J. V. Pancho, J. P. Herberger, and D. L. Plucknett. 1979. A Geographical Atlas of World Weeds. New York: John Wiley & Sons.

Horton, D. E., and H. Fano. 1985. Potato Atlas. Lima, Peru: Centro Internacional de la Papa.

Hoyt, E. 1988. Conserving the Wild Relatives of Crops. Rome: International Board for Plant Genetic Resources.

Huaman, Z. 1986. Conservation of potato genetic resources at CIP. CIP Circ. 14:1–7.

Inoue, A., and K. Horikoshi. 1989. A pseudomonas thrives in high concentrations of toluene. Nature 338:264–266.

International Board for Plant Genetic Resources (IBPGR). 1982. Barley Descriptors. AGPG:IBPGR/82/49. Rome: International Board for Plant Genetic Resources Secretariat.

IBPGR. 1984a. IBPGR Regional Committee for Southeast Asia, Five-Year Plan of Action, 1985–1989. Rome: International Board for Plant Genetic Resources.

IBPGR. 1984b. Exchange of Information. IBPGR/84/157. Rome: International Board for Plant Genetic Resources.

IBPGR. 1984c. Documentation in Genebanks: A Status Report. Rome: International Board for Plant Genetic Resources Secretariat.

IBPGR. 1985a. Ecogeographical Surveying and In Situ Conservation of Crop Relatives. Rome: International Board for Plant Genetic Resources.

IBPGR. 1985b. Report of a Working Group on Forages (Second Meeting). Rome: International Board for Plant Genetic Resources.

IBPGR. 1986. Report of a Working Group on Barley (Second Meeting). Rome: International Board for Plant Genetic Resources.

IBPGR. 1987. Annual Report 1986. Rome: International Board for Plant Genetic Resources.

IBPGR. 1988a. Report of a Working Group on *Allium* (Third Meeting). Rome: International Board for Plant Genetic Resources.

IBPGR. 1988b. Conservation and Movement of Vegetatively Propagated Germplasm: In Vitro Culture and Disease Aspects. Report of the IBPGR Advisory Committee on In Vitro Storage. Rome: International Board for Plant Genetic Resources.

IBPGR. 1988c. Annual Report 1987. Rome: International Board for Plant Genetic Resources.

IBPGR. 1989a. Annual Report 1988. Rome: International Board for Plant Genetic Resources.

IBPGR. 1989b. Report of an International *Beta* Genetic Resources Workshop. Rome: International Board for Plant Genetic Resources.

IBPGR. 1990. Annual Report 1989. Rome: International Board for Plant Genetic Resources.

IBPGR. 1991. Annual Report 1990. Rome: International Board for Plant Genetic Resources.

International Board for Plant Genetic Resources-International Rice Research Institute Rice Advisory Committee. 1980. Descriptors for Rice, *Oryza sativa* L. Manila, Philippines: International Rice Research Institute.

International Board for Plant Genetic Resources, International Center for Agricultural Research in Dry Areas, and International Crops Research Institute for the Semi-Arid Tropics. 1985. Chickpea Descriptors. AGPG:IBPGR 85/35. Rome: International Board for Plant Genetic Resources Secretariat.

International Rice Research Institute (IRRI). 1975. Standard Evaluation System for Rice. Los Baños, Philippines: International Rice Research Institute.

IRRI. 1985. IRRI Highlights 1984. Los Baños, Philippines: International Rice Research Institute.

IRRI. 1991. Rice Germplasm: Collecting, Preservation, Use. Los Baños, Philippines: International Rice Research Institute.

International Rice Research Institute and International Board for Plant Genetic Resources. 1980. Descriptors for Rice. Manila, Philippines: International Rice Research Institute.

International Union for the Conservation of Nature and Natural Resources (IUCN). 1978. Categories, Objectives, and Criteria for Protected Areas. Gland, Switzerland: International Union for the Conservation of Nature and Natural Resources.

IUCN. 1987. Centres of Plant Diversity. A Guide and Strategy for Their Conservation. Kew, U.K.: International Union for the Conservation of Nature and Natural Resources.

International Union for the Conservation of Nature and Natural Resources, United Nations Environment Program, and World Wildlife Fund. 1980. World Conservation Strategy: Living Resource Conservation for Sustainable Development. Gland, Switzerland: International Union for the Conservation of Nature and Natural Resources.

Jain, S. K. 1975a. Genetic reserves. Pp. 379–398 in Crop Genetic Resources for Today and Tomorrow, O. H. Frankel and J. G. Hawkes, eds. New York: Cambridge University Press.

Jain, S. K. 1975b. Patterns of survival and evolution in plant populations. Pp. 49–89 in Population Genetics and Ecology, S. Karlin and E. Nevo, eds. New York: Academic Press.

James, E. 1972. Organization of the United States National Storage Laboratory. In Viability of Seeds, E. H. Roberts, ed. London: Chapman & Hall.

Janzen, D. H. 1973. Tropical agroecosystems. Science 182:1212.

Johnson, J. 1987. Universities propose new release policies. Seedsmen's Dig. 38(4):16–18.

Jondle, R. J. 1989. Overview and status of plant proprietary rights. Pp. 5–15 in Intellectual Property Rights Associated with Plants, B. E. Caldwell and J. A. Schillinger, eds. ASA Special Publication 52. Madison, Wis.: American Society for Agronomy.

Kabuye, C. H. S., G. M. Mungai, and J. G. Mutangah. 1986. Flora of Kora National Reserve. Pp. 57–104 in Kora: An Ecological Inventory of the Kora National Reserve, Kenya, M. Coe and N. M. Collins, eds. London: Royal Geographical Society.

Kahler, A. L., R. W. Allard, M. Krazakowa, C. F. Wehrhahm, and E. Nevo. 1980. Associations between enzyme phenotypes and environment in the slender wild oat (*Avena barbata*) in Israel. Theor. Appl. Genet. 56:31–47.

Kahn, R. P. 1977. Plant Quarantine: Principles, Methodology, and Suggested Approaches. Chapter 25 in Transfer of Genetic Resources, W. B. Hewitt and L. Chiarappa, eds. Cleveland: CRC Press.

Kahn, R. P. 1979. A concept of pest risk analysis. EPPO Bull. 9:119–130.

Kahn, R. P. 1982. The host as a vector: Exclusion as a control. Chapter 7 in Pathogens, Vectors, and Plant Diseases, K. F. Harris and K. Maramorosch, eds. New York: Academic Press.

Kahn, R. P. 1983. A model plant quarantine station: Principles, concepts, and requirements. Pp. 303–333 in Exotic Plant Quarantine Pests and Procedures for Introduction of Plant Materials, K. G. Singh, ed. Serdang, Selangor, Malaysia: Plant Quarantine Center and Training Institute, Association of South East Asian Nations.

Kahn, R. P. 1985. Technologies to Maintain Biological Diversity: Assessment of Plant Quarantine Practices. Report submitted to the Office of Technology Assessment, U.S. Congress, October 16, 1985.

Kahn, R. P. 1988. The importance of seed health in international exchange seed exchange. Pp. 7–20 in Rice Seed Health. Manila, Philippines: International Rice Research Institute.

Kahn, R. P. 1989a. Plant Protection and Quarantine: Biological Concepts. Boca Raton, Fla.: CRC Press.

Kahn, R. P. 1989b. Plant quarantine as a disease control strategy in tropical countries. Rev. Trop. Plant Pathol. 6:151–180.

Kalton, R. R., P. A. Richardson, and N. M. Frey. 1989. Inputs in private sector plant breeding and biotechnology research programs in the United States. Diversity 5(4):22–25.

Kaniewski, W. K., and P. E. Thomas. 1988. A two-step ELISA for rapid, reliable detection of potato viruses. Amer. Pot. J. 65:561–571.

Karpati, J. F. 1981. Post-entry quarantine and plant introduction. Pp. 39–41 in Plant Production and Protection Paper No. 27. Rome: Food and Agriculture Organization of the United Nations.

Karpati, J. F. 1983. Plant quarantine on a global basis. Seed Sci. Technol. 11:1145–1157.

Keystone Center. 1988. Final Report of the Keystone International Dialogue Series on Plant Genetic Resources. Session I: Ex Situ Conservation of Plant Genetic Resources. Washington, D.C.: Genetic Resources Communications Systems.

Keystone Center. 1990. Final Consensus Report of the Keystone International Dialogue Series on Plant Genetic Resources. Madras Plenary Session, January 29 to February 2, 1990. Washington, D.C.: Genetic Resources Communications Systems.

Keystone Center. 1991. Final Consensus Report: Global Initiative for the Security and Sustainable Use of Plant Genetic Resources. Keystone International Dialogue Series on Plant Genetic Resources. Oslo Plenary Session, May 31–June 4, 1991. Washington, D.C.: Genetic Resources Communications Systems.

Kimura, M. 1968. Evolutionary rate at the molecular level. Nature 217:624–626.

Kimura, M., and J. F. Crow. 1964. The number of alleles that can be maintained in a finite population. Genetics 49:725–738.

Kimura, K., H. Fujimaki, M. Sakejima, K. Matsuba, H. Yoshida, and K. Mori. 1986. Analyses of agronomic traits of introduced rice genetic resources. Res. Rep. Agric. Dev. Hokuriku Area 13:1–93.

Kirsop, B. E., ed. 1980. The Stability of Industrial Organisms. Kew, U.K.: Commonwealth Mycological Institute.

Kirsop, B. E. 1988a. Microbial Strain Data Network: A service to biotechnology. Int. Ind. Biotech. 8:24–27.

Kirsop, B. E. 1988b. Resources centres. Pp. 1–35 in Living Resources for Biotechnology: Yeasts, B. E. Kirsop and C. P. Kurtzman, eds. New York: Cambridge University Press.

Kirsop, B. E. 1988c. Culture and preservation. Pp. 74–98 in Living Resources for Biotechnology: Yeasts, B. E. Kirsop and C. P. Kurtzman, eds. New York: Cambridge University Press.

Kirsop, B. E., and E. J. DaSilva. 1988. Organisation of resource centers. Pp. 173–187 in Living Resources for Biotechnology: Yeasts, B. E. Kirsop and C. P. Kurtzman, eds. New York: Cambridge University Press.

Kirsop, B. E., and C. P. Kurtzman, eds. 1988. Living Resources for Biotechnology: Yeasts. New York: Cambridge University Press.

Kirsop, B. E., and J. J. S. Snell, eds. 1984. Maintenance of Microorganisms: A Manual of Laboratory Methods. New York: Academic Press.

Kiyosawa, S. 1982. Genetics and epidemiological modeling of breakdown of plant disease resistance. Annu. Rev. Phytopathol. 20:93–117.

Kloppenburg, J., Jr., ed. 1988. Seeds and Sovereignty: The Use and Control of Plant Genetic Resources. Durham, N.C.: Duke University Press.

Kloppenburg, J., Jr., and D. L. Kleinman. 1987a. The plant germplasm controversy: Analyzing empirically the distribution of the world's plant genetic resources. BioScience 37/3(March):190–198.

Kloppenburg, J., Jr., and D. L. Kleinman. 1987b. Seeds of struggle: The geopolitics of genetic resources. Tech. Rev. 90 (February-March):46–53.

Kloppenburg, J. R., Jr., and D. L. Kleinman. 1988. Seeds of controversy: National property versus common heritage. Pp. 172–203 in Seeds and Sovereignty: The

Use and Control of Plant Genetic Resources, J. R. Kloppenburg, Jr., ed. Durham, N.C.: Duke University Press.

Knight, T. A. 1799. Philosophical Transactions, Royal Society, London:192.

Knorr, L. C. 1977. Citrus. Pp. 111–115 in Plant Health and Quarantine in International Transfer of Genetic Resources, W. B. Hewitt and L. Chiarappa, eds. Cleveland: CRC Press

Knuepffer, H. 1983. Computer in Genbanken—Eine Uebersicht. Kulturpflanze 31:77–143.

Komagata, K. 1987. Relocation of the World Data Center. MIRCEN J. 3:337–342.

Konopka, J., and J. Hanson, eds. 1985. Documentation of Genetic Resources: Information Handling Systems for Genebank Management. Rome: International Board for Plant Genetic Resources.

Korovina, O. N. 1980. Organization of reserves and protected areas in the USSR—The basis of conserving populations of wild relatives of cultivated plants. Tr. Prikl. Bot., Genet. Selek. 68(3):3–14.

Krasner, S. D. 1985. Structural Conflict: The Third World Against Global Liberalism. Berkeley: University of California Press.

Krichevsky, M. I., B.-O. Fabricus, and H. Sugawara. 1988. Information resources. Pp. 31–53 in Living Resources for Biotechnology: Filamentous Fungi, B. E. Kirsop and D. L. Hawksworth, eds. New York: Cambridge University Press.

Kristiansen, B., and J. D. Bu'Lock. 1980. Developments in industrial fungal biotechnology. Pp. 203–223 in Fungal Biotechnology, J. E. Smith, D. R. Berry, and B. Kristiansen, eds. London: Academic Press.

Kyomo, M. L. 1991. Conservation and utilization of genetic resources in the Southern African Development Coordination Conference (SADCC) region: A collaborative approach. Pp. 277–286 in Crop Genetic Resources of Africa, Volume 1, F. Attere, H. Zedan, N. Q. Ng, and P. Perrino, eds. Rome: International Board for Plant Genetic Resources, United Nations Environment Program, and International Institute for Tropical Agriculture.

Ladizinsky, G. 1989. Ecological and genetic considerations in collecting and using wild relatives. Pp. 297–305 in The Use of Plant Genetic Resources, A. H. D. Brown, O. H. Frankel, D. R. Marshall, and J. T. Williams, eds. New York: Cambridge University Press.

Lambat, A. K., R. Nath, P. M. Mukewar, A. Majumdar, I. Rani, P. Kaur, J. L. Varshney, P. C. Agarwal, R. K. Khetarpal, and U. Dev. 1983. International spread of Karnal bunt of wheat. Phytopath. Medit. 22:213–214.

Lawrence, Z. 1982. The identity and typification of *Aspergillus parasiticus*. Mycotaxon 15:293–305.

Lawrence, T., J. Toll, and D. H. van Sloten. 1986. Directory of Germplasm Collections. 2. Root and Tuber Crops. Rome: International Board for Plant Genetic Resources.

Lesser, W. 1987. Plant breeders' rights and monopoly myths. Nature 330:215.

Li, Z., and Y. Zhu. 1989. Rice male sterile cytoplasm and fertility restoration. Pp. 85–102 in Hybrid Rice. Manila, Philippines: International Rice Research Institute.

Lindlow, S. E. 1983. Methods of preventing frost injury caused by epiphytic ice-nucleation bacteria. Plant Dis. 67:327–333.

Linnington, S., and R. D. Smith. 1987. Deferred regeneration: A manpower-efficient technique for germplasm conservation. Plant Genet. Resources Newsl. 70:2–12.

Lipton, M., and R. Longhurst. 1989. New Seeds and Poor People. Baltimore: Johns Hopkins University Press.

Lyman, J. M. 1984. Progress and planning for germplasm conservation of major food crops. Plant Genet. Resources Newsl. 60:3–21.

MacGregor, R. C. 1973. The Emigrant Pests. A report prepared for the Animal and Plant Health Inspection Service, U.S. Department of Agriculture, Washington, D.C.

Malik, K. A., and D. Claus. 1987. Bacterial culture collections: Their importance to biotechnology and microbiology. Biotech. Genet. Eng. Rev. 5:137–197.

Mangelsdorf, P. C. 1974. Corn: Its Origin, Evolution and Improvement. Cambridge, Mass.: Harvard University Press.

Mantell, S. 1989. Recent advances in plant biotechnology for third world countries. Pp. 27–40 in Agricultural Biotechnology: Prospects for the Third World, J. Farringdon, ed. London: Overseas Development Institute.

Manwan, I., S. Sama, and S. Rizui. 1985. Use of varietal rotation in the management of tungro diseases in Indonesia. Indonesian Agric. Res. Develop. J. 7:43–48.

Mapletoft, R. J. 1987. The technology of embryo transfer. Pp. 2–40 in Proceedings of the International Symposium Embryo Movement, D. Hare and G. Seidel, eds. Ottawa, Ontario, Canada: Lowe-Martin.

Marshall, D. R. 1989. Limitations to the use of germplasm collections. Pp. 105–120 in The Use of Plant Genetic Resources, A. H. D. Brown, O. Frankel, D. R. Marshall, and J. T. Williams, eds. New York: Cambridge University Press.

Marshall, D. R., and A. H. D. Brown. 1975. Optimum sampling strategies in genetic conservation. Pp. 53–80 in Crop Genetic Resources for Today and Tomorrow, O. H. Frankel and J. G. Hawkes, eds. New York: Cambridge University Press.

Marshall, D. R., and A. J. Pryor. 1978. Multiline varieties and disease control. I. The "dirty crop" approach with each component carrying a unique resistance gene. Theor. Appl. Genet. 51(4):177–184.

Marth, E. H. 1982. Cheese. Pp. 65–112 in Prescott & Dunn's Industrial Microbiology, 4th ed., G. Reed, ed. Westport, Conn.: AVI Publishing.

Martignoni, M. E., and P. J. Iwai. 1981. A catalogue of viral diseases of insects, mites and ticks. Pp. 897–911 in Microbial Control of Pests and Plant Diseases 1970–1980, H. D. Burges, ed. London: Academic Press.

Martyn, E. B. 1968. Plant virus names. Phytopathol. Paper 9:1–204.

Martyn, E. B. 1971. Plant virus names. Phytopathol. Paper 9(Suppl.1):1–41.

Mathys, G. 1977. Phytosanitary regulations and the transfer of genetic resources. Pp. 327–331 in Plant Health and Quarantine in International Transfer of Genetic Resources, W. B. Hewitt and L. Chiarappa, eds. Cleveland: CRC Press.

Maunder, A. B. 1992. Identification of useful germplasm for practical plant breeding programs. Pp. 147–169 in Plant Breeding in the 1990s, H. T. Stalker and J. P. Murphy, eds. Wallingford, U.K.: CAB International.

McGowan, V. F., and V. B. D. Skerman. 1986. World Catalogue of Rhizobium Collections, 3d ed. St. Lucia, Queensland, Australia: World Data Center for Microorganisms.

McMullen, N. 1987. Seeds and World Agricultural Progress. Washington, D.C.: National Planning Association.

Meers, J. L., and P. E. Milsom. 1987. Organic acids and amino acids. Pp. 359–383 in Basic Biotechnology, J. Bu'Lock and B. Kristiansen, eds. Orlando, Fla.: Academic Press.

Moo-Young, M., S. Hasnain, and J. Lamptey, eds. 1986. Biotechnology and Renewable Energy. London: Elsevier Applied Science Publishers.

Mooney, P. R. 1979. Seeds of the Earth: A Private or Public Resource? Ottawa, Ontario, Canada: Inter Pares.

Mooney, P. R. 1983. The Law of the Seed. Development Dialogue, Volumes 1 and 2. Uppsala, Sweden: Dag Hammarskjöld Foundation.

Mooney, P. R. 1985. The law of the lamb. Dev. Dial. 1:103–108.

Morel, G. 1975. Meristem culture techniques for the long-term storage of cultivated plants. Pp. 327–332 in Crop Genetic Resources for Today and Tomorrow, O. H. Frankel and J. G. Hawkes, eds. New York: Cambridge University Press.

Morris, G. J., D. Smith, and G. E. Coulson. 1988. A comparative study of the changes in the morphology of hyphae during freezing and viability upon thawing for twenty species of fungi. J. Gen. Microbiol. 134:2897–2916.

Nabhan, G. P. 1985. Native American crop diversity, genetic resource conservation, and the policy of neglect. Agric. Human Val. II(3):14–17.

Nabhan, G. P. 1989. Enduring Seeds. Native American Agriculture and Wild Plant Conservation. San Francisco: North Point Press.

Nakayama, K. 1982. Amino acids. Pp. 748–801 in Prescott & Dunn's Industrial Microbiology, 4th ed., G. Reed, ed. Westport, Conn.: AVI Publishing.

National Research Council (NRC). 1972. Genetic Vulnerability of Major Crops. Washington, D.C.: National Academy of Sciences.

NRC. 1978. Conservation of Germplasm Resources: An Imperative. Washington, D.C.: National Academy of Sciences.

NRC. 1989a. Alternative Agriculture. Washington, D.C.: National Academy Press.

NRC. 1989b. Lost Crops of the Incas. Washington, D.C.: National Academy Press.

NRC. 1991a. Managing Global Genetic Resources: The U.S. National Plant Germplasm System. Washington, D.C.: National Academy Press.

NRC. 1991b. Managing Global Genetic Resources: Forest Trees. Washington, D.C.: National Academy Press.

NRC. 1993. Managing Global Genetic Resources: Livestock. Washington, D.C.: National Academy Press.

National Institutes of Health (NIH). 1984. NIH Policy Relating to Reporting and Distribution of Unique Biological Materials Produced with NIH Funding. 13 NIH Guide for Grants and Contracts. Bethesda, Md.: National Institutes of Health.

Nault, L. R., and W. R. Findley. 1981. *Zea diploperennis:* A primitive relative offers new traits to improve corn. Desert Plants 2(4):203–205.

ne Nsaku, N. 1991. La conservation des ressources phytogénétiques des pays des grands lacs (CEPGL: Burundi, Rwanda et Zaïre). Pp. 269–276 in Crop Genetic Resources of Africa, Volume 1, F. Attere, H. Zedan, N. Q. Ng, and P. Perrino, eds. Rome: International Board for Plant Genetic Resources, United Nations Environment Program, and International Institute for Tropical Agriculture.

Neergaard, P. 1977. Methods for detection and control of seed-borne fungi and bacteria. Pp. 33–38 in Plant Health and Quarantine in International Transfer of Genetic Resources, W. B. Hewitt and L. Chiarappa, eds. Cleveland: CRC Press.

Neergaard, P. 1984. Seed health in relation to the exchange of germplasm. Pp. 2–21 in Seed Management Techniques for Genebanks, J. B. Dickie, S. Linington, and J. T. Williams, eds. Cleveland: CRC Press.

Ng, N. Q., P. Perrino, F. Attere, and H. Zedan, eds. 1991. Crop Genetic Resources of Africa, Volume 2. Rome: International Board for Plant Genetic Resources, United Nations Environment Program, and International Institute for Tropical Agriculture.

Noy-Meir, Y. Anikster, M. Waldman, and A. Ashri. 1989. Population dynamics research for in situ conservation: Wild wheat in Israel. Plant Genet. Resources Newsl. 75/76:9–11.

Office of Technology Assessment (OTA). 1984. Commercial Biotechnology: An International Analysis. OTA-BA-218. Washington, D.C.: U.S. Government Printing Office.

OTA. 1986. A Review of U.S. Competitiveness in Agricultural Trade—A Technical Memorandum. Report 60 OTA-TM-TET-29. Washington, D.C.: U.S. Government Printing Office.

OTA. 1987a. Technologies to Maintain Biological Diversity. Report OTA-F-330. Washington, D.C.: U.S. Government Printing Office.

OTA. 1987b. New Developments in Biotechnology: Ownership of Human Tissues and Cells. Special Report OTA-BA-337. Washington, D.C.: U.S. Government Printing Office.

OTA. 1989. New Developments in Biotechnology: Patenting Life. Special Report OTA-BA-370. Washington, D.C.: U.S. Government Printing Office.

Oldfield, M. L. 1984. The Value of Conserving Genetic Resources. Washington, D.C.: U.S. Department of the Interior.

Oldfield, M. L., and J. B. Alcorn. 1987. Conservation of traditional agroecosystems. BioScience 37:199–208.

Olsen, J., and K. Allermann. 1987. Microbial biomass as a protein source. Pp. 285–308 in Basic Biotechnology, J. Bu'Lock and B. Kristiansen, eds. Orlando, Fla.: Academic Press.

Paterson, A. H., E. S. Lander, J. D. Hewitt, S. Peterson, S. E. Lincoln, and S. D. Tanksley. 1988. Resolution of quantitative traits into Mendelian factors by using a complete linkage map of restriction fragment length polymorphisms. Nature 335:721–726.

Peacock, W. J. 1984. The impact of molecular biology on genetic resources. Pp. 268–276 in Crop Genetic Resources: Conservation and Evaluation, J. H. W. Holden and J. T. Williams, eds. London: George Allen & Unwin.

Peberdy, J. F. 1987. Developments in protoplast fusion in fungi. Microbiol. Sci. 4:108–114.

Peeters, J. P., and N. W. Galwey. 1988. Germplasm collections and breeding needs in Europe. Econ. Bot. 42:503–521.

Peeters, J. P., and J. T. Williams. 1984. Towards better use of genebanks with special reference to information. Plant Genet. Resources Newsl. 60:22–32.

Perret, P. 1989. The role of networks of dispersed collections. Pp. 157–170 in The Use of Plant Genetic Resources, A. H. D. Brown, O. H. Frankel, D. R. Marshall, and T. J. Williams, eds. New York: Cambridge University Press.

Perrin, R. K., K. A. Kunnings, and L. A. Ihnen. 1983. Some Effects of the U.S. Plant Variety Protection Act of 1970. Economics Research Report No. 46. Raleigh, N.C.: North Carolina State University.

Perry, M., A. K. Stoner, and J. D. Mowder. 1988. Plant germplasm information management system: Germplasm Resources Information Network. Hortscience 23(1):57–60.

Plant Health Division, Agriculture Canada. 1978–1986. Intercepted Plant Pests. Ottawa, Ontario: Plant Health Division, Agriculture Canada.

Plant Protection and Quarantine. 1960–1988. Foreign Import Regulations. Washington, D.C.: U.S. Department of Agriculture.

Plant Protection and Quarantine. 1980–1986. List of Intercepted Pests. Washington, D.C.: U.S. Department of Agriculture.

Plucknett, D. L. 1985. "The law of the seed" and the CGIAR. Dev. Dial. 1:97–102.

Plucknett D. L., and N. J. H. Smith. 1987. Sustaining agricultural yields. BioScience 36(1):40–45.

Plucknett, D. L., and N. J. H. Smith. 1988. Plant Quarantine and the International Transfer of Germplasm. Study Paper No. 25. Washington, D.C.: Consultative Group on International Agricultural Research, World Bank.

Plucknett, D. L., N. J. H. Smith, J. T. Williams, and N. M. Anishetty. 1987. Gene Banks and the World's Food. Princeton, N.J.: Princeton University Press.

Ponte, J. G., and G. Reed. 1982. Bakery foods. Pp. 246–292 in Prescott & Dunn's Industrial Microbiology, 4th ed., G. Reed, ed. Westport, Conn.: AVI Publishing.

Porter, W. M., and D. H. Smith, Jr. 1982. Computer assisted management of the USDA Small Grain Collection. Plant Dis. 66(5):435–438.

Prescott-Allen, R., and C. Prescott-Allen. 1983. Genes from the Wild: Using Wild Genetic Resources for Food and Raw Materials. London: Earthscan.

Prescott-Allen, R., and C. Prescott-Allen. 1984. Canada's National Parks as In Situ Genebanks. A Report to Parks Canada. Victoria, British Columbia, Canada: Padata.

Prescott-Allen, C., and R. Prescott-Allen. 1986. The First Resource: Wild Species in the North American Economy. New Haven, Conn.: Yale University Press.

Prescott-Allen, R., and C. Prescott-Allen. 1990. How many plants feed the world? Conserv. Biol. 4(4):365–374.

Primrose, S. B. 1986. The application of genetically engineered microorganisms in the production of drugs. J. Appl. Bacteriol. 61:91–116.

Pundir, R. P. S., K. N. Reddy, and M. H. Mengesha. 1988. ICRISAT Chickpea Germplasm Catalog: Evaluation and Analysis. Patancheru, India: International Crop Research Institute for the Semi-Arid Tropics.

Purchase, H. G. 1986. Future applications of biotechnology in poultry. Avian Dis. 30(1):47–59.

Qi, C., and F. Yunliu. 1988. Molecular cloning and expression of *Bacillus thuringiensis* subsp. *galleriae insecticidal* crystal protein gene in *Escherichia coli*. Pp. 183–191 in Recent Advances in Biotechnology and Applied Biology, S.-T. Chang, K.-Y. Chen, and N. Y. S. Woo, eds. Hong Kong: Chinese University of Hong Kong Press.

Reckhaus, P. M., and I. Adamou. 1986. Rice diseases and their economic importance in the Niger. Plant Protection Bull. 34(2):77–82.

Reed, G. 1982. Microbial biomass, single cell protein, and other microbial products. Pp. 541–592 in Prescott & Dunn's Industrial Microbiology, 4th ed., G. Reed, ed. Westport, Conn.: AVI Publishing.

Rice Research Board. 1986. Seventeenth Annual Report to the California Rice Growers. Yuba City, Calif.: Rice Research Board.

Rice Research Board. 1987. Eighteenth Annual Report to the California Rice Growers. Yuba City, Calif.: Rice Research Board.

Rick, C. M. 1987. Genetic resources in *Lycopersicon*. Pp. 17–26 in Tomato Biotechnology: Proceedings of a Symposium Held at the University of California, Davis, August 20–22, 1986, D. J. Nevins and R. A. Jones, eds. New York: A. R. Liss.

Rick, C. M., J. F. Fobes, and M. Holle. 1977. Genetic variation in *Lycopersicon pimpinellifolium*: Evidence of evolutionary change in mating systems. Plant Syst. Evolut. 127:139–170.

Roberts, L. 1987. Who owns the human genome? Science 237:358–361.

Robertson, D. W., and D. von Wettstein. 1971. Report of the Committee on Genetic Marker Stocks, Nomenclature, and Gene Symbols. P. 618 in Barley Genetics. II. Proceedings of the Second International Barley Genetics Symposium, R. A. Nilan, ed. Pullman: Washington State University Press.

Robinson, P. K., J. O. Reeve, and K. H. Goulding. 1988. The biotechnological potential of immobilized microalgal cells. Pp. 193–204 in Recent Advances in Biotechnology and Applied Biology, S.-T. Chang, K.-Y. Chen, and N. Y. S. Woo, eds. Hong Kong: Chinese University of Hong Kong Press.

Rogers, S. O., and A. J. Bendich. 1985. Extraction of DNA from milligram amounts of fresh, herbarium and mummified plant tissues. Plant Mol. Biol. 5:69–76.

Rogosa, M., M. I. Krichevsky, and R. R. Colwell. 1986. Coding Microbiological Data for Computers. New York: Springer.

Rolz, C. 1984. Bioconversion of lignocellulose. Pp. 43–48 in Biotechnology in the Americas: Prospects for Developing Countries, W. D. Sawyer, ed. Washington, D.C.: Interciencia Association.

Rote Taube. 1991. English translation. Intl. Rev. Ind. Property Copyright Law 22(1):136–142.

Rural Development Administration. 1985. Rice Varietal Improvement in Korea. Suweon, Republic of Korea: Rural Development Administration.

Russell, N. C. 1989. Guadalajara hosts international symposium on *Zea diploperennis* and the conservation of plant genetic resources. Diversity 5(1):8–9.

Rutger, J. N. 1986. Needed applications of biotechnology in rice genetics and breeding. Pp. 31–36 in World Biotech Report 1986, Vol. 2, Part 7. London: Online.

Ruttan, V. W. 1982. Agricultural Research Policy. Minneapolis: University of Minnesota Press.

Sahasrabudhe, S. R., and V. V. Modi. 1987. Microbial degradation of chlorinated aromatic compounds. Microbiol. Sci. 4:300–303.

Sakai, A., S. Kobayashi, and I. Oiyama. 1990. Cryopreservation of nucellar cells of navel orange (*Citrus sinensis* Osb. var. *Brasiliensis tanaka*) by vitrification. Plant Cell Rep. 9:30–33.

Salhuana, W., Q. Jones, and R. Sevilla. 1991. The Latin American Maize Project model for rescue of irreplaceable germplasm. Diversity 7(1):40–42.

Sastrapradja, D. S., S. Sastrapradja, M. A. Rifai, and B. H. Siwi. 1985. A government role for the activities on plant genetic resources: Indonesia. Pp. 77–85 in Proceedings of the International Symposium on South East Asian Plant Genetic Resources, K. L. Mehra and S. Sastrapradja, eds. Bogor, Indonesia: Lembaga Biologi Nasional.

Sauer, C. O. 1938. Theme of plant and animal destruction in economic history. J. Farm Econ. 20:765–775.

Schlosser, S. 1985. The use of nature reserves for in situ conservation. Plant Genet. Resources Newsl. 61:23–25.

Schoulties, C. L., C. P. Seymour, and J. W. Miller. 1983. Where are the exotic disease threats? Chapter 6 in Exotic Plant Pests and North American Agriculture, C. L. Wilson and C. L. Graham, eds. New York: Academic Press.

Schutz, W. M., and C. A. Brim. 1971. Inter-genotypic competition in soybeans. III. An evaluation of stability in multiline mixtures. Crop Sci. 11:684–689.

Scott, P. R., P. W. Benedikz, and C. J. Cox. 1982. A genetic study of the relationship between height, time of ear emergence, and resistance to *Septoria nodorum* in wheat. Plant Pathol. 31:45–60.

Scott, P. R., P. W. Benedikz, H. G. Jones, and M. A. Ford. 1985. Some effects of canopy structure and microclimate on infection of tall and short wheats by *Septoria nodorum*. Plant Pathol. 34:578–593.

Seaward, M. R. D. 1982. Principles and priorities of lichen conservation. J. Hattori Bot. Lab. 52:401–406.

Seidewitz, L. 1974–1976. Thesaurus for the International Standardization of Gene Bank Documentation. Parts I, II, III, IV, and V. Braunschweig, Germany: Institut für Pflanzenbau und Saatgutforschung der Forschungsanstalt für Landwirtschaft.

Seshu, D. V., K. Alluri, F. Cuevas-Perez, S. W. Ahn, M. Akbar, K. G. Pillai, and M. J. Rosero. 1989. Report of Activities and Accomplishments of the International Rice Testing Program, 1985–89. Los Baños, Philippines: International Rice Research Institute.

Sgaravatti, E. 1986. World List of Seed Sources. Rome: Food and Agriculture Organization of the United Nations.

Shaffer, M. 1987. Minimum viable populations: Coping with uncertainty. Pp. 69–86 in Viable Populations for Conservation, M. E. Soulé, ed. New York: Cambridge University Press.

Shands, H. L., P. J. Fitzgerald, and S. A. Eberhart. 1989. Program for plant germplasm preservation in the United States: The U.S. National Plant Germplasm System. Pp. 97–115 in Biotic Diversity and Germplasm Preservation, Global Imperatives, L. Knutson and A. K. Stoner, eds. Dordrecht, Netherlands: Kluwer Academic Publishers.

Silvey, V. 1986. The contribution of new varieties to cereal yields in England between 1947 and 1983. J. Nat. Inst. Agric. Bot. 17:155–168.

Simmonds, N. W. 1962. Variability in crop plants: Its use and conservation. Biol. Rev. 37:422–465.

Simmonds, N. W. 1979. Principles of Crop Improvement. New York: Longman.

Simmonds, N. W. 1991. Genetics of horizontal resistance to diseases of crops. Biol. Rev. 66:189–241.

Simmonds, N. W., and S. Rajaram, eds. 1988. Breeding Strategies for Resistance to the Rusts of Wheat. Mexico City: Centro International de Mejoramiento de Maíz y Trigo.

Singh, E. L. 1988. Embryo transfer: Possibilities for disease transmission. Paper prepared for the Committee on Managing Global Genetic Resources: Agricultural Imperatives, National Research Council, Washington, D.C.

Singh, V. K., P. C. Verna, and B. M. Khannah. 1979. Fungicidal spray schedule for the control of *Alternaria* blight of wheat. Indian J. Mycol. Plant Pathol. 9(2):200–204.

Smith, J. S. C. 1988a. Diversity of United States hybrid maize germplasm; isozymic and chromatographic evidence. Crop Sci. 28:63–69.

Smith, D. 1988b. Culture and preservation. Pp. 75–99 in Living Resources for Biotechnology: Filamentous Fungi, D. L. Hawksworth and B. E. Kirsop, eds. New York: Cambridge University Press.

Smith, D., and A. H. S. Onions. 1983. The Preservation and Maintenance of Living Fungi. Kew, U.K.: Commonwealth Mycological Institute.

Sossou, J., S. Karunaratne, and A. Kovoor. 1987. Collecting palm: In vitro explanting in the field. Plant Genet. Resources Newsl. 69:7–18.

Soulé, M. E., ed. 1987. Viable Populations for Conservation. New York: Cambridge University Press.

Sprague, G. F., and M. T. Jenkins. 1943. A comparison of synthetic varieties, multiple crosses, and double crosses in corn. J. Am. Soc. Agron. 35:137–147.

Staines, J. E., V. F. McGowan, and V. B. D. Skerman. 1986. World Directory of Collections of Cultures of Microorganisms, 3d ed. Brisbane, Queensland, Australia: World Data Center.

Stavely, J. R. 1983. The 1982 Bean Rust Nurseries in the United States. Geneva, N.Y.: Bean Improvement Cooperative.

Stavely, J. R. 1988. Bean rust in the United States in 1987. Annu. Rep. Bean Improv. Coop. 31:130–131.

Steinkraus, K. H., ed. 1983. Fermented Foods. New York: Marcel Dekker.

Strickberger, M. W. 1976. Genetics. New York: Macmillan.

Templeton, G. E., D. O. TeBeest, and R. J. Smith. 1979. Biological control with mycoherbicides. Annu. Rev. Phytopathol. 17:301–310.

Thitai, G. N. W. 1991. Countries' activities summaries. Pp. 247–262 in Crop Genetic Resources of Africa, Volume 1, F. Attere, H. Zedan, N. Q. Ng, and P. Perrino, eds.

Rome: International Board for Plant Genetic Resources, United Nations Environment Program, and International Institute for Tropical Agriculture.

Timothy, D. H., P. H. Harvey, and C. Dowswell. 1988. Development and Spread of Improved Maize Varieties and Hybrids in Developing Countries. Washington, D.C.: U.S. Agency for International Development.

Turner, P. D. 1977. Rubber (*Hevea brasiliensis* M.). Pp. 241–252 in Plant Health and Quarantine in International Transfer of Genetic Resources, W. B. Hewitt and L. Chiarappa, eds. Cleveland: CRC Press.

United Nations. 1973. Report of the United Nations Conference on the Human Environment, Stockholm, June 5–16, 1972. New York: United Nations.

United Nations. 1992. Article 15 in the Convention on Biological Diversity, United Nations Conference on the Environment and Development, Rio de Janeiro, Brazil, June 5, 1992.

United Nations Development Program and International Board for Plant Genetic Resources. 1986. Directory of European Institutions Holding Crop Genetic Resources Collections, E. Bettencourt and P. M. Perret, eds. Rome: International Board for Plant Genetic Resources.

Upshall, A., A. A. Kumar, M. C. Bailey, M. D. Parker, M. A. Favreau, K. P. Lewison, M. L. Joseph, J. M. Maraganore, and G. L. McKnight. 1987. Secretion of active human tissue plasminogen activator from the filamentous fungus *Aspergillus nidulans*. Bio/Technology 5:1301–1304.

U.S. Department of Agriculture. 1981. The National Plant Germplasm System: Current Status (1980), Strengths and Weaknesses, Long-Range Plan (1983–1997). Washington, D.C.: Science and Education, U.S. Department of Agriculture.

van Arsdell, J. N., S. Kwok, V. L. Schweickart, M. B. Ladner, D. H. Gelfand, and M. A. Innis. 1987. Cloning, characterization, and expression in *Saccharomyces cerevisiae* of endogluconase I from *Trichoderma reesei*. Bio/Technology 5:60–64.

van Hintum, T. J. L. 1988. GENIS: A fourth generation information system for the database management of genebanks. Plant Genet. Resources Newsl. 75/76:13–15.

van Sloten, D. H. 1990a. IBPGR and the challenges of the 1990s: A personal point of view. Diversity 6(2):36–39.

van Sloten, D. H. 1990b. Facing the Future: IBPGR in the 1990s. Paper presented at the International Centers Week, October 29 to November 2, 1990, World Bank, Washington, D.C.

Vanderborcht, T. 1988. A centralized database for the common bean and its use in diversity analysis. Pp. 51–65 in Genetic Resources of *Phaseolus* Beans, P. Gepts, ed. Dordrecht, Netherlands: Kluwer Academic Publishers.

Vaughn, R. H. 1982. Lactic acid fermentation of cabbage, cucumbers, olives, and other products. Pp. 185–236 in Prescott & Dunn's Industrial Microbiology, 4th ed., G. Reed, ed. Westport, Conn.: AVI Publishing.

Vavilov, N. I. 1951. The origin, variation, immunity and breeding of cultivated plants. Chronica Bot. 13(1/6):1–364.

Venette, J. R., and D. A. Jones. 1982. Tilt (propiconazol) controls rust on dry edible beans in North Dakota. Annu. Rep. Bean Improv. Coop. 25:19–20.

Vitkovskij, V. L., and S. V. Kuznetsov. 1990. The N. I. Vavilov All Union Research Institute of Plant Industry. Diversity 6(1):15–18.

Wang, H. L., and C. W. Hesseltine. 1982. Oriental fermented foods. Pp. 492–538 in Prescott & Dunn's Industrial Microbiology, 4th ed., G. Reed, ed. Westport, Conn.: AVI Publishing.

Wellhausen, E. J., L. M. Roberts, and E. X. Hermandez (in collaboration with P. C.

Mangelsdorf). 1952. Races of Maize in Mexico. Cambridge, Mass.: Bussey Institute, Harvard University.

Welsh, J., and M. McClelland. 1990. Fingerprinting genomes using PCR with arbitrary primers. Nucl. Acids Res. 18:7213–7218.

Whealy, K. 1988. Garden Seed Inventory: Second Edition. Decorah, Iowa: Seed Saver Publications.

White, G. A., S. A. Eberhart, P. A. Miller, and J. D. Mowder. 1988. Plant materials registered by *Crop Science* incorporated into National Plant Germplasm System. Crop Sci. 28:716–717.

Whyte, R. O., and G. Julén, eds. 1963. Proceedings of a technical meeting on plant exploration and introduction. Genetica Agraria 17:1–573.

Widrlechner, M. P. 1987. Variation in the breeding system of *Lycopersicon pimpinellifolium:* Implications for germplasm maintenance. Plant Genet. Resources Newsl. 70:38–43.

Wilkes, H. G. 1967. Teosinte: The Closest Relative of Maize. Cambridge, Mass.: Bussey Institute, Harvard University.

Wilkes, H. G. 1977. The world's crop plant germplasm—An endangered resource. Bull. At. Sci. 33:8–16.

Wilkes, G. 1989a. Germplasm preservation: Objectives and needs. Pp. 13–41 in Biotic Diversity and Germplasm Preservation, Global Imperatives, L. Knutson and A. K. Stoner, eds. Boston: Kluwer Academic Publishers.

Wilkes, H. G. 1989b. Teosinte in Mexico as a model for in situ conservation: The challenge. University of Massachusetts, Boston. Unpublished paper.

Wilkes, H. G., and S. Wilkes. 1972. The green revolution. Environment 14:32–39.

Williams, J. T. 1984. A decade of crop genetic resources research. Pp. 1–17 in Crop Genetic Resources: Conservation and Evaluation, J. H. W. Holden and J. T. Williams, eds. London: George Allen & Unwin.

Williams, J. T. 1988. Recent changes in emphasis in international genetic resources work. Pp. 7–12 in Crop Genetic Resources of East Asia, S. Suzuki, ed. Rome: International Board for Plant Genetic Resources.

Williams, J. T. 1989a. Plant germplasm preservation: A global perspective. Pp. 81–96 in Biotic Diversity and Germplasm Preservation, Global Imperatives, L. Knutson and A. K. Stoner, eds. Dordrecht, Netherlands: Kluwer Academic Publishers.

Williams, J. T. 1989b. Practical considerations relevant to effective evaluation. Pp. 235–244 in The Use of Plant Genetic Resources, A. H. D. Brown, O. H. Frankel, D. R. Marshall, and J. T. Williams, eds. New York: Cambridge University Press.

Williams, J. K., A. R. Kubelik, K. J. Livak, J. A. Rafalski, and S. V. Tingey. 1990. DNA polymorphisms amplified by arbitrary primers are useful as genetic markers. Nucl. Acids Res. 18:6531–6535.

Wilson, E. O., and F. M. Peter, eds. 1988. Biodiversity. Washington, D.C.: National Academy Press.

Winter, D. L. 1985. The promise of cyclosporine. Pp. 160–173 in 1986 Yearbook of Science and the Future. Chicago: Encyclopedia Britannica.

Withers, L. A. 1985. Cryopreservation of cultured cells and meristems. Pp. 253–315 in Cell Culture and Somatic Cell Genetics of Plants, Volume 2: Growth, Nutrition, Differentiation and Preservation, I. K. Vasil, ed. Orlando, Fla.: Academic Press.

Withers, L. A. 1986. In vitro approaches to the conservation of plant genetic resources. Pp. 261–276 in Plant Tissue Culture and Its Agricultural Applications, L. A. Withers and P. G. Alderson, eds. London: Butterworths.

Withers, L. A. 1987. In vitro methods for collecting germplasm in the field. Plant Genet. Resources Newsl. 69:2–6.

Withers, L. A. 1991a. In vitro conservation. Biol. J. Linnean Soc. 43:31–42.

Withers, L. A. 1991b. Preservation of plant tissue cultures. Pp. 243–267 in Maintenance of Microorganisms and Cultured Cells, B. E. Kirsop and J. J. S. Snell, eds. London: Academic Press.

Withers, L. A., and S. K. Wheelans. 1986. IBPGR in vitro conservation databases—A report. Pp. 13–14 in IBPGR Advisory Committee on In Vitro Storage: Report of the Third Meeting. Rome: International Board for Plant Genetic Resources.

Witt, S. C. 1985. BriefBook: Biotechnology and Genetic Diversity. San Francisco: California Agricultural Lands Project.

Wolf, E. C. 1987. On the Brink of Extinction: Conserving the Diversity of Life. Washington, D.C.: Worldwatch Institute.

Wolfe, M. S. 1985. The current status and prospects of multiline cultivars and variety mixtures for disease resistance. Annu. Rev. Phytopathol. 23:251–273.

Wolfe, M. S. 1988. The use of variety mixtures to control diseases and stabilize yield. Pp. 91–100 in Breeding Strategies for Resistance to the Rusts of Wheat, N. W. Simmonds and S. Rajaram, eds. Mexico City: Centro Internacional de Mejoramiento de Maíz y Trigo.

Wood, R. K. S., and M. J. Way, eds. 1988. Biological control of pests, pathogens, and weeds: Developments and prospects. Phil. Trans. Roy. Soc. London B318:109–376.

World Intellectual Property Organization, Committee of Experts on Biotechnological Inventions on Industrial Property. 1987. Industrial Property Protection of Biotechnological Inventions. BIOT/CE/III/2. Geneva: World Intellectual Property Organization.

Wu, L.-C. 1987. Strategies for conservation of genetic resources. Pp. 183–211 in Cultivating Edible Fungi, P. J. Wuest, D. J. Royse, and R. B. Beelman, eds. Amsterdam: Elsevier.

Yarwood, C. E. 1983. History of plant pathogen introductions. Chapter 3 in Exotic Plant Pests and North American Agriculture, C. L. Wilson and C. L. Graham, eds. New York: Academic Press.

Yuan, R. T., and Y. Lin. 1989. Biotechnology in the People's Republic of China and Hong Kong. International Trade Administration 37. Washington, D.C.: U.S. Department of Commerce.

Zedan, H., and R. Olembo. 1988. UNEP activities aimed at the promotion of applied environmental microbiology in developing countries. Pp. 75–88 in Recent Advances in Biotechnology and Applied Microbiology, S.-T. Chang, J.-Y. Chan, and N. Y. S. Woo, eds. Hong Kong: Chinese University of Hong Kong Press.

Zimdahl, R. L. 1983. Where are the principal exotic weed pests? Chapter 7 in Exotic Plant Pests and North American Agriculture, C. L. Wilson and C. L. Graham, eds. New York: Academic Press.

Zuber, M. S. 1975. Corn germplasm base in the U.S.: Is it narrowing, widening, or static? Pp. 277–286 in Proceedings of the 30th Annual Corn and Sorghum Industry Research Conference. Washington, D.C.: American Seed Trade Association.

Zuber, M. S., and L. L. Darrah. 1981. 1979 U.S. corn germplasm base. Pp. 234–249 in Proceedings of the 35th Annual Corn and Sorghum Industry Research Conference. Washington, D.C.: American Seed Trade Association.

Glossary

accession A distinct, uniquely identified sample of seeds, plants, or other germplasm materials that is maintained as an integral part of a germplasm collection.

active collection Germplasm accessions that are maintained under conditions of short- and medium-term storage for the purpose of study, distribution, or use.

allele A contraction of allelomorph; any of a number of alternative forms of a gene. The forms differ in DNA (deoxyribonucleic acid) sequence and affect the functioning of a single gene product (RNA [ribonucleic acid] or protein). New alleles arise from existing ones through mutation. The alleles of a gene occupy the same site or locus on a chromosome.

aneuploid An unbalanced genome with extra or missing chromosomes, having more or less than an integral multiple of the haploid number of chromosomes.

anther culture The culturing of the part of the stamen that bears the pollen grains (anther) or of a single pollen grain, as a method of producing haploids, homozygotes, or all-male plants.

apomixis Asexual reproduction in plants through the formation of seeds without fertilization (agamospermy), or the formation of a new individual from a group of cells without the production of an embryo or seed (vegetable reproduction).

base collection A comprehensive collection of germplasm accessions held for the purpose of long-term conservation; stocks from the base collection replenish exhausted or expired stocks in the active collection.

biological diversity The variety and variability among living organisms and the ecological complexes in which they occur.

bisexual Having both sexes.

breeding line A genetic group that has been selected and bred for their special combinations of traits.

characterization Assessment of the presence, absence, or degree of specific traits that are largely morphological and little influenced in their expression by varying environmental conditions.

chromosome The structure in the cell nucleus on which genes are located.

clonal propagation The reproduction of plants through asexual means, such as cuttings, grafts, or tissue culture.

clone A group of genetically identical individuals that result from asexual, vegetative multiplication; any plant that is propagated vegetatively and that is therefore a genetic duplicate of its parent.

collection A sample (for example, variety, strain, population) maintained at a genetic resources center for the purposes of conservation or use (an accession); a group of collected samples.

cultivar A contraction of cultivated variety. *See also* variety.

cytogenetics The combined study of cells and genes at the chromosome or cytoplasmic level.

cytoplasmic (mitochondrial) gene The mitochondrial DNA that is found outside of the nucleus in small, oblate bodies called mitochondria.

deoxyribonucleic acid (DNA) The basis of heredity; the substance of the genetic code.

distribution To supply adequate samples of genetic resource stocks to breeders and other users.

effective population size The equivalent number of parents if all contributed the same number of progeny to the next generation.

enhancement The process of improving a germplasm accession by breeding while retaining the important genetic contributions of the accession. This process may entail simple selection.

epistasis The interaction of genes at different loci. The situation in which one gene affects the expression of another.

evaluation The assessment of plants in a germplasm collection for potentially useful genetic traits, many of which may be environmentally variable (for example, pest or disease resistance, fruit quality, flavor, yield).

exploration The search for materials in the field.

ex situ conservation Maintenance or management of an organism away from its native environment. For crop germplasm, this term typically refers to maintenance in seed banks or repositories.

folk variety Local varieties of cultivated plants developed by indigenous farmers in traditional agricultural systems. By modern standards, such varieties are often highly variable genetically.

gene The basic functional unit of inheritance.

gene pool All the genes within a population.

genetic diversity In a group such as a population or species, the possession of a variety of genetic traits and alleles that frequently result in differing expressions in different individuals.

genetic drift The random fluctuations of gene frequencies due to sampling effects or unintended selection; although genetic drift occurs in all populations, its effects are most evident in small populations.

genetic resources In the context of this report, the term is synonymous with germplasm. *See* germplasm.

genetic stocks Accessions that typically possess one or more special genetic traits that make them of interest for research.

genetic vulnerability The condition that results when a crop or a plant species is genetically and uniformly susceptible to a pest, pathogen, or environmental hazard.

genome The complete set of chromosomes found in each cell nucleus of an individual. The haploid nucleus of gametes (sperm or egg) contains one genome. The diploid cells that make up the bulk of the living tissue of the plant body contain two sets.

genotype In the context of this report, plants with a specific complement of genes.

germplasm Seeds, plants, or plant parts that are useful in crop breeding, research, or conservation because of their genetic attributes. Plants, seed, or cultures that are maintained for the purposes of studying, managing, or using the genetic information they possess.

green revolution Increased production from the introduction of high-yielding varieties of major grain crops that were also aided by the more intensive use of fertilizers and irrigation.

haploid A cell or organism with a single genome.

heterozygous Having different alleles at a locus.

hybrid The progeny of a cross between two different species, races, cultivars, or breeding lines.

hybridization The process of crossing individuals that possess different genetic makeups.

inbreeding The breeding of individuals that are related.

in situ conservation Maintenance or management of an organism within its native environment. For landraces, this term includes maintenance in traditional agricultural systems.

isoallele An allele whose effect can only be distinguished from that of the normal allele by special tests.

isoenzyme (isozyme) Different chemical forms of the same enzyme that can generally be distinguished from one another by electrophoresis.

isogenic Genetically identical individuals or lines except for one or a few genes.

landrace A population of plants, typically genetically heterogeneous, commonly developed in traditional agriculture from many years of farmer-directed selection, and which is specifically adapted to local conditions.

legume Any member of the pea or bean family (Leguminosae or alternately, Fabaceae), for example, beans, peanuts, and alfalfa.

multiallelic genetic stock For cultivated plants, includes multiple gene marker stocks that are useful for linkage studies; also stocks with special combinations of loci necessary for the expression of a single trait.

neutral allele Those alleles whose differential contributions to fitness are so small that their frequencies change more owing to drift than to natural selection.

nominal population size Number of individuals per generation.

null hypothesis The standard hypothesis used in testing the statistical significance of the difference between the means of samples drawn from two populations. It states there is no difference between the populations from which the samples are drawn. If the probability is 0.05 or less, the null hypothesis is rejected and the difference is said to be significant.

obsolete variety Plant varieties that are no longer grown commercially; such varieties may be maintained in collections for use in breeding programs.

outbreeding The crossing of genetically unrelated plants or animals; crossbreeding.

passport data Information about a sample or specimen and the collection site, recorded at the time of collection. This information should include time of collection, exact location, identifying characteristics, ecological condition of location, names and numbers assigned to the accession by the collector, and any other relevant observations.

perennial crops Crop plants for which individuals are productive over several years. They include herbaceous perennials that die back annually, such as asparagus, and woody perennials with stems that may live for many years, such as apples, citrus crops, or mangos.

phenotype The combined expression of the environmental and genetic (hereditary) influences on an organism; the visible characteristics of an organism.

plant breeders' lines Unreleased lines or parents of hybrids maintained by breeders as part of their working stocks. Breeders usually develop and carry many lines of which only a small number are ever released into commercial production.

pleiotropy The phenomenon of a single gene being responsible for more than one phenotypic effect.

polymorphic In the context of this report, plants with several to many variable phenotypic or genetic forms.

population A group of individuals of the same species that occupy a particular geographic area or region. In general, individuals within a population potentially interbreed with one another.

preservation Storage of materials in collections under conditions that promote long-term survival and the use of propagation methods that protect genetic integrity during regeneration.

primary gene pool For plants, a cultivated species and its wild relatives that are readily intercrossed so that gene transfer is relatively simple.

primitive variety *See* folk variety.

pyramiding A strategy for developing long-lasting genetic resistance in a cultivar by combining two or more different genes that confer similar resistance into the same breeding line.

quarantine Measures that isolate introduced stocks to ensure that they do not carry diseases or pests injurious to the stocks in the importing country.

rare Small populations that are not currently endangered, but that are at some risk of loss.

regeneration Grow-out of a seed accession for the purpose of obtaining a fresh sample with high viability and adequate numbers of seeds.

restriction fragment length polymorphisms (RFLP) Variation that occurs in the length of DNA fragments resulting from digestion of the extracted DNA with one of several restriction enzymes that cleave DNA at specific recognition sites. Changes in the genetic composition result in fragments of altered length.

secondary gene pool For plants, all the biological species that can be crossed with a cultivated species, but where hybrids are usually sterile. Gene transfer is difficult but not impossible.

seed viability The ability of a seed to germinate under appropriate conditions.

single-gene or single-trait variants Lines that carry variants for qualitative characters. Examples include differences in morphological and physiological characters, electrophoretic variation in proteins, and fragment length variation in DNA generated by restriction enzymes.

species A taxonomic subdivision; a group of morphologically similar organisms that actually or potentially interbreed and are reproductively isolated from other such groups.

tertiary gene pool For plants, those species that can be crossed with a cultivated species only with difficulty and where gene transfer is usually only possible with radical techniques. Biotechnology has, at least in theory, greatly enlarged this pool because transformation (a radical technique) makes possible the introduction of DNA from any species.

tissue culture A technique for cultivating cells, tissues, or organs of plants in a sterile, synthetic medium; includes the tissues excised from a plant and the culture of pollen or seeds.

tuber A thickened, compressed, fleshy, usually underground stem that may function as a storage organ for food (starch) or water, or for propagation.

variety A plant type within a cultivated species that is distinguishable by one or more characters. When reproduced from seeds or by asexual means (for example, cuttings) its distinguishing characters are retained. The term is generally considered to be synonymous with cultivar.

weed A plant growing where it is not wanted by humans.

wild and weedy relatives For plants, those species that share a common ancestry and ecogeographic area with a crop species but that have not been domesticated. Most crops have wild or weedy relatives that differ in their degree of relationship to the crop. The ease with which genes can be transferred from them to the crop varies.

wild species Species that have not been subject to breeding to alter them from their (wild) state.

Abbreviations

ARS	Agricultural Research Service (U.S. Department of Agriculture), United States
AVRDC	Asian Vegetable Research and Development Center, Taiwan
BGASC	Barley Genetic and Aneuploid Stock Collection, United States
CAC	crop advisory committee
CATIE	Centro Agronómico Tropical de Investigación y Enseñanza (Tropical Agriculture Research and Training Center), Costa Rica
CENARGEN	Centro Nacional de Recursos Genéticos e Biotecnologia (National Center for Genetic Resources and Biotechnology), Brazil
CGIAR	Consultative Group on International Agricultural, United States
CIAT	Centro Internacional de Agricultura Tropical (International Center for Tropical Agriculture), Colombia
CIMMYT	Centro Internacional de Mejoramiento de Maíz y Trigo (International Maize and Wheat Improvement Center), Mexico
CIP	Centro Internacional de la Papa (International Potato Center), Peru

CMGS	Committee for the Maintenance of Genetic Stocks, United States
CRRI	Central Rice Research Institute, India
CSIRO	Commonwealth Scientific and Industrial Research Organization, Australia
DNA	deoxyribonucleic acid
EC	European Community (collective name given to the consolidation of the European coal and steel community, the Common Market, and the European Atomic Energy Community), Belgium
ECP/GR	European Cooperative Program for Crop Genetic Resources Networks
EEC	European Economic Community (Common Market), Belgium
EUCARPIA	European Association for Research on Plant Breeding, The Netherlands
FAO	Food and Agriculture Organization of the United Nations, Italy
FAST	Forecasting and Assessment in the Field of Science and Technology, Belgium
GED	Global Environment Facility (implemented jointly by the World Bank, the United Nations Environment Program, and the United Nations Development Program)
GRIN	Germplasm Resources Information Network (U.S. Department of Agriculture), United States
HYV	high-yielding variety
IARC	international agricultural research center
IARI	Indian Agricultural Research Institute, India
IBPGR	International Board for Plant Genetic Resources, Italy
ICDA	International Coalition for Development Action, Belgium
ICTA	Instituto de Ciencia y Tecnologi Agricola, Guatemala
IITA	International Institute of Tropical Agriculture, Nigeria
IRRI	International Rice Research Institute, The Philippines
IUCN	International Union for Conservation of Nature and

Natural Resources (currently known as the World
Conservation Union), Switzerland

LAMP   Latin American Maize Project

MiCIS   Microbial Culture Information Service, United
                Kingdom
MINE   Microbial Information Network Europe (European
                Economic Community), Belgium
MIRCENS   microbial resources centers
MSDN   Microbial Strain Data Network, United Kingdom

NPGS   National Plant Germplasm System (U.S. Department
                of Agriculture), United States

OECD   Organization for Economic Cooperation and
                Development, France

PCR   polymerase chain reaction
PI   Plant Introduction Number
PORIM   Palm Oil Research Institute of Malaysia
PRECODEPA   Programa Regional Cooperativa de Papa
                (Cooperative Regional Potato Program)
PVP   plant variety protection
PVPA   U.S. Plant Variety Protection Act
PVPC   Plant Variety Protection Certificate

RFLP   restriction fragment length polymorphism

TGSC   Charles M. Rick Tomato Genetics Resource Center,
                United States

UPOV   Union for the Protection of New Varieties; also,
                Union Internationale pour la Protection des
                Obtentions Vegetales (International Union for the
                Protection of New Varieties of Plants), Switzerland
USDA   U.S. Department of Agriculture

WARDA   West Africa Rice Development Association, Ivory
                Coast
WDC   World Data Center for Microorganisms, United States
WFCC   World Federation for Culture Collections, United
                Kingdom

Authors

ROBERT W. ALLARD (*Subcommittee Chairman*) is emeritus professor of genetics at the University of California, Davis. He has a Ph.D. degree in genetics from the University of Wisconsin. His areas of research include plant population genetics, gene resource conservation, and plant breeding. He is a member of the National Academy of Sciences.

PAULO DE T. ALVIM Since 1963, Alvim has been the scientific director for the Comissão Executiva do Plano da Lavoura Cacaueira (Executive Commission of the Program for Strengthening Cacao Production), Brazil. He earned a Ph.D. degree from Cornell University with specialization in plant physiology, tropical agriculture, and ecology.

JOHN H. BARTON Since 1975, Barton has been a professor of law and director of the International Center on Law and Technology at Stanford University, where he earned his law degree. He is an expert on international trade law and high-technology law and has written and consulted extensively on the legal aspects of biotechnology, including both biosafety and intellectual property issues.

FREDERICK H. BUTTEL is professor of rural sociology at the University of Wisconsin, where he earned a Ph.D. degree in sociology. He is also adjunct professor of rural sociology at Cornell University. His areas of interest are in technology and social change, particularly in relation to agricultural research, biotechnology, and the environment.

TE-TZU CHANG In 1991, Chang retired as head of the International Rice Germplasm Center at the International Rice Research Institute, where he was also a principal scientist. He has been a visiting professor at the University of the Philippines, Los Baños, since 1962. He earned a Ph.D. degree in plant genetics and breeding from the University of Minnesota. He had a vital role in the Green Revolution in rice. Chang has broad experience in managing and designing plant gene banks.

PETER R. DAY (*Committee Chairman*) Before joining Rutgers University as director of the Center for Agricultural Molecular Biology in 1987, Day was the director of the Plant Breeding Institute, Cambridge, United Kingdom. He has a Ph.D. degree from the University of London, and is a leader in the field of biotechnology and its application to agriculture.

ROBERT E. EVENSON Since 1977, Evenson has been a professor of economics at Yale University. He has a Ph.D. degree in economics from the University of Chicago. His research interests include agricultural development policy with a special interest in the economics of agricultural research.

HENRY A. FITZHUGH (*Subcommittee Chairman*) is deputy director general for research at the International Livestock Center for Africa, Ethiopia. He received a Ph.D. degree in animal breeding from Texas A&M University. His field of research is the development and testing of biological and socioeconomic interventions to improve the productivity of livestock in agricultural production systems.

MAJOR M. GOODMAN is professor of crop science, statistics, genetics, and botany at North Carolina State University (NCSU) where he has been employed since 1967. He has a Ph.D. degree in genetics from NCSU, and his areas of research are plant breeding, germplasm conservation and utilization, numerical taxonomy, history and evolution of maize, and applied multivariate statistics. Goodman is a member of the National Academy of Sciences.

JAAP J. HARDON In 1985, Hardon became the director of the Center for Genetics Resources, The Netherlands. He has a Ph.D. degree in plant genetics from the University of California. His specialty is plant breeding and genetics.

DONALD R. MARSHALL In 1991, Marshall left the Waite Agricultural Research Institute at the University of Adelaide, Australia, to accept a position as professor of plant breeding at the University of Sydney. He has a Ph.D. degree in genetics from the University of California, Davis. His professional interests are population genetics, plant breeding, host-parasite interactions, and genetic resources conservation.

SETIJATI SASTRAPRADJA is affiliated with the National Center for Research in Biotechnology at the Indonesian Institute of Science. She has a Ph.D. degree in botany from the University of Hawaii.

CHARLES SMITH is a professor of animal breeding strategies at the University of Guelph, Canada. He has a Ph.D. degree in animal breeding from Iowa State University. His research area is in animal breeding strategies, including genetic conservation, and he has been involved in international efforts to conserve domestic animal germplasm.

JOHN A. SPENCE In 1989, Spence was appointed head of the Cocoa Research Unit at the University of the West Indies, Trinidad and Tobago. He has a Ph.D. degree from the University of Bristol, United Kingdom. His research interests are cocoa tissue culture and cryopreservation as alternatives to holding field germplasm collections.

Index

Descriptors and descriptor states, 162,
165, 181, 205, 206–207, 209
Developing countries
crop genetic resources, 6, 15, 40
environmental funding, 27
exploitation of, 92, 323–324
genetic diversity of crops, 16, 27, 69,
72, 74, 75, 76, 80–81, 114
germplasm banks in, 42, 87, 90–91, 92
maize germplasm trade, 291
microorganism applications in, 252
MIRCEN-sponsored collections, 253
national programs, 81, 92–93
ownership issues, 13, 24–25, 27, 101,
279, 282, 290, 299–300
potato breeding programs, 294
plant protection legislation, 282
private sector role in, 112–113, 115
public sector role in, 113
quarantine policies, 275–276
recommended roles, 26–27, 129, 301,
345–346
seed organizations, 113
seed policies, 113–114, 115
use of germplasm resources, 180
yield increases, 92, 188
see also Conflicts over genetic
resources
Diagnostic probes, patents, 286
Differential survival, 157–158
Dioecious species, 159
Disease
transmission by embryo transfer, 196–
197 *see also* Epidemics; Pathogens;
individual pathogens
DNA
artificial chromosomes, 198
defined, 408
introduction from other species, 31
markers, 121, 199, 201–202, 224
measures of variation, 121
plasmid libraries, 23
probes, 10–11, 23, 52, 199, 200–202,
204, 220, 224
randomly amplified polymorphic, 202
recombinant technology, 189, 203
transformation technique, 11, 31, 38,
197, 202
DNA sequences
data banks, 23, 38, 197–199, 203
from dead materials, 38–39, 198, 201

as genetic resources, 38, 197–198
linkage maps, 197, 199
restriction fragment length
polymorphisms, 23, 67–68, 121,
199–202
Documentation
compatibility between systems, 209–
210, 211–212
computerization, 210–213, 216
deficiencies in, 140–141
descriptive information, 8–11, 206–207
environmental conditions, 208–209
of evaluation efforts, 165–166
of genetic resources, 205–217
importance, 38
and management of collections, 205
recommendations, 21, 187, 216–217
status report, 213–214
technology development and, 215–216
and use of systems, 181, 212–213
see also Information on germplasm
collections
Domesticated species
evolution of, 33–34, 129
in situ conservation, 127–128
plant-pathogen relationships, 60
Dot blot tests, 11
Double cropping, 59
Drosophila spp. (Fruit flies), 37, 138, 219,
220
Drought, 51, 57
Dutch elm disease (*Ceratocystis ulmi*), 58
Dwarf bunt (*Tilletia controversa*), 58, 70

E

Earworms (*Heliothis zea*), 73
Eastern Europe
genetic resources work, 350, 352
security of collections in, 42
Ecogeographical surveys, 124–125
Economic assessment of genetic
resources
contributions of international
collections, 317–318
Hedonic pricing method, 308–309
incorporation into advanced cultivars
and, 307–308
institutional releases and, 312
landrace appearance and, 312–313
markets, 304–306

T

Recent Publications of the Board on Agriculture

Policy and Resources

Pesticides in the Diets of Infants and Children (1993), 408 pp., ISBN 0-309-04875-3.

Managing Global Genetic Resources: Livestock (1993), 294 pp., ISBN 0-309-04394-8.

Sustainable Agriculture and the Environment in the Humid Tropics (1993), 720 pp., ISBN 0-309-04749-8.

Agriculture and the Undergraduate: Proceedings (1992), 296 pp., ISBN 0-309-04682-3.

Water Transfers in the West: Efficiency, Equity, and the Environment (1992), 320 pp., ISBN 0-309-04528-2.

Managing Global Genetic Resources: Forest Trees (1991), 244 pp., ISBN 0-309-04034-5.

Managing Global Genetic Resources: The U.S. National Plant Germplasm System (1991), 198 pp., ISBN 0-309-04390-5.

Sustainable Agriculture Research and Education in the Field: A Proceedings (1991), 448 pp., ISBN 0-309-04578-9.

Toward Sustainability: A Plan for Collaborative Research on Agriculture and Natural Resource Management (1991), 164 pp., ISBN 0-309-04540-1.

Investing in Research: A Proposal to Strengthen the Agricultural, Food, and Environmental System (1989), 156 pp., ISBN 0-309-04127-9.

Alternative Agriculture (1989), 464 pp., ISBN 0-309-03985-1.

Understanding Agriculture: New Directions for Education (1988), 80 pp., ISBN 0-309-03936-3.

Designing Foods: Animal Product Options in the Marketplace (1988), 394 pp., ISBN 0-309-03798-0; ISBN 0-309-03795-6 (pbk).

Agricultural Biotechnology: Strategies for National Competitiveness (1987), 224 pp., ISBN 0-309-03745-X.

Regulating Pesticides in Food: The Delaney Paradox (1987), 288 pp., ISBN 0-309-03746-8.

Pesticide Resistance: Strategies and Tactics for Management (1986), 480 pp., ISBN 0-309-03627-5.

Pesticides and Groundwater Quality: Issues and Problems in Four States (1986), 136 pp., ISBN 0-309-03676-3.

Soil Conservation: Assessing the National Resources Inventory, Volume 1 (1986), 134 pp., ISBN 0-309-03649-9; Volume 2 (1986), 314 pp., ISBN 0-309-03675-5.

New Directions for Biosciences Research in Agriculture: High-Reward Opportunities (1985), 122 pp., ISBN 0-309-03542-2.

Genetic Engineering of Plants: Agricultural Research Opportunities and Policy Concerns (1984), 96 pp., ISBN 0-309-03434-5.

Nutrient Requirements of Domestic Animals Series and Related Titles

Nutrient Requirements of Horses, Fifth Revised Edition (1989), 128 pp., ISBN 0-309-03989-4; diskette included.

Nutrient Requirements of Dairy Cattle, Sixth Revised Edition, Update 1989 (1989), 168 pp., ISBN 0-309-03826-X; diskette included.

Nutrient Requirements of Swine, Ninth Revised Edition (1988), 96 pp., ISBN 0-309-03779-4.

Vitamin Tolerance of Animals (1987), 105 pp., ISBN 0-309-03728-X.

Predicting Feed Intake of Food-Producing Animals (1986), 95 pp., ISBN 0-309-03695-X.

Nutrient Requirements of Cats, Revised Edition (1986), 87 pp., ISBN 0-309-03682-8.

Nutrient Requirements of Dogs, Revised Edition (1985), 79 pp., ISBN 0-309-03496-5.

Nutrient Requirements of Sheep, Sixth Revised Edition (1985), 106 pp., ISBN 0-309-03596-1.

Nutrient Requirements of Beef Cattle, Sixth Revised Edition (1984), 90 pp., ISBN 0-309-03447-7.

Nutrient Requirements of Poultry, Eighth Revised Edition (1984), 71 pp., ISBN 0-309-03486-8.

Further information, additional titles (prior to 1984), and prices are available from the National Academy Press, 2101 Constitution Avenue, NW, Washington, DC 20418, 202/334-3313 (information only); 800/624-6242 (orders only); 202/334-2451 (fax).